Electrical Safety of Low-Voltage Systems

Dr. Massimo A. G. Mitolo
Professional Engineer

New York Chicago San Francisco
Lisbon London Madrid Mexico City
Milan New Delhi San Juan
Seoul Singapore Sydney Toronto

The McGraw-Hill Companies

Cataloging-in-Publication Data is on file with the Library of Congress

1 2 3 4 5 6 7 8 9 0 DOC/DOC 0 1 5 4 3 2 1 0 9

ISBN 978-0-07-150818-6
MHID 0-07-150818-X

Sponsoring Editor
Stephen S. Chapman

Editing Supervisor
Stephen M. Smith

Production Supervisor
Pamela A. Pelton

Project Manager
Sonia Taneja, Aptara, Inc.

Copy Editor
Sunil Kumar Ojha, Aptara, Inc.

Proofreader
Prakash Sharma

Indexer
Slavka Zlatkova

Art Director, Cover
Jeff Weeks

Composition
Aptara, Inc.

To my parents, Domenico and Fernanda,
true examples of unconditional love and boundless support,
to whom I owe everything.

To my wife, Jennifer,
my partner in the Quest.
To my daughter, Alessandra,
may she always live in a safe and interesting world.

Felix qui potuit rerum cognoscere causas.
Happy is he who was able to know the causes of things.

VIRGIL (70–19 BC)
Georgics, bk. 2, l. 490 (19 BC)

Old friend, what are you looking for?
After those many years abroad you come with
images you tended under foreign skies
far away from your own land.

GEORGE SEFERIS

Contents

Preface

Electrical safety may be perceived only as a list of prudent actions to or not to undertake in the presence of energized objects, constituting the defense against direct contact with live parts. However, the safety of persons also depends on their exposure to indirect contact, that is, contact with parts normally not in tension, but likely to become energized due to faults. Thus, the attitude toward live parts is not the only key in preventing accidents.

This book, prompted by this concept, is an attempt, from the academic point of view, to bridge the existing gap between life-safety electrical issues in low-voltage systems (i.e., not exceeding 1 kV) and their proper comprehension and design solution, in light of applicable IEC and IEEE standards. We assume, in fact, that we can analytically quantify the hazards caused by indirect contact, thereby promoting a proper design for the electrical system and minimizing the related risk.

The book, based on my 20-year-long experience as a professor and as a professional engineer, provides an explanation of the fault-loops in different types of grounding systems (i.e., TT, TN, and IT) and of the faults occurring on both sides of the supply (i.e., the primary and secondary of substation transformers). The crucial role played by the state of the neutral is deeply examined, thereby allowing the comprehension of the reasons behind the methodologies of protection against electric shock, which are required by current standards and codes.

The book's audience consists of electrical engineering students who need to know the principles of electrical safety as well as professional engineers who are involved in the bonding and grounding of power systems. Background requirements include a knowledge of a.c. electric circuits, algebra, complex numbers, and basic calculus.

Each chapter is arranged in a format that is aimed at promoting the reader's understanding by providing many figures and equivalent circuits to clarify, both visually and analytically, the concepts discussed, such as the determination of fault currents and touch voltages. Several chapters also have a section of frequently asked questions at the end, with relative answers based on the actual inputs of students and professionals.

The first three chapters explain the fundamental principles of electrical safety, providing the basic concepts of protection against direct contact and indirect contact as well as the mathematical interpretation of safety and risk of standard protective measures.

Chapter 4 discusses the role of the earth as an available return path to the supply source of fault currents, thus analyzing the theory of ground potentials and ground resistances of electrodes.

Chapter 5 describes the effects of currents passing through the human body as interfering with the body's own electricity as well as causing thermal stress to its tissues. This chapter also explains the concepts of permissible body current and permissible touch voltage as used in IEC and IEEE technical standards.

Chapters 6 through 9 explain the protection against indirect contact in different grounding systems, such as TT, TN, PME, and IT, and detail voltage exposures and protective issues in each of them.

Chapter 10 is devoted to the extra-low-voltage systems and describes the safety issues arising under fault conditions.

Chapter 11 describes the fundamental components of earthing arrangements, explains their functions, and provides minimum acceptable sizes following applicable technical standards. An analytical method to determine the minimum cross-sectional area of protective conductors, assuming an adiabatic thermal process during faults, is also offered.

Chapter 12 discusses the effects of overvoltages, in particular the temporary ones, within different types of grounding systems as well as the stress voltages that may arise under fault conditions, possibly causing the breakdown of the basic insulation of equipment.

Chapter 13 examines the safety issues caused by static electricity and residual voltages, eventually present on de-energized items. The energy stored in charged objects is calculated and the mitigation strategies to reduce it are described.

Chapter 14 discusses the methodologies of measurement employed during the design phase (e.g., soil resistivity test) and after the installation of the electrical system as well as prior to putting it into service (e.g., earth resistance test).

The final chapter analyzes the safety requirements against indirect contact employed in special installations or locations, where environmental conditions may increase the risk of indirect contact (i.e., marinas, train stations, swimming pools, surgery rooms, etc.).

The three appendices discuss the basic concepts of sinusoids and phasors, the fundamental conventions, and the network theorems that are extensively used throughout the text. Their purpose is to give the reader a basic theoretical support for the comprehension of the technical methodologies profusely applied in the book.

Writing this book has been a formidable journey through the core of the bonding and grounding of electrical systems, and I do hope that it will shed some light on some of the concepts commonly accepted by the community of the practitioner engineers, but perhaps not completely understood.

My sincere thanks go to all my colleagues and friends from industry and academia alike for their constructive critiques during the drafting of the manuscript. Last, but not least, many thanks for continuous and effective support to my wife Jennifer, a "solidly" grounded person, who lights up my life.

Dr. Massimo A. G. Mitolo

About the Author

Massimo A. G. Mitolo, educated in Italy, received the doctoral degree in electrical engineering from the University of Naples "Federico II." His field of research is the analysis and grounding of power systems. Dr. Mitolo has been a registered Professional Engineer in Italy since 1991 and is currently working as an Associate Electrical Engineer at Chu & Gassman Consulting Engineers in New York. He is an IEEE Senior Member and is very active within the IEEE IAS Industrial & Commercial Power Systems Department, where he is currently the Vice Chair of the Power Systems Engineering Main Technical Committee, the Chair of the Papers Review Subcommittee, the Chair of the Power Systems Analysis Subcommittee, and the Vice Chair of the Power Systems Grounding Subcommittee. Dr. Mitolo is also an Associate Editor of the IEEE Manuscript Central.

CHAPTER 1

Basic Definitions and Nomenclature

Defendit numerous.
In numbers safety.

IUVENALIS (circa 60–120 BC)

1.1 Introduction

Any discipline has its own language, basic definitions, and nomenclature. These elements are crucial for a deep understanding of the core of the discipline itself. With this in mind, the following text describes both fundamental terms and schematics that will be intensively used throughout this book. They constitute a very effective tool for comprehending the influence of the state of the neutral (i.e., system grounding) on electrical safety of low-voltage systems.

1.2 Basic Definitions and Nomenclature

1.2.1 Basic Insulation

The insulation applied to live parts and necessary to provide basic protection against electric shock.

1.2.2 Class 0 Equipment

Equipment outfitted only with the basic insulation and no bonding terminals.

1.2.3 Class I Equipment

Equipment outfitted with basic insulation and bonding terminals; automatic disconnection of supply can be carried out as protection against electric shock in the case of failure of the basic insulation.

1.2.4 Class II Equipment

Equipment outfitted with a double insulation, consisting of basic insulation plus supplementary insulation, or a reinforced insulation.

1.2.5 Class III Equipment

Equipment in which protection against electric shock solely relies on supply at safety extra-low voltage. Thus, the extra safety features built into Class I and Class II appliances are not required.

1.2.6 Direct Contact

Contact with parts of the installation normally live.

1.2.7 Indirect Contact

Contact with metal parts not normally live (e.g., exposed-conductive-parts), but energized under fault conditions. (The basic difference between the definitions of direct and indirect contact is the presence, between the live part and the person, of a metal enclosure.)

1.2.8 Disconnection of Supply

Protection against indirect contact may be carried out by automatic disconnection of supply. A protective device shall automatically disconnect the supply to the faulty circuit or equipment so that a prospective touch voltage exceeding 50 V a.c. r.m.s. (or 120 V ripple-free d.c.) does not persist for a time sufficient to cause a risk of harmful physiological effect in a person.

1.2.9 Exposed-Conductive-Part (ECP)

ECP is a conductive part, forming part of electrical equipment, which can be touched (even if out of reach), and which is not live, but which may become live when basic insulation fails. A conductive part that can be energized just because it is in touch with an ECP shall not be considered an ECP. Sometimes ECPs are referred to as *noncurrent-carrying metal parts*.

1.2.10 Extra-Low Voltage

Voltage supplied from a source that does not exceed 50 V between conductors and between conductors and earth.

1.2.11 Extraneous-Conductive-Part (EXCP)

EXCP is a conductive part, not forming part of the electrical system, which can be touched, and is liable to introduce a "zero" potential (i.e., earth potential) or an arbitrary potential. Both of these potentials are dangerous.

Examples of EXCPs are the metalwork for gas, water, and heating systems, the metallic frame of a building, conductive floors, walls, etc.

1.2.12 Functional Insulation

It is the insulation between conductive parts at different potentials that is necessary only for the proper functioning of the appliance.

1.2.13 Ground

The earth, that is to say, a conductive mass whose potential is conventionally considered as zero.

1.2.14 IT Grounding System

Power system isolated from earth (ungrounded) (Fig. 1.1) or high-resistance grounded (HRG) (Fig. 1.2); ECPs are independently grounded from the power source. The neutral may be distributed, even though it is advisable not to ship it in order to facilitate its insulation from ground.

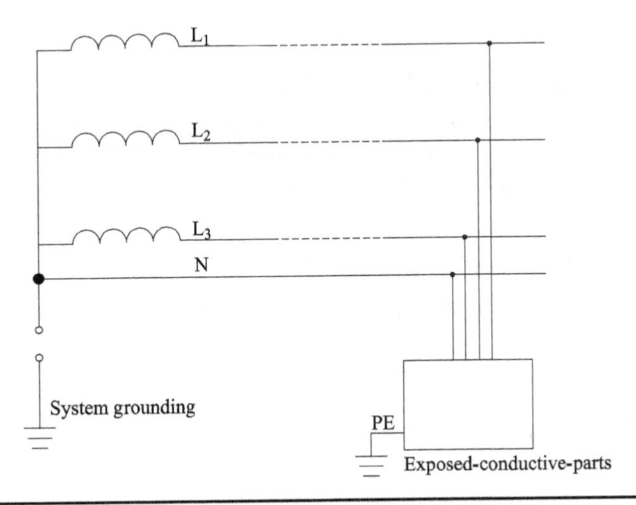

FIGURE 1.1 Power system isolated from earth (ungrounded).

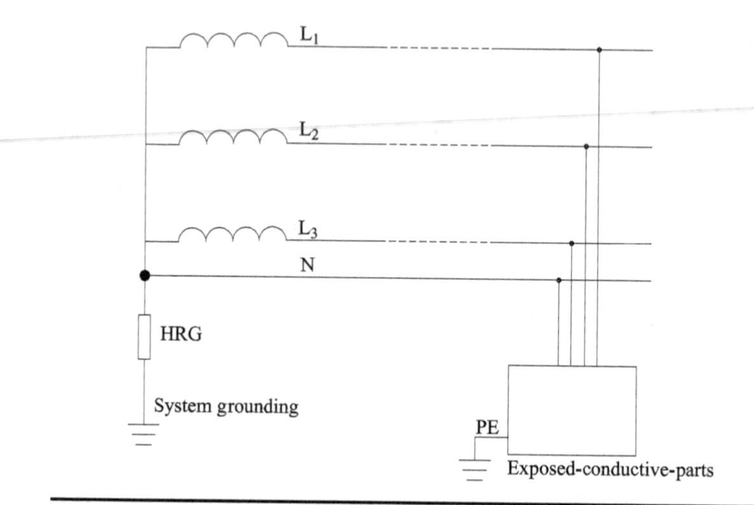

Figure 1.2 Power system high-resistance grounded (HRG).

1.2.15 Neutral-Protective Conductor (PEN)

The PEN conductor combines the functions of both a protective conductor and a neutral conductor.

1.2.16 Protective Bonding Conductor

Conductor provided for protective-equipotential bonding. Its purpose is to guarantee the same potential between metal parts possibly at dangerous different potential upon fault. It is also referred to as *bonding jumper*.

1.2.17 Protective Conductor (PE)

Conductor provided for the purpose of safety against electric shock. Its function is to safely drain the ground-fault current to the source. It is also referred to as *equipment grounding conductor*.

1.2.18 Remote or Zero Potential

Potential of a point conventionally assumed as at the infinite (see Fig. 1.3 for symbol).

Figure 1.3 Symbol of zero potential.

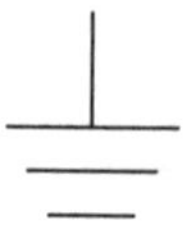

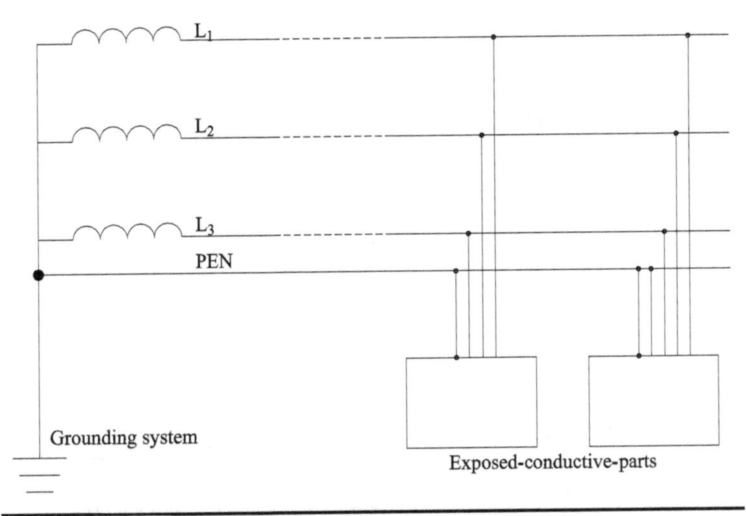

FIGURE 1.4 TN-C grounding system. Neutral and protective functions are combined in a single conductor throughout the electrical system.

1.2.19 TN Grounding System

A solidly grounded power system; the system has one point directly grounded and the ECPs directly connected to that point by protective conductors.

1.2.20 TN-C Grounding System

Same definition as that for TN grounding system; neutral and protective functions are combined in a single conductor (i.e., PEN conductor) throughout the electrical system (Fig. 1.4).

1.2.21 TN-C-S Grounding System

Same definition as that for TN grounding system; neutral and protective functions are combined in a single conductor (i.e., PEN conductor) in a part of the electrical system (Fig. 1.5).

1.2.22 TN-S Grounding System

Same definition as that for TN grounding system; separate neutral and protective conductors are used throughout the system (Fig. 1.6).

1.2.23 TT Grounding System

A solidly grounded power system; ECPs are directly connected to the ground, independently of the grounding of any point of the power supply system (Fig. 1.7).

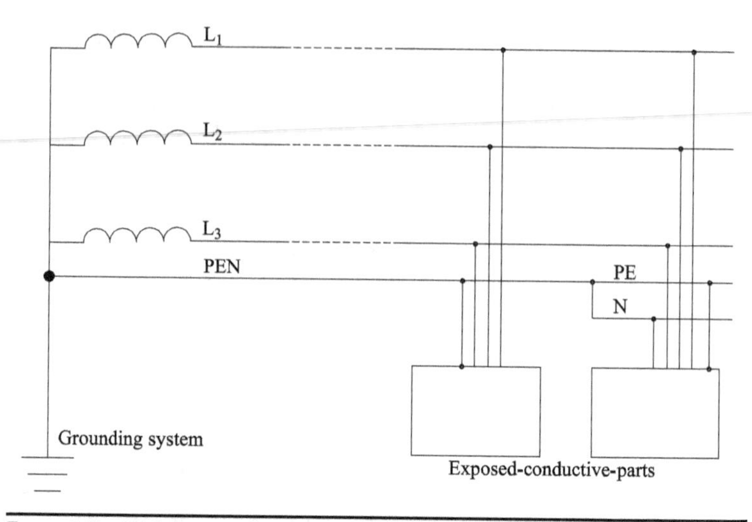

FIGURE 1.5 TN-C-S grounding system. Neutral and protective functions are combined in a single conductor in a part of the electrical system.

1.2.24 Prospective Touch Voltage

The prospective touch voltage V_{ST} is defined as the potential difference between the faulted ECP and the earth occupied by the person, at the distance of 1 m from the ECP, when the ECP is not being touched by the person.

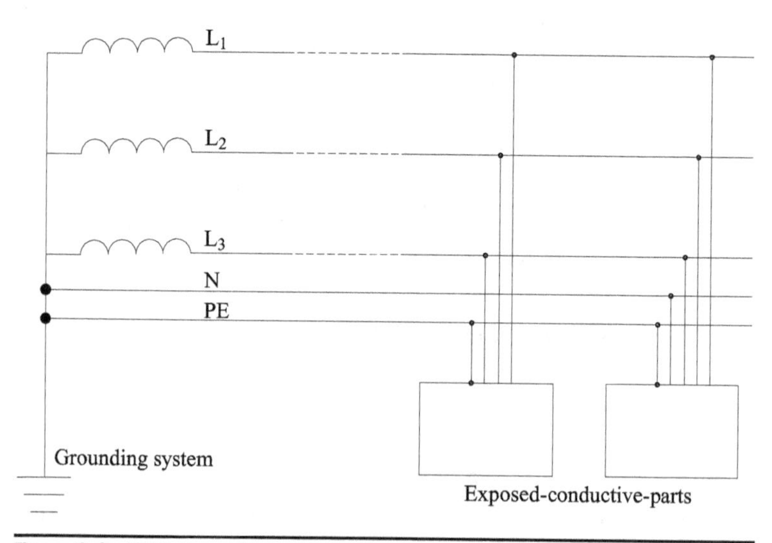

FIGURE 1.6 TN-S grounding system. Separate neutral and protective conductors are used throughout the electrical system.

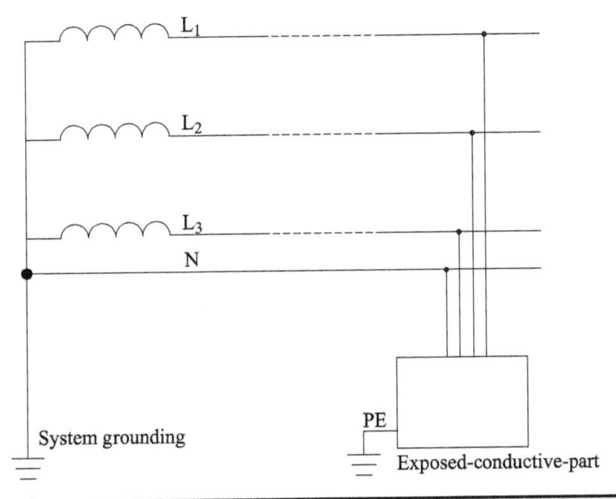

FIGURE 1.7 In TT systems, ECPs are directly connected to the ground, independently of the grounding of any point of the power supply system.

1.2.25 Touch Voltage

The touch voltage V_T is defined as the voltage differential, which a person may be subject to, between both hands and both feet.

1.2.26 Prospective Step Voltage

The prospective step voltage V_{SS} is defined as the potential difference between two points on the surface of the earth, displaced by the distance of 1 m, when the earth is not being touched by the person.

1.2.27 Step Voltage

The step voltage V_S is defined as the potential difference, which a person may be subject to, between the two feet, conventionally displaced by 1 m.

CHAPTER 2

Fundamentals of Electrical Safety

Mysterious affair, electricity.
SAMUEL BECKETT (1906–1989)

2.1 Introduction

Electrical safety is not exclusively defined by the prudent conduct of individuals in the presence of energized objects. A sensible attitude toward electrical equipment may only prevent direct contact, that is, an accidental contact with parts normally live (e.g., energized conductors, terminals, bus bars inside of equipment, etc.).

Persons are also exposed to the risk of indirect contact, that is, contact with faulty *exposed-conductive-parts* (ECPs). ECPs are items supplied by the electrical systems that are not normally live, but that are accidentally energized due to failure of the basic insulation (Fig. 2.1).

Indirect contact is more insidious than direct contact, as it may occur even during the reasonable use of electrical equipment. Safety is carried out by systematically applying measures of protection against both types of contacts, which might occur during the common interaction between a person and an electrical equipment. Protection against direct contact, also referred to as *basic protection*, is achieved with effective separation of persons from live parts, whereas protection against indirect contact, also referred to as *fault protection*, is accomplished by automatically disconnecting the supply. In some specific situations, discussed later in this chapter, fault protection can also be carried out without disconnection of supply.[1]

It is important to note that all electrical systems must be properly maintained, so as to reasonably prevent danger of electric contacts.

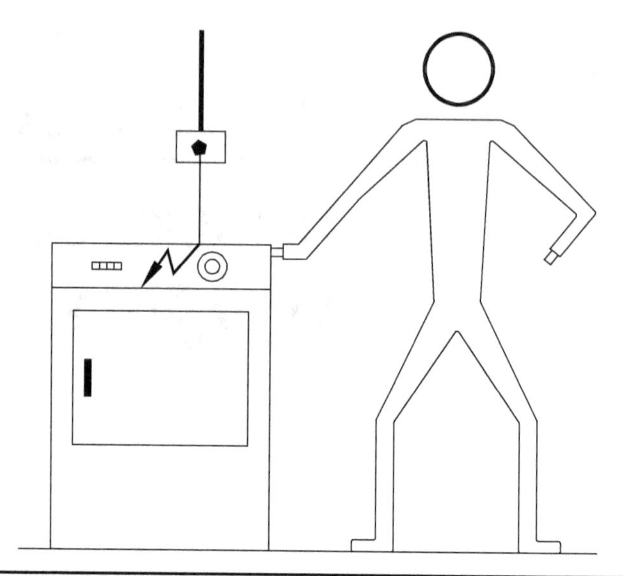

FIGURE 2.1 Indirect contact.

2.2 Protection Against Direct Contact

It is understood that all electrical equipment must have provisions to guarantee protection against direct contact. In the following sections, the fundamental strategies of basic protection are examined.

2.2.1 Insulation of Live Parts

In order to operate, electric equipment contains parts at different potentials, which must be properly insulated from each other and from their enclosure through the functional insulation.

The basic insulation prevents persons from coming in contact with live parts and is the fundamental protection against direct contact. To be effective as a protection, the insulation material must completely cover the live parts and should be removable only by destruction (Fig. 2.2).

The basic insulation must be capable of withstanding the possible stresses during the functioning of the equipment without losing its integrity. Electric fields, mechanical collisions, high temperatures, and the aging of the insulating material are the possible causes of failure of the basic insulation. It is essential, then, that the basic insulation has sufficient mechanical strength to withstand the stress caused by the normal operation of equipment. As a consequence, insulating paints, and similar products, cannot be considered suitable for the basic insulation; however, they can be used as the functional insulation (e.g., insulation between windings of transformers or motors).

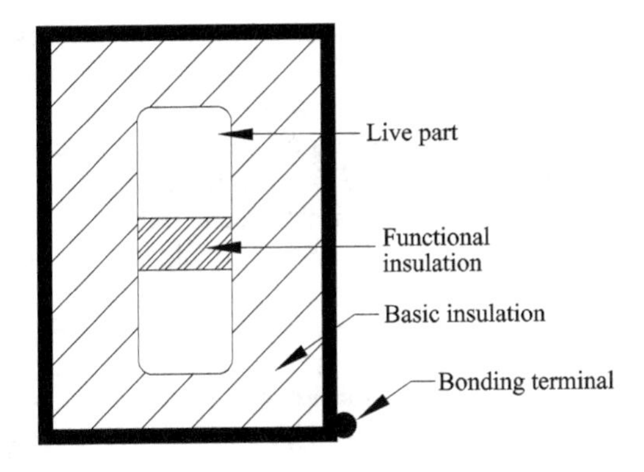

FIGURE 2.2 Diagrammatic representation of functional and basic insulations in Class I equipment.

Figure 2.2 shows a piece of Class I equipment, that is, an ECP outfitted with a bonding terminal to allow the grounding of the enclosure.

2.2.2 Enclosures and Barriers

Both enclosures and barriers are constructions, firmly held in their positions, intended to prevent persons from intentionally, or accidentally, touching live parts without the aid of tools.

As the term suggests, enclosures provide protection in any approaching direction to the equipment by "enclosing" it. Live parts are inside the protective construction. Barriers, instead, may offer the same defined degree of protection against direct contact, but only in a limited number of approaching "routes" to the equipment. Safety is equally achieved if live parts are kept "behind" barriers, instead of inside of an enclosure.

For instance, barriers may be used around an open-type piece of equipment when, due to its height, the access from above is naturally precluded to persons. The "top" is, therefore, deemed unnecessary for safety and the enclosure is not strictly required.

Removal of barriers, or opening of enclosures, must be possible only by using keys or tools so as to prevent the accidental elimination of the fundamental protection against direct contact. The necessity of keys or tools as a "rule of engagement" to the equipment can be waived if removal/opening of protection can occur only after the supply is disconnected.

The minimum insulation requirement for enclosures and barriers is that live parts be inaccessible to a person's finger. This requirement limits the size of openings in equipment, for example, vents.

The IEC *International Protection Code*[2] has standardized designations composed of the letters IP followed by two characteristic numerals, which describe the degree of protection offered by different types of enclosures and barriers. The first characteristic numeral (0 to

6) indicates the degree of protection against access of person's finger/back of hand to hazardous parts as well as against ingress of solid foreign objects. The second numeral (0 to 8) designates the degree of protection against ingress of water through enclosures and barriers. An optional letter (A to D) designates, just like the first numeral, the degree of protection against direct contact. A brief description of the characteristic numerals and optional letters can be found in Fig. 2.3.

1st Numeral	Protection of Equipment Against Solid Particles	Against Person's Access With
0	Nonprotected	Nonprotected
1	> 50 mm diam.	Back of hand
2	> 12.5 mm diam.	Finger
3	> 2.5 mm diam.	Tool
4	> 1 mm diam.	Wire
5	Dust	Wire
6	Dust proof	Wire

2nd Numeral	Protection of Equipment Against Ingress of Water
0	Nonprotected
1	Vertical dripping
2	Dripping (15° tilted)
3	Rain (spraying water at an angle up to 60° on either side of the vertical)
4	Splashes from any direction
5	Jets from any direction
6	Powerful jets from any direction (flow rate > 12.5 dm^3/min)
7	Temporary immersion
8	Continuous immersion

Optional Letter	Protection Against Person's Access With
A	Back of hand
B	Finger
C	Tool
D	Wire

FIGURE 2.3 Brief description of the IP designations.

Each numeral requires different tests be applied to equipment to obtain the IP rating. The jointed test finger, the rigid sphere, and the test wire are the standard rating tools.

To guarantee safety, enclosures and barriers are required by international standards to have at least a degree of protection of IPXXB, which does not allow access to a person's finger. The symbol X means there are no requirements for that specific characteristic numeral. The IP2X degree of insulation is not equivalent to IPXXB, but better. An IP2X enclosure, or barrier, in fact, must pass the following two tests:

1. The standard jointed finger (length 80 mm and diameter 12 mm), applied with a test force[3] of 10 N to all sides and openings of the enclosure, must not touch any live parts in every possible position of its two joints.
2. A 12.5-mm-diameter rigid sphere must not entirely pass through any opening (test force of 30 N).

An IPXXB enclosure, instead, must pass only the above first test to provide the same degree of safety against electrocution. However, IPXXB enclosures, although safe for persons, may allow the ingress of foreign objects of 12.5 mm diameter, or smaller, into the equipment, and, therefore, might not be suitable in certain locations.

Let us examine the case in Fig. 2.4. The enclosure is "permeable" to the test sphere, which can penetrate inside, and thus cannot be classified as IP2X; at the same time, though the enclosure does not allow contact with live parts, as the jointed finger cannot touch any live part, ergo its rating is IPXXB.

If enclosures or barriers have readily accessible horizontal top surfaces (e.g., height less than 2.5 m), a more stringent insulation is required. To prevent the additional risk of direct contact due to small

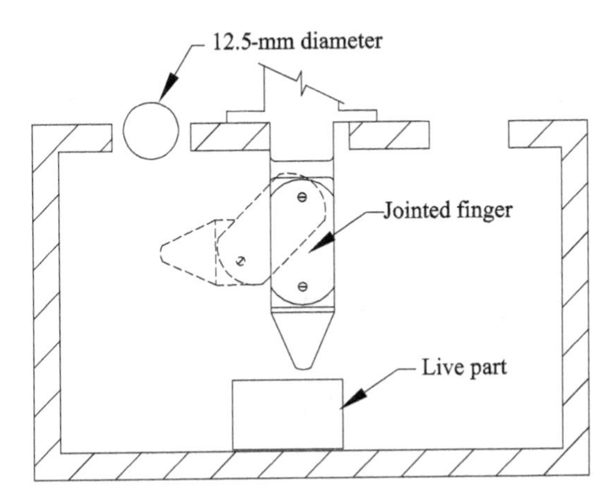

FIGURE 2.4
Enclosure IP1XB.

metal objects, which falling through openings may bridge the gap between persons and live parts, the degree of protection IPXXD or IP4X is necessary. These two designations maintain the same previously exemplified logic, with the only difference being the use of the test wire (length 100 mm and diameter 1 mm) instead of the jointed finger.

It must be clear that the judgment of the electrical engineer is necessary to establish the optimum degree of insulation of equipment, in light of both the actual environmental conditions of the location and its normal operations. It is also important to note that a too severe degree of insulation, if unnecessary, can damage the equipment by limiting its ventilation and, thereby, raising its internal temperature beyond safe limits.

2.2.2.1 Enclosures and Mechanical Impacts

A serious hazard for persons is the accidental rupture of enclosures due to external mechanical impacts, which can expose live parts and trigger explosive atmospheres. Enclosures, therefore, must have the capability to protect their own contents. Such ability is specified by the international IK code,[4] which indicates the degree of protection against harmful impacts. The IK code rates enclosures through the code letters IK followed by the characteristic group numeral (00 to 10), indicating an impact energy value in joules (see Table 2.1).

The IK code contemplates the maximum value of impact energy of 20 J; when higher impact energy is required, the IK code recommends a value of 50 J.

2.2.3 Protection by Obstacles

Obstacles are elements placed between exposed live parts and persons (e.g., fence, handrail, mesh, screen, etc.). They prevent direct contacts by increasing the distance from energized parts, which, otherwise, would be accessible. Safety is, therefore, assured by keeping exposed live parts out of reach. Unlike enclosures and barriers, obstacles could be intentionally circumvented, as, by definition, they may not be firmly held in their positions; therefore, obstacles offer only a limited degree of protection and that too only for accidental touch. This protective measure, consequently, should be exclusively adopted in areas accessible to skilled personnel in the field of electricity.

	IK01	IK02	IK03	IK04	IK05	IK06	IK07	IK08	IK09	IK10
Impact energy (J)	0.15	0.2	0.35	0.5	0.7	1	2	5	10	20

Table 2.1 Relation Between IK Code and Impact Energy

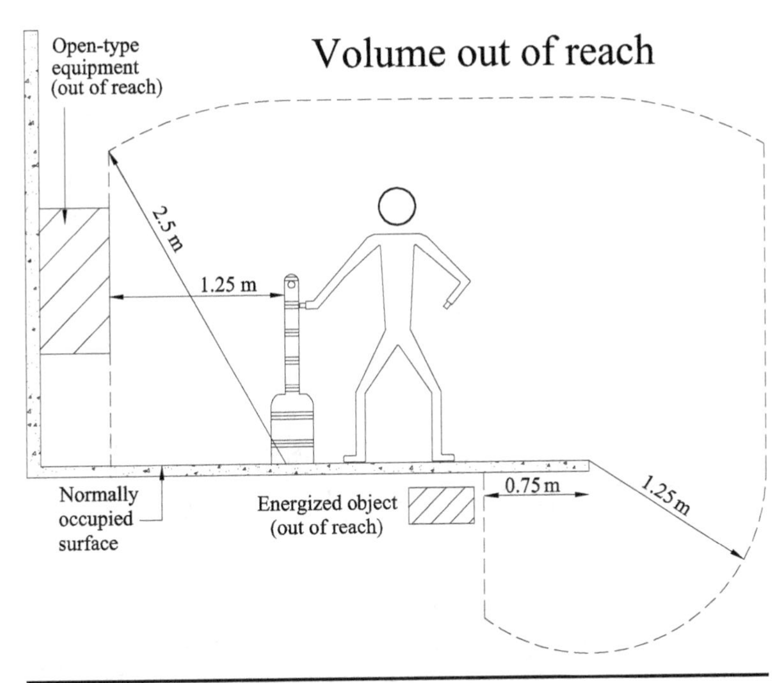

FIGURE 2.5 Volume out of reach.

We conventionally deem out-of-reach energized objects placed outside of the volume defined by the reach of the person's arm. The horizontal arm's extent is conventionally assumed to be 1.25 m, but as the contact can also occur in the overhead direction, the average height of persons must be included. Therefore, the conventional length of 2.5 m from the floor is also considered arm's reach. The extent of arm's reach is to be measured from the obstacle (Fig. 2.5).

Skilled persons are deemed safe as long as exposed energized parts are in the volume out of reach (i.e., outside of the dotted line).

If persons normally handle long conductive items (e.g., tools, ladders, etc.), larger clearance distances must be considered to take into account the additional risk due to their length so as to provide the same level of safety.

2.2.4 Additional Protection by Residual Current Devices

Residual current devices (RCDs) are also referred to as *residual current operated circuit-breakers* (RCCBs) or *ground-fault circuit interrupters* (GFCIs). RCDs with operating current I_{dn} not exceeding 30 mA are additional means of protection against direct contact. When they are used in households and similar environments, nontrained people should be able to easily operate them.

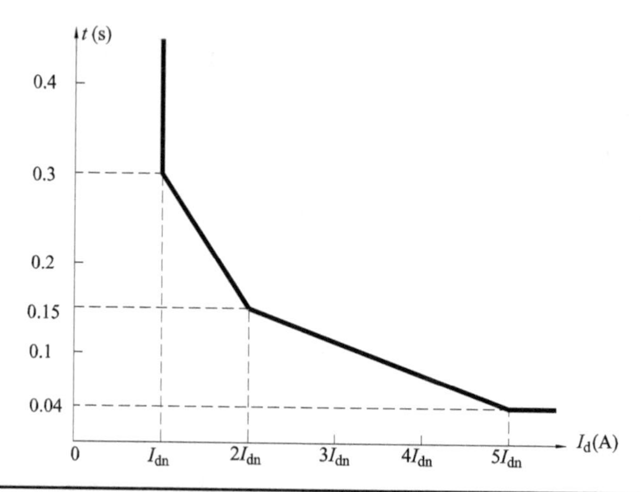

FIGURE 2.6 Permissible operating time as a function of the ground-fault current.

The term residual current[5] I_d indicates the vector sum of all alternating currents flowing through a circuit's wires, single-phase or three-phase,[6] including the neutral conductor, and is expressed in terms of the root mean square (r.m.s.) value. The RCD executes this sum, which is zero in normal conditions. Should a fault occur, I_d becomes greater than zero and is equal to the r.m.s. of the ground-fault current I_G. The RCD compares this nonzero value to its rated operating current I_{dn} and if $I_d > I_{dn}$ disconnects the supply to the faulty circuit. The clearing time will occur within a conventional safe time as established by applicable standards. RCDs, in fact, do not limit the magnitude of the ground-fault current, but only the time this current circulates to ground. Figure 2.6 shows the permissible operating times[7] not to be exceeded by general purpose RCDs as a function of the residual current I_d, usually expressed as a multiple of the rated operating current I_{dn}.

Besides the residual operating current, the RCD is characterized by another important parameter: the residual nonoperating current I_{dNO}, which represents the maximum r.m.s value of the residual current that does not cause its operation. Standard value for I_{dNO} is $0.5I_{dn}$ and therefore the RCD does not operate for $I_d < 0.5I_{dn}$; it might operate in the range $0.5I_{dn} < I_d \leq I_{dn}$ and must surely operate for $I_d < I_{dn}$.

For a better understanding of the functioning of the residual current devices, let us examine Fig. 2.7, which shows a single-phase RCD.

In the absence of ground faults, we have

$$\underline{I}_{Ph} = \underline{I}_N \Rightarrow \underline{I}_{Ph} - \underline{I}_N = 0 \tag{2.1}$$

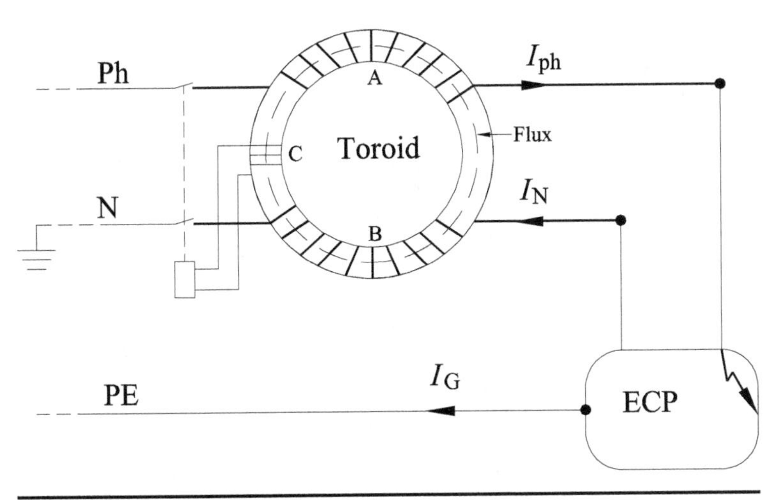

Figure 2.7 Single-phase RCD.

If a fault puts in contact the phase conductor with the enclosure, a current $\underline{I}_G$ will flow through the protective conductor,[8] causing phase and the neutral currents to differ. If we consider the point of contact with the enclosure as a "generalized" node, we can apply the first Kirchoff's principle:

$$\underline{I}_{Ph} = \underline{I}_N + \underline{I}_G \Rightarrow \underline{I}_{Ph} - \underline{I}_N = \underline{I}_G \neq 0 \tag{2.2}$$

As a consequence, the resulting magnetic flux $\underline{\Phi}$ along the RCD's toroid, which is proportional to the net current $\underline{I}_G$ flowing through the windings A and B, is no longer zero. Thus, an electromotive force is generated within the dedicated coil C, which will quickly activate the circuit breaker if $|\underline{I}_G| > I_{dn}$ and disconnect the supply.

The same protective residual logic can be applied to three-phase systems (Fig. 2.8).

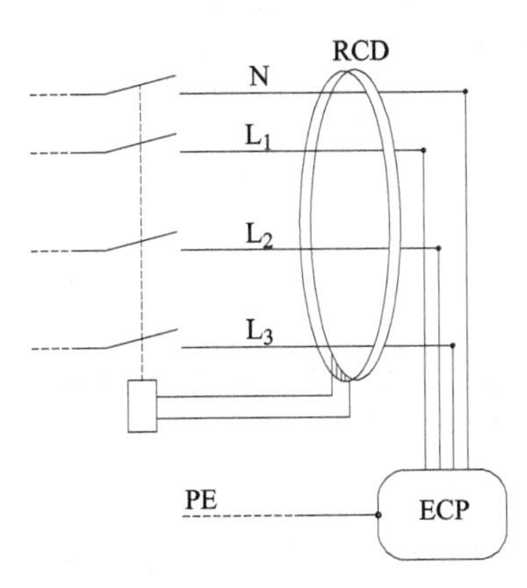

Figure 2.8 A residual current device in a three-phase circuit.

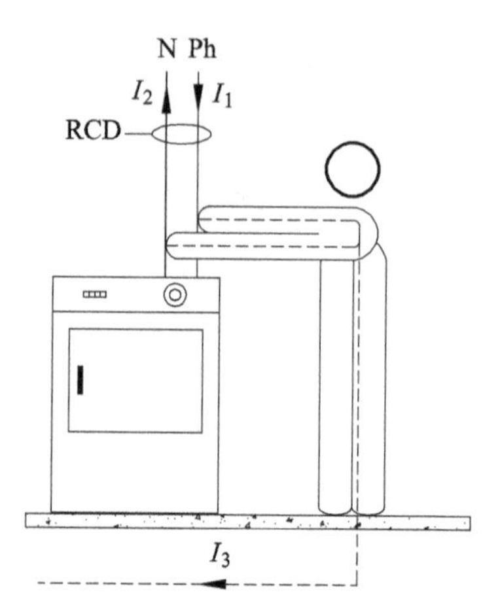

FIGURE 2.9 Direct contact phase-to-neutral.

The three-phase RCD is a transformer whose primary winding is constituted by the line conductors themselves. The vector sum of the line currents and the neutral current in healthy three-phase circuits is always zero, and therefore, in the secondary winding, which has the task of switching off the supply, no current will circulate. If a fault occurs, the vector sum becomes nonzero due to the current leaving the system through the PE not passing through the toroid. The RCD, then, activated by its secondary winding, will trip the circuit breaker.

RCDs must be considered as an additional means of protection and do not substitute for the other fundamental protective measures against direct contact previously examined. RCDs, in fact, can protect persons by disconnecting the supply only in the case of contact between energized objects and the ground. They can sense only fault currents not returning to the source through the legitimate path. Consequently, direct contact between the phase and the neutral conductors may not activate the RCD, as there may not be enough ground current circulation for it to sense (Fig. 2.9).

The RCD will only sense the component $\underline{I}_3$, while the larger current $\underline{I}_1$ will circulate through the person's body. $\underline{I}_3$ may not be large enough to exceed the RCD's operating threshold, which cannot disconnect the supply.

2.3 Protection Against Indirect Contact

The failure of the basic insulation may cause electrocution owing to the accidental presence of voltage-to-ground over metal parts not normally live (Fig. 2.1). This condition is particularly dangerous as it is not under a person's control despite any prudent conduct. Protective measures, active and passive, against this type of fault situation must be considered.

2.3.1 Protection by Automatic Disconnection of Supply

The automatic disconnection of the faulty circuit from its source is an active protective measure aimed to limit the persistence of prospective touch voltages on an ECP to a time that the human body can withstand without incurring harmful physiological effects. The protective device must promptly trip in accordance with the magnitude of the touch voltage: the higher this value, the faster it must trip. As explained in Chap. 5, the time–voltage safety curve describes the permissible prospective touch voltage for persons as a function of the contact duration in any type of earthing system (e.g., 50 V a.c. can be withstood for no more than 5 s).

As later shown, maximum disconnection times of protective devices have been elaborated as a function of the nominal voltage and the type grounding of the electrical system, rather than of the perspective touch voltage.

Disconnection of supply upon faults is a measure that requires an efficient bonding of the ECPs to the earthing system so that protective devices, by sensing the leakage to earth, can intervene at the inception of the ground fault even before a person comes in contact with energized objects. This protection is suitable if electrical items are equipped with bonding terminals (i.e., Class I equipment of Fig. 2.2). Under this point of view, the disconnection of supply can be considered as a preventive approach to safety.

If the ground-fault current is high enough (e.g., TN systems), automatic circuit breakers can be employed to switch off the faulty circuit. RCDs may also be used to disconnect the voltage source upon ground faults (especially in TT systems). In this regard, it is important to underline the importance of the inclusion of the neutral conductor through the RCD's toroid (Fig. 2.8).

If the neutral is not included, any unbalanced load could cause nuisance tripping of the device. In fact, the vector sum of the phase currents circulating through the toroid would not be compensated by the neutral current, causing a nonzero result. If the system does not carry the neutral conductor, the vector sum of the line currents through the toroid is normally zero, even if the load is unbalanced, but becomes nonzero in the case of a ground fault, allowing the operation of the device.

On the other hand, the protective conductor PE must be excluded from the RCD, otherwise the device would never trip. The fault currents over the PE, in fact, would return to the source passing through the toroid, thereby causing the vector sum of the currents to be zero, despite the presence of the fault.

As discussed later, it is important that protective devices are coordinated with the value of the earthing system resistance so as to prevent the persistence of fault potentials on accessible parts for a dangerous amount of time.

2.3.1.1 Nuisance Trippings of RCDs

Nuisance trippings of RCDs disconnect the supply in the absence of any actual danger for persons, thereby causing an unnecessary loss of service. Typical reasons of nuisance tripping of RCDs are the overvoltages resulting from both switching transients and lightning.[9] Surge voltages momentarily overstress the capacitance-to-ground of cables and equipment, forcing the circulation of leakage current to earth for a few microseconds. RCDs may, then, react and initiate the parting of the breaker's contacts, which can take few milliseconds. As a result, the supply will be disconnected when the overvoltage is already expired.

RCDs may also trip upon starting of three-phase motors. High inrush currents, in fact, may not be perfectly balanced among the phases and therefore cause the tripping of the RCD.

Another cause of nuisance trippings may be the leakage currents inevitably flowing through the insulation of equipment during its normal operations. The issue of high leakage currents in equipment will be discussed in Chap. 15.

2.3.2 Protection Without Automatic Disconnection of Supply

Passive means of protection, that is, not involving disconnection of supply, may be used to prevent the occurrence of hazardous situations in case of failure of the basic insulation. The continuity of the service is, then, preserved, which is particularly important in installations where the loss of energy can be detrimental to safety. Such protective measure is typically used when skilled and instructed persons strictly supervise the installation.

2.3.2.1 Protection by Use of Class II Equipment or Equivalent Insulation

If the basic insulation fails, in order to prevent the appearance of potentials on the exposed parts of electrical items, a supplementary and independent layer of insulation material may be added to safeguard persons against indirect contact (Fig. 2.10).

Basic insulation plus supplementary insulation form a double insulated, or Class II, piece of equipment, which is identified by the symbol in Fig. 2.11.

To reduce the probability of simultaneous failure of the two insulations, manufacturers must install (and test) them in a way that

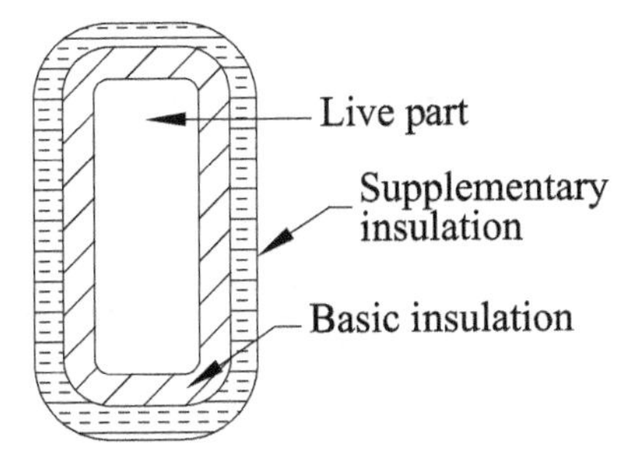

Figure 2.10 Diagrammatic representation of Class II equipment.

allows a degree of independence from the same deteriorating factors. The complete independence between the two layers of insulation is extremely difficult to achieve as they are essentially subject to the same stress factors.

In lieu of two independent layers of insulating material, a single insulating stratum can be applied to live parts as long as the same degree of protection against electric shock as the double insulation is provided. This equivalent measure is defined as *reinforced insulation.*

Enclosures of Class II equipment can be either conductive or insulating.

2.3.2.2 Protection by Nonconducting Locations

This measure is intended to prevent, through the nonconductive nature of the location itself, the exposure to dangerous potential differences between simultaneously accessible parts and between live parts and the earth. The insulation of the location, in fact, avoids, or drastically limits, the circulation of current through a person's body in case of contact with faulty equipment. Thus, nonconducting locations must have insulating floor and walls, characterized by a resistance to ground of at least 50 kΩ, if the nominal voltage of the installation is less than 500 V, and of at least 100 kΩ, if the nominal voltage exceeds 500 V. Also this measure, like Class I and Class II equipment, relies on two layers of protection to ensure safety[10]: basic insulation of equipment and of location.

Equipment in nonconducting locations must not be connected to earthing systems. The connection to ground, in fact, would introduce into the premises a zero potential, thereby defeating the purpose of having a location insulated from ground. Therefore, Class 0 equipment, that is, items with basic insulation and without bonding

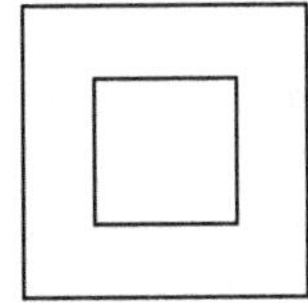

Figure 2.11 Symbol of Class II equipment.

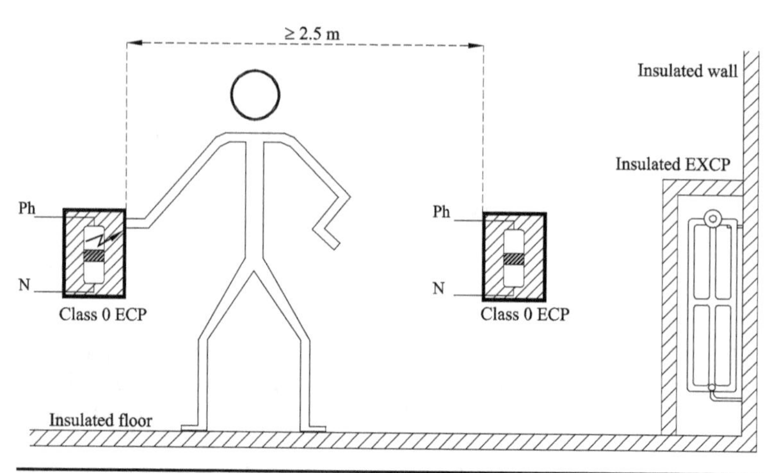

Figure 2.12 Protection by nonconducting location.

terminal, must be used in nonconducting locations. In addition, arrangements to also keep insulated *extraneous-conductive-parts* (EXCPs), such as radiators, must be in place.

Even though the failure of an ECP is not dangerous, a second fault to ground involving a different, and simultaneously accessible, ECP can expose persons to dangerous potential differences. All the ECPs, thus, must be properly separated from each others by spacing not less than 2.5 m. The interposition of obstacles between ECPs that cannot be spaced by 2.5 m is a possible equivalent solution.

In the above conditions, indirect contact caused by breakdown of the basic insulation of equipment is not dangerous, since, within nonconducting locations, persons cannot be exposed to any potential difference (Fig. 2.12).

Because of its delicate nature, protection by nonconducting location is suitable only in installations strictly supervised by qualified persons.[11] The key feature of this protection is, in fact, the absence of any earth reference, which might be introduced into the premises by unaware persons, via portable grounded equipment supplied by extension cords and/or EXCPs.

2.3.2.3 Protection by Earth-Free Local Equipotential Bonding

In earth-free locations, the appearance of dangerous touch voltages is prevented by means of local equipotential bonding conductors. In the case of failure of the basic insulation, in fact, such conductors, by connecting together both simultaneously accessible Class I equipment supplied by different phase conductors and the floor (if conductive), can prevent, or reduce, the appearance of dangerous potential difference in the installation being protected. The equipotential bonding,

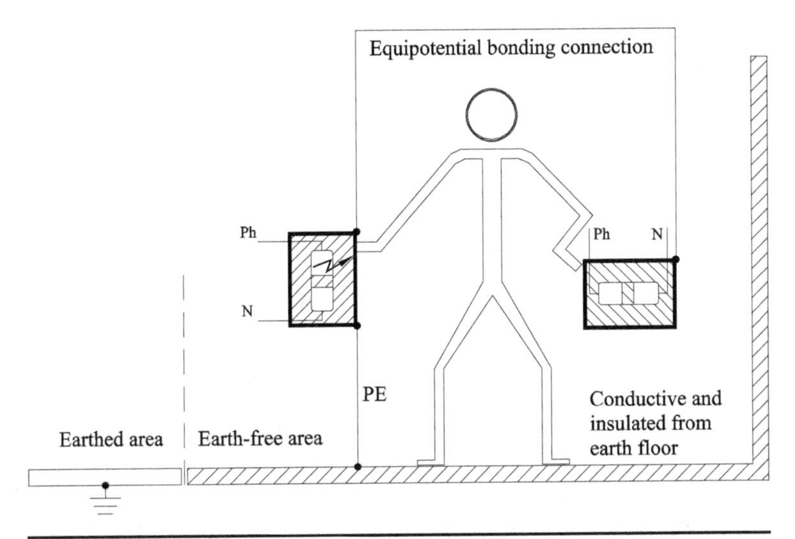

FIGURE 2.13 Earth-free local equipotential bonding between Class I equipment.

in fact, makes the location a Faraday cage, which becomes equipotentially energized upon faults (Fig. 2.13).

In the above conditions, persons cannot undergo any potential gradients, even if different ECPs are within reach.

Similar to nonconducting locations, in the case of faults the supply is not disconnected and the equipment may stay energized for an unknown period of time. This is why the local equipotential bonding must not be linked to the grounding system of the building, which may be present as a protective measure in other areas protected by disconnection of supply. Such a link, in fact, could energize the grounding system and thereby transfer unresolved faults that occurred in the earth-free area to grounded ECPs elsewhere in the same structure.

A major hazard of this protection, suitable only in strictly supervised installations, is at the interface with adjacent rooms, whose floor is earthed. A person standing over both floors at the same time is exposed to potential differences upon faults in the earth-free location. A solution to this hazard is the interposition between the two locations of a sufficiently wide insulated floor section.

2.3.2.4 Protection by Electrical Separation

As per IEC, electrically separated systems use isolating transformers, with the same value of primary and secondary voltages not exceeding 500 V. Such transformers isolate persons from ground, and from other circuits, thereby preventing the circulation of earth currents upon faults. In this arrangement, the ECPs must not be grounded.

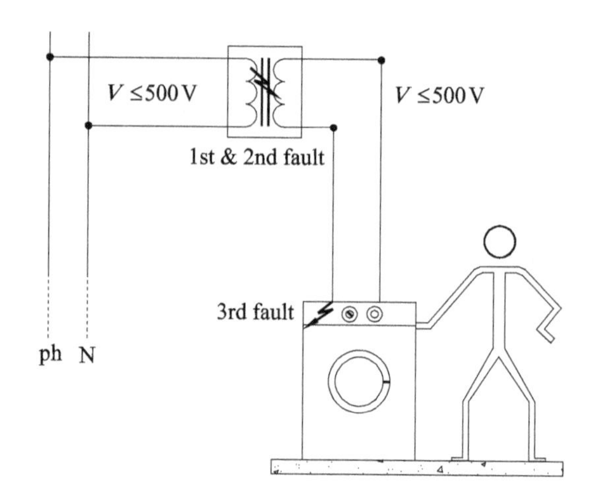

FIGURE 2.14 Person in contact with the primary voltage in separated systems.

Between the primary and secondary windings of an isolating transformer is a double or reinforced insulation or a grounded metallic screen or sheath. The double insulation lowers the probability of persons coming into contact with the primary voltage, which is dangerous as it is ground-referenced (Fig. 2.14).

Persons are exposed to the danger of indirect contact if three concurrent faults, not necessarily simultaneous, occur: failure of the first layer of the double insulation, failure of the second layer of the double insulation, and failure of the basic insulation of the ECP. Thus, three levels of protection must fail to determine a hazardous situation.

However, according to international standards,[12] the isolating transformer, with the double insulation, is no longer required, and an ordinary transformer, that is, with only basic insulation between the windings, may be used in separated systems. With an ordinary transformer, persons are in danger if two concurrent faults take place: failure of the basic insulations of the transformer and of the appliance. This solution is consistent with the other standard protective measures against indirect contact, which, as already said, are characterized by two "layers" of protection.

As anticipated, owing to the high impedance to ground of the separated system, in the case of failure of the basic insulation of an ECP, the ground current cannot flow. Thus, even if the ECP is energized, persons are not in danger, as the fault-loop cannot be established through the earth.

In reality, no electrical system is truly isolated from ground, even when there is no intentional earthing connection of the source. Every circuit is, in fact, "coupled" to earth through a "virtual" capacitor, whose two armatures are the circuit wires and the actual earth; the dielectric is the means interposed between the armatures (e.g., the air).

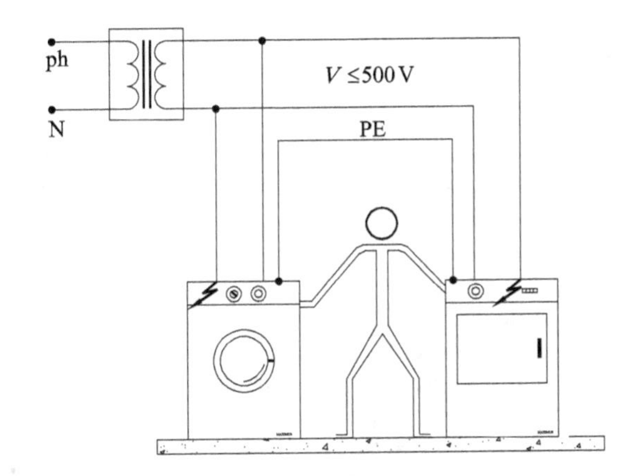

Figure 2.15 Double ground fault in electrically separated systems.

This virtual capacitor has a capacitance proportional to the length of the cables (i.e., the armatures) and its presence can introduce an unwanted connection to earth in separated systems. In order to safely limit the capacitive currents to ground, international standards recommend that the product of the nominal voltage of the separated circuit (in volts) and its length (in meters) should not exceed 10^5 V·m, and the length of the wiring system should not exceed 500 m. These two conditions actually define the electrically separated systems and, thereby, the number of transformers necessary to fulfill it.[13]

As shown in Fig. 2.15, a hazardous situation can be determined by a double ground fault involving simultaneously both poles of the separated system, when a single transformer supplies more than one piece of equipment.

Persons, then, can be exposed to a dangerous potential difference while in simultaneous contact with two faulted ECPs. In these conditions, the person would close the fault-loop, allowing circulation of current through his/her body. This hazardous condition can be avoided by means of nonearthed equipotential bonding conductors connecting together ECPs of the same separated circuit. The equipotential connection, while cancelling, or drastically reducing, the potential difference between the enclosures, "converts" the double ground fault into a short circuit, which can promptly be cleared by overcurrent devices.

It is important to note that isolating transformers cannot be considered, per se, an effective protection against direct contact, but must be coupled with the basic insulation of components (e.g., wires). The absence of the basic insulation, in fact, would expose persons to the risk of touching simultaneously bare parts connected to different poles of the transformer, with lethal consequences.

FAQs

Q. Regarding the IP degree of protection that enclosures must provide, what does it mean that we should consider actual environmental conditions during the normal operations of equipment?

A. In the decision-making process to establish the optimum IP degree of protection, we must not consider improbable events that possibly might occur to the equipment. For instance, an outlet in the backyard of a dwelling unit is legitimately expected to be subject not only to rain but also to water splashes from any direction; therefore, the right rating of its enclosure is IPX4 and not IPX3. On the other hand, an outlet in the living room of the same dwelling unit will not be subject to rain, or splashes, during its usual and normal operations. We must not consider, therefore, the presence, for example, of possible floods in the house, and accordingly overrate the receptacle.

Q. What is the difference between functional and basic insulations?

A. The functional insulation allows the functioning of the equipment by insulating parts at different potentials, whereas the basic insulation protects against direct and indirect contact. The two insulations, therefore, are accordingly tested with different methods, as they must satisfy different requirements.

Q. What is the difference between direct and indirect contact?

A. Indirect contact occurs through metal enclosures, which are energized due to the failure of the basic insulation of live parts inside it. Direct contact occurs by "directly" touching live parts, which were erroneously deemed harmless, for example, during maintenance of equipment.

Q. Is protection against direct contact by obstacles really safe?

A. Protection against direct contact of open-type equipment by obstacles is actually realized by two "layers" of protection: The first one is the distance from live parts, as "marked" by obstacles, which prevent accidental contacts, and the second one is the technical competence of qualified persons interacting with the open-type equipment.

In essence, obstacles, combined with technical skills, provide a degree of protection against direct contact, but they are not supposed to prevent intentional contact with energized parts. Under the above conditions, protection against direct contact by obstacles can be considered safe.

Endnotes

1. See, for reference, IEC 60364-4-41: 2005, "*Low-Voltage Electrical Installations, Part 4-41: Protection for Safety—Protection Against Electric Shock*," 5th ed.
2. IEC 60529: 2001, "*Degrees of Protection Provided by Enclosures (IP Code).*"
3. The test is carried out with the aid of a push-type dynamometer.
4. IEC 62262, "*Degrees of Protection Provided by Enclosures for Electrical Equipment Against External Mechanical Impacts (IK code)*," 2002-02-12.

5. The residual current is also referred to as *zero-sequence current*.
6. Electrical energy is mostly transported and supplied by using a so-called three-phase AC system. In normal situations, voltages, or currents, have a relative phase angle of 120°.
7. As per IEC 61008-1: 1996-12, + A1:2002, + A2:2006, "*Residual Current Operated Circuit-Breakers Without Integral Overcurrent Protection for Household and Similar Uses (RCCBs).*"
8. The magnitude of $\underline{I}_G$ depends on the grounding system adopted, as its actual path back to the source may, or may not, include the earth. This analysis will be performed in the next chapters.
9. See Chap. 12.
10. The "layers" of protection of Class I equipment are constituted by the basic insulation and the disconnection of supply, as better explained in Chap. 3.
11. By the term *qualified* we intend persons responsible, capable, and trained to perform electrical tasks.
12. IEC 61558-1: 2005-09, "*Safety of Power Transformers, Power Supplies, Reactors and Similar Products—Part 1: General Requirements and Tests.*"
13. In case the system does not comply with the above two conditions, the system becomes an *IT* system, as explained in Chap. 9.

CHAPTER 3

Mathematical Principles of Electrical Safety

Do not worry about your difficulties in Mathematics.
I can assure you mine are still greater.
ALBERT EINSTEIN (1879–1955)

3.1 Introduction

To prevent damage to persons, electrical equipment are manufactured with "built-in" protective features (e.g., basic insulation, double insulation, bonding provisions, etc.), and after their installation, more standard protective measures (PMs) against direct/indirect contact may be added as per the electrical design (e.g., protective devices such as circuit breakers, residual current devices, etc.).

This practice lowers the risk of electric shock below a specific threshold considered acceptable by codes and standards.

Protective measures, like any other manufactured item, can fail, though, and the electrical equipment may expose persons to the risk of electric shock. As the failure of the PMs, like any other item, can be statistically predicted, in this chapter we will examine the possibility to quantify such a risk.

3.2 Mathematical Definition of Safety

The electrical safety of a piece of equipment against the appearance of dangerous voltages on its enclosure is a parameter that can be thought

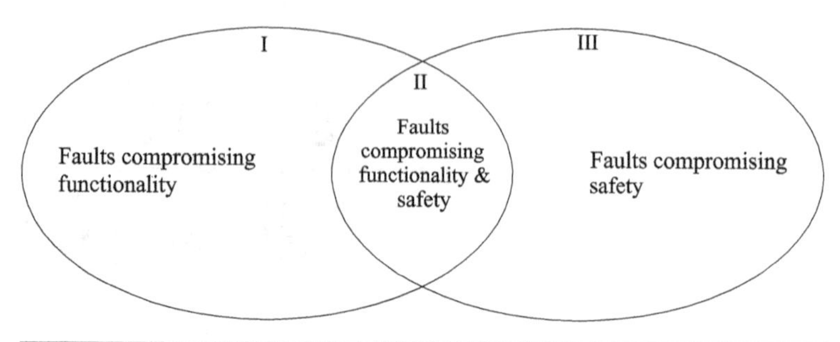

FIGURE 3.1 Faults compromising safety.

of as a function of time. We can define safety at time t as the probability (i.e., quantity comprised between 0 and 1) that the item will not cause a dangerous voltage exposure as a consequence of faults. Electrical safety of an ECP must be referred to as the absence of "superficial" dangerous potentials on its enclosure and must not be confused with its functionality. Some faults, in fact, may compromise safety against electric shock but not the operation of the equipment, which may keep working. This concept is shown in the Venn diagram in Fig. 3.1.

Faults falling in set III, but not in set II, create the most hazardous situation, as the lack of safety is not revealed by the loss of functionality of the equipment.

In formulas:

$$S(t) = \frac{N - F(t)}{N} \tag{3.1}$$

where N denotes the total number of identical items, while $F(t)$ is the number of equipment among N, whose enclosure became "hot" after the time t. The numerator of Eq. (3.1) represents the number of "safe" items against electric shock after the cumulative time t during which items have been functioning. As the exposure time t to risk progresses, the number of items becoming "live" will increase and safety asymptotically will approach zero. Hypothetically speaking, after infinite time, electrical accident will surely happen, as the basic insulation as well as other deployed PMs will no longer carry out their protective functions because of their inevitable aging. On the other hand, safety is at its maximum value (i.e., unity) when either the item is not energized or its failure cannot cause any hazardous situations (e.g., the item functions at extremely low voltages[1]).

We can link safety to the failure rate of the single PM deployed on an item (e.g., Class I equipment). If the PM malfunctions (e.g., the basic insulation fails), the system being protected becomes unsafe. The reliability of the protective measure equates to the safety against indirect contact of the entire equipment.

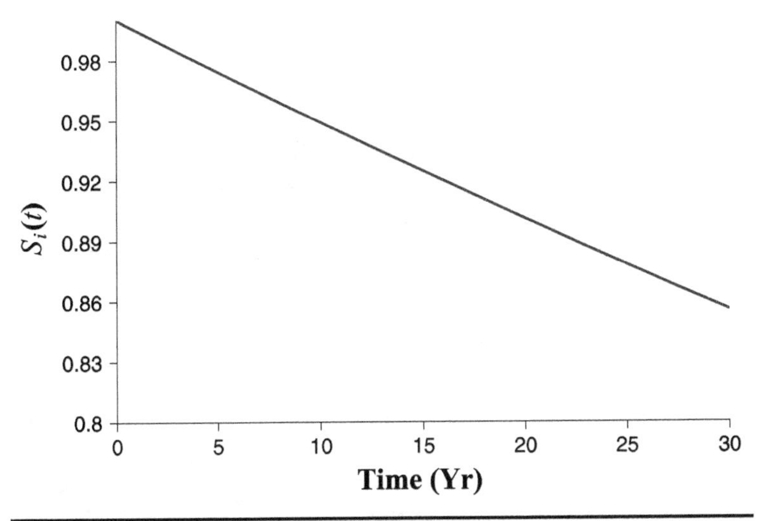

FIGURE 3.2 Safety offered by a circuit breaker.

We can then use the *negative exponential distribution* to quantify the safety of the generic ith protective measure as follows:

$$S_i(t) = \frac{N - F(t)}{N} = e^{-\lambda_i t} \tag{3.2}$$

where λ_i represents the failure rate, defined as the mean number of failures per unit-time, for example, years, of the ith protective measure (e.g., the failure rate for a circuit breaker[2] is 0.0052 failure per year). We will, herein, assume a constant failure rate, that is, the failure associated with the steady-state period of the life of the protective component. The failures, therefore, will be considered as due to random causes and not due to infant mortality or deterioration caused by the age of the PM.

Safety as offered by a circuit breaker to a piece of equipment is shown in Fig. 3.2.

If n protective measures are simultaneously deployed, we need to superpose their protective effects in order to calculate the total safety achieved by the item against electric shock. If all the n measures must simultaneously operate to ensure safety, the protection system is defined as "serial." If, on the contrary, all PMs must fail in order for safety to fail, the system is defined as *parallel* or *redundant*.

Safety for serial and parallel systems can be, respectively, evaluated through Eqs. (3.3) and (3.4):

$$S_S(t) = \prod_1^n S_i(t) = S_1(t) \ldots S_{n-1}(t) S_n(t) \tag{3.3}$$

$$S_P(t) = 1 - \prod_1^n [1 - S_i(t)] = 1 - [1 - S_1(t)] \ldots [1 - S_n(t)] \tag{3.4}$$

where the subscript i indicates the ith PM and n indicates the total number of PMs.

3.3 Risk of Indirect and Direct Contact

How can we evaluate the risk a person is subject to when interacting with electrical items? Basically, three simultaneous adverse events must occur to expose a person to damage (i.e., physical injuries, death) caused by indirect contact:

1. A surface potential must appear on the equipment enclosure
2. Person must touch the enclosure
3. The surface potential's magnitude must exceed the safe limits

The above conditions are tied together by a logic "AND" and so removing just one of them makes the hazard disappear.

The logic AND corresponds to the algebraic multiplication sign; hence, the probability that the above-described events occur, and therefore the magnitude of the residual risk $r(t)$ at any given time after PMs have been applied can be evaluated by Eq. (3.5).

$$r(t) = [1 - S(t)]k(t)v(t) \tag{3.5}$$

where $[1-S(t)]$ is the probability that the enclosure is energized due to an internal fault, which "perforates" the basic insulation. This term is also referred to as *insecurity*.

$k(t)$ is the probability that a person touches the faulted enclosure. For example, hand-held devices can perform their function only if held, and therefore, $k(t) = 1$; the same applies to restrictive locations (e.g., metal tanks), where workers may be in permanent bodily contact with electrical equipment due to limited freedom of movement.[3] On the contrary, appliances in ordinary locations (e.g., dishwashers) can operate even in the absence of persons, ergo the probability that they are touched during a fault is very low, and $k(t)$ is well below 1.

$v(t)$ is the probability that the touch voltage exceeds the dangerous values and/or the maximum durations established by technical standards in reference to a person's body resistance. $v(t)$ depends on the magnitude of the fault potential, which may reach the same value as the system nominal voltage, if the grounding/bonding is not effective or is missing.

To reduce $r(t)$, at least one of the factors in Eq. (3.5) must be kept as close to zero as possible. When $k(t) < 1$, the preferred approach is to lower $v(t)$ by limiting the time the fault potential persists on the enclosure by prompt automatic disconnection of supply. If $k(t) = 1$, Class II equipment, which lowers the factor $[1-S(t)]$, may be the best choice.

For a practical understanding of the previous definition of risk, let us consider the following example:

A simultaneous failure of the supports of a bridge, during rush hour, can cause significant damage to persons (i.e., loss of human life; $v(t)$ is high). In addition, the probability of commuters transiting over the bridge is high (i.e., $k(t)$ is high). On the other hand, however, the probability that the bridge collapses should be remote (i.e., $[1-S(t)]$ is very low), thanks to the redundancy in the bridge's supports. Ergo, the resulting risk is low, although not zero, and deemed acceptable.

To quantify the residual risk $r(t)$ for direct contact, we can still apply Eq. (3.5). In this case, $v(t)$ has the same value as in indirect contact because its value depends on the maximum permissible voltage, which is common for both cases; also $k(t)$ has equal value, as live parts, erroneously considered harmless, have the same probability to be touched just as an ECP.

The major difference in the two expressions of the residual risk is the value of the insecurity $[1-S(t)]$: in the case of direct contact, the probability that the part is energized equals, of course, 100%, whereas in indirect contact such probability is much lower because of the protective measures. As a result, in correspondence of the very same maximum permissible voltage, the residual risk for direct contact is greater than the residual risk for indirect contact.

3.4 The Acceptable Residual Risk

In reality, the risk against electric shock can be reduced, but not completely eliminated, if not at unsustainable expenses. For example, a protective device can fail, but in general we do not, nor are we required to, install multiple identical devices in series as a redundant protection[4] because this practice would be cost-prohibitive. Thus, the residual risk must be "acceptable" with regard to electric shocks as a compromise between achievable safety and its cost. What is the acceptable risk then?

The residual risk is deemed acceptable after the application of standard protective measures, if its probabilistic value calculated in Eq. (3.5) falls below an arbitrary threshold as basically established by:

- Up-to-date applicable technical standards and codes, indicating minimum safety requirements
- *Authorities Having Jurisdiction's* dictates, which may provide technical and "legal" interpretation of the aforementioned minimum safety requirements
- Economic resources available to increase safety beyond the minimum aforementioned requirements

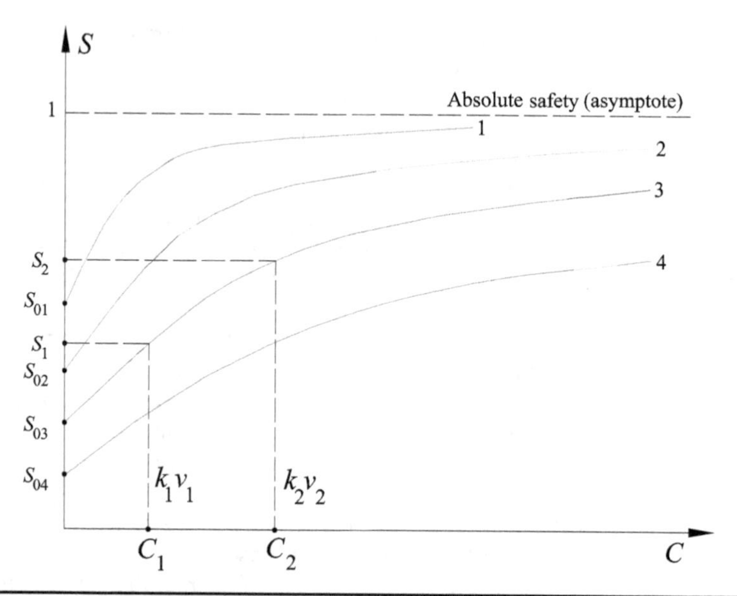

FIGURE 3.3 Safety–cost curves.

Without any doubt we could lower the residual risk by increasing the financial investment in protective measures. Safety would accordingly increase as a function of the cost C of the PMs, following the safety–cost curves diagrammatically shown in Fig. 3.3, relative to different pieces of equipment (i.e., equipment 1 to 4).

It can be noticed that if we do not apply any protective measure to the electrical item (i.e., $C = 0$), the resulting safety against electric shock is at its minimum value S_0, which may be unacceptable (e.g., S_{04} for the item of curve 4 in Fig. 3.3). The safety–cost curves asymptotically approach the absolute safety, equal to 1, when the cost of protective measures approaches infinity. In reality, the cost of safety must be a finite value, and criteria are necessary to determine the optimum investment to realize it.

One criterion consists of determining the maximum cost C_1 for the PMs that we are willing to tolerate to face the value $k_1 v_1$ characteristic of the equipment being analyzed. We can then find on the curve the resultant value of S_1 that the item has achieved (e.g., Fig. 3.3; vertical line to curve 3). If the magnitude of S_1 satisfactorily lowers $r(t)$, we deem the equipment safe.

It is important to note that the product $k_1 v_1$ for a piece of equipment may increase with time to the value $k_2 v_2 > k_1 v_1$, for example, if we move it from ordinary locations to restrictive locations. In that case, a superior cost C_2 may be necessary to compensate the larger value $k_2 v_2$ through a higher S_2 (Fig. 3.3).

Alternatively, one might establish the desired level of safety S_1 and accordingly determine the relative cost C_1. If the magnitude of the cost C_1 falls within the allocated budget, we achieve the desired safety.

With the above methodologies, results may be unacceptable to the designer: in the first case, S_1 may be too low and one must increase the cost; in the latter case, the cost C_1 may exceed the available budget, which forces the designer to lower the desired value S_1. In other cases, it can be realized that small increases in cost can remarkably raise safety and, vice versa, minimum decreases in safety may allow substantial reductions in costs.

As shown in Fig. 3.3, the safety curves almost saturate when a certain cost is exceeded and show almost negligible improvements even if C is very much increased. Thus, a different approach consisting of evaluating the resulting increments in safety ΔS_i from successive unit increments in cost (i.e., $\Delta C = 1$) can be carried out. The values ΔS_i are then compared with an acceptable minimum value ΔS_0, which is established as a function of the product kv. ΔS_0 represents the value beyond which investing in lowering safety is deemed no longer cost-effective (Fig. 3.4). In this case, in fact, each additional unit of cost yields less and less additional safety, or, conversely, obtaining one more unit of safety costs more and more.

The optimum cost for safety in Fig. 3.4 is C_1. Should kv increase, in order to lower the risk, we must accept higher costs for the PMs (e.g.,

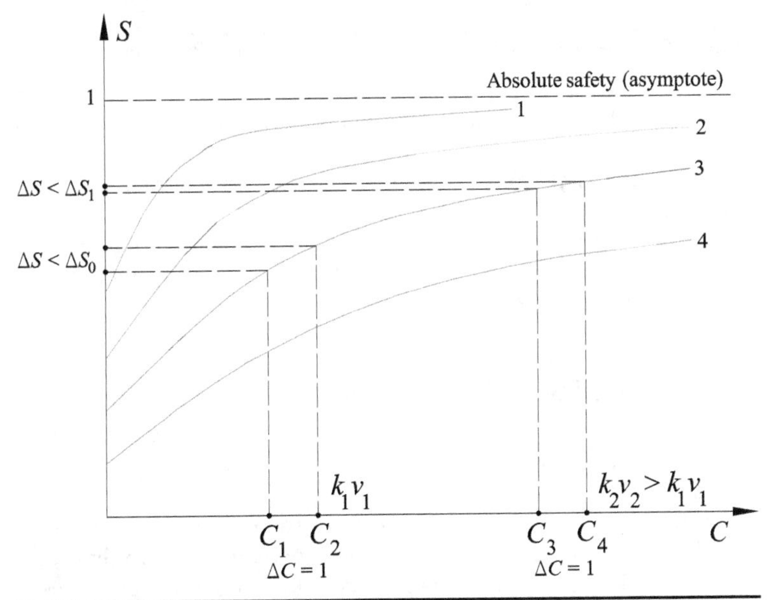

FIGURE 3.4 Increment in safety caused by unitary increment of cost.

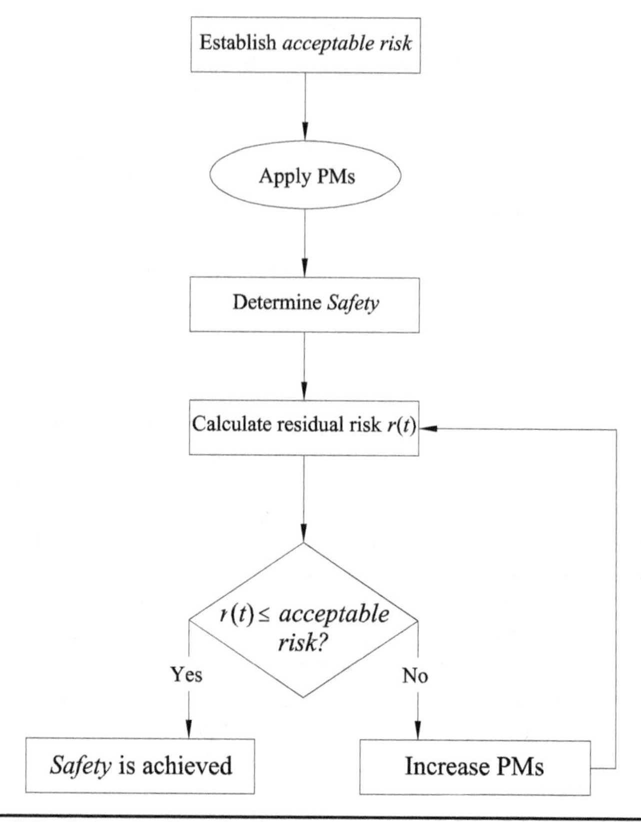

FIGURE 3.5 Determination of safety as the result of an iterative process.

C_3 in Fig. 3.4) and lower acceptable values for ΔS_0 (i.e., $\Delta S_1 < \Delta S_0$ in Fig. 3.4).

Achieving safety can be thought as the result of an iterative process as described in the flow chart of Fig. 3.5.

In following sections, we will examine the standard protective measures, already introduced in Chap. 2, by analyzing their safety and risk against indirect contact.

3.5 Safety and Risk of Basic Insulation

Let us calculate safety, and the resulting risk, of an electrical item with no conductive enclosure, whose live parts have basic insulation (e.g., cables in air) (Fig. 3.6).

In the absence of a conductive enclosure, persons can only be exposed to direct contact, should the basic insulation fail. In that case, electric shock occurs only if the person touches the point of the insulation's failure, which exposes the live part.

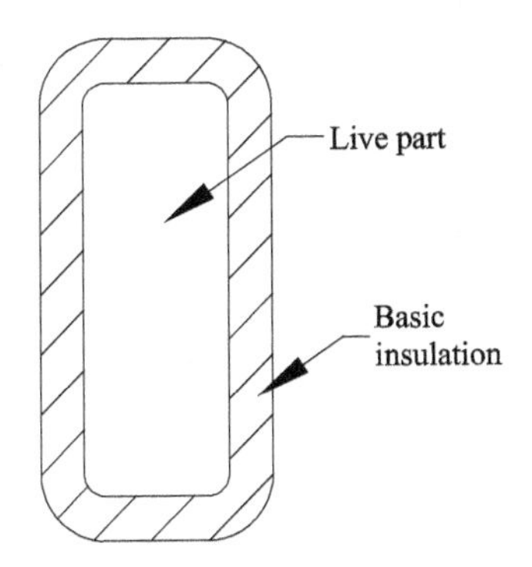

Figure 3.6 Electrical item with no enclosure and basic insulation.

The safety of an item, at any time, as offered by the *basic insulation*, is estimated as shown in Eq. (3.6):

$$S_{BI}(t) = e^{-\lambda_{BI} t} \tag{3.6}$$

where λ_{BI} is the failure rate of the basic insulation. The related risk is indicated as $r_{BI}(t)$.

3.6 Safety and Risk of Class 0 Equipment

Let us consider a Class 0 piece of equipment, that is, an electrical item with basic insulation in an enclosure without a bonding terminal.[5]

If the basic insulation fails and persons touch the faulted enclosure, indirect contact would occur. The safety S_0 of this configuration coincides with the previous one examined in Sec. 3.4. In fact, the failure rate of the basic insulation does not change, and therefore

$$S_0(t) = S_{BI}(t) \tag{3.7}$$

The fault potential, though, appears over the whole metal enclosure, increasing the probability that persons are subject to a touch potential. This causes $k_0(t) > k_{BI}(t)$. Hence, the risk of electric shock caused by indirect contacts becomes greater by adding the enclosure, even though safety is the same.

In formulas:

$$r_0(t) > r_{BI}(t) \tag{3.8}$$

In ordinary locations, the risk $r_0(t)$ is not considered acceptable by any standards or codes, which require Class I equipment (i.e., outfitted with bonding terminals) to be used.[6]

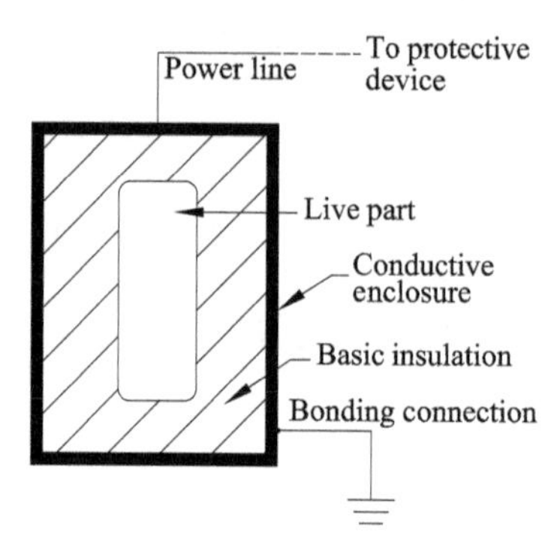

FIGURE 3.7 Class I equipment in conjunction with protective device.

3.7 Safety and Risk of Class I Equipment

Class I equipment must be used in conjunction with overcurrent and/or residual protective devices, which allow the prompt disconnection of supply upon faults (Fig. 3.7).

It is, in fact, an international requirement that at least two levels of protection be present against indirect contact. In our case, the basic insulation is the first one and the bonding of the enclosure, in conjunction with the protective device, is the second.

In order for a person touching the ECP to be shocked, three events must occur: failures of the basic insulation, failure of the bonding/grounding connection,[7] and failure of the protective device. The equipment, therefore, is protected by a "redundant" system because even though the basic insulation fails, the protective device, due to the bonding connection, can sense the fault current and clear it. Vice versa, the failure of the protective device, and/or the bonding/grounding connection, does not immediately expose persons to live potentials, in the presence of a sound basic insulation.

Let S_{BGC} and S_{PD}, respectively, be safety of the bonding/grounding connection and of the protective device. The serial safety of this combined protective measure is as per Eq. (3.9):

$$S_{BGCPD}(t) = S_{BGC}(t)S_{PD}(t) = e^{-(\lambda_{BGC}+\lambda_{PD})t} \tag{3.9}$$

We have assumed λ_{BGC} and λ_{PD}, respectively, as the failure rates of the bonding/grounding connection and of the protective device.

The related total safety S_I, as offered by Class I equipment in Fig. 3.7, is expressed in Eq. (3.10), in light of Eq. (3.4).

$$\begin{aligned} S_I(t) &= 1 - [1 - S_{BI}(t)][1 - S_{BGCPD}(t)] \\ &= e^{-\lambda_{BI}t} + e^{-\lambda_{BGCPD}t} - e^{-(\lambda_{BI}+\lambda_{BGCPD})t} \end{aligned} \tag{3.10}$$

A correct comparison between S_I and S_{BI} can be performed only if we take into consideration the possibility that Class I equipment's

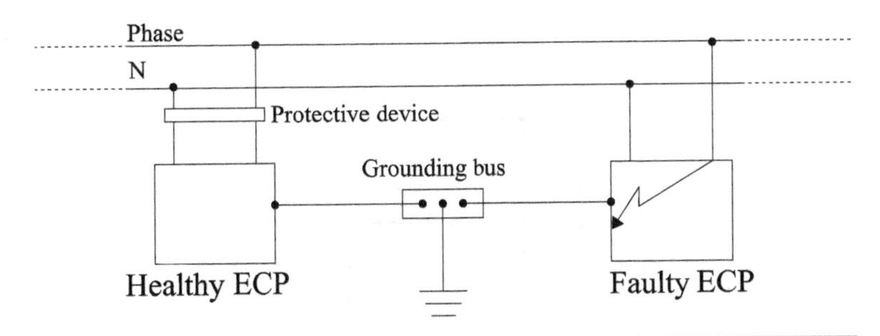

Figure 3.8 The fault potential is transferred from faulty ECP to healthy ECP.

bonding connection can energize its enclosure independently of any fault within it. In fact, ground faults occurring within other ECPs, which are bonded to the same grounding system, can cause potentials to be transferred to healthy equipment (Fig. 3.8).

This fault condition, which the healthy ECP's protective device is unable to clear, decreases $S_I(t)$ by a factor $F_{TP}(t)$, defined as the probability that the bonding connection energizes the healthy enclosure due to transferred voltages.

In formulas:

$$S_{ITOT}(t) = S_I(t) - F_{TP}(t) \tag{3.11}$$

Consequently, Class I equipment is "safer" than an electrical item equipped with only basic insulation, when

$$S_{ITOT}(t) > S_{BI}(t) \tag{3.12}$$

However, even by assuming true the inequality (3.12), the residual risk $r_{ITOT}(t)$ of Class I equipment is not necessarily less than $r_{BI}(t)$. As previously explained, in fact, the risk also depends on the probability that persons will be in contact with the fault potential and the metal enclosure of the Class I item elevates such a risk. Therefore, even though $S_{ITOT}(t) > S_{BI}(t)$, it is not automatically true that $r_{ITOT}(t) < r_{BI}(t)$.

3.8 Safety and Risk of Class II Equipment

Class II equipment is diagrammatically shown in Fig. 2.11. These items generally have no conductive enclosure (e.g., cables, drills, hairdryers, etc.) and therefore electric shock can be caused only by direct contact.

Safety $S_{II}(t)$ is given by Eq. (3.13), where λ_{SI} is the failure rate of the supplementary insulation. This is a parallel system and thus we need to take into account both the contribution of basic insulation (S_{BI}) and

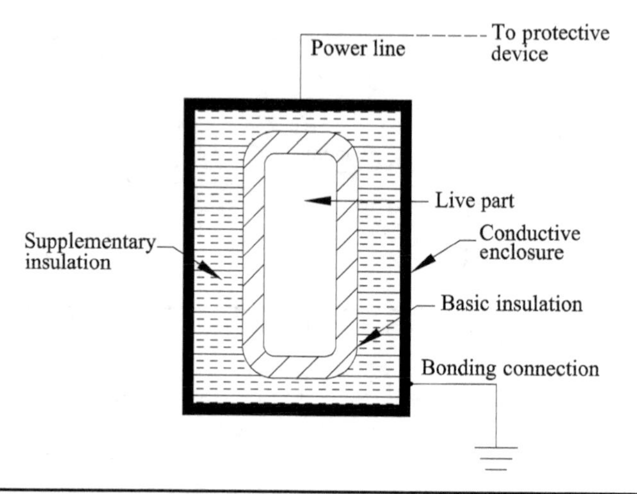

Figure 3.9 Diagrammatic representation of Class II equipment in bonded conductive enclosure.

supplementary insulation (S_{SI}). Eq. (3.3) applies:

$$S_{II}(t) = 1 - [1 - S_{BI}(t)][1 - S_{SI}(t)] = e^{-\lambda_{BI}t} + e^{-\lambda_{SI}t} - e^{-(\lambda_{BI}+\lambda_{SI})t} \quad (3.13)$$

Let us compare S_{II} and S_{BI}. Since $\lambda_{SI} < \lambda_{BGCPD}$, we obtain:

$$S_{II}(t) > S_I(t) \quad (3.14)$$

In addition, in the absence of a conductive enclosure, the probability that persons can touch a fault potential caused by the failure of both insulation layers is much lower than in the case of Class I equipment (i.e., $k_{II} < k_I$), hence, $r_{II}(t) < r_I(t)$.

What would happen if Class II equipment were installed in a metal enclosure that is bonded (Fig. 3.9)?

We can reasonably assume that the enclosure is more likely to be energized due to voltages transferred by the bonding connection than due to failure of its own double insulation. In fact, the probability of failure of Class II equipment is considered very low when compared to the probability $F_{TP}(t)$ that the enclosure becomes live due to transferred voltages. Thus, international standards prohibit the bonding of Class II equipment.[8]

We can express safety of Class II equipment in bonded enclosure $S_{IIBE}(t)$ as

$$S_{IIBE}(t) = 1 - [1 - S_I(t)][1 - S_{SI}(t)] - F_{TP}(t) \quad (3.15)$$

Therefore, $S_{IIBE}(t) < S_{II}(t)$ and the presence of a metal enclosure causes $k_{IIBE} > k_{II}$ and $r_{IIBE}(t) > r_{II}(t)$.

3.9 Safety and Risk of Electrical Separation

The protective measure by electrical separation has been examined in Chap. 2. In this section, let us assume an ordinary separation transformer (i.e., no double insulation between the primary and secondary windings). Persons are at risk of electric shock if the basic insulations of both the separation transformer and the ECP fail (Fig. 3.10).

Safety $S_{ES}(t)$ is given by Eq. (3.16):

$$S_{ES}(t) = 1 - \left[1 - S_{BI}^{TR}(t)\right]\left[1 - S_{BI}^{ECP}(t)\right] = e^{-\lambda_{BI}^{TR}t} + e^{-\lambda_{BI}^{ECP}t} - e^{-(\lambda_{BI}^{TR}+\lambda_{BI}^{ECP})t} \tag{3.16}$$

where λ_{BI}^{TR} and λ_{BI}^{ECP}, respectively, indicate the failure rate of the basic insulations of the separation transformer and of the appliance.

$S_{ES}(t)$ is of the same magnitude as $S_{II}(t)$ given in Eq. (3.13). However, the risk of touch voltages is greater for the electrical separation because of the presence of the metal enclosure of the ECP, which increases the probability of contact, and therefore $r_{ES}(t) > r_{II}(t)$.

Also for the electrical separation, the bonding of the enclosures is not permitted, because the probability that the transformer and the ECP fail is less than the probability that the bonding connection dangerously energizes the ECP.

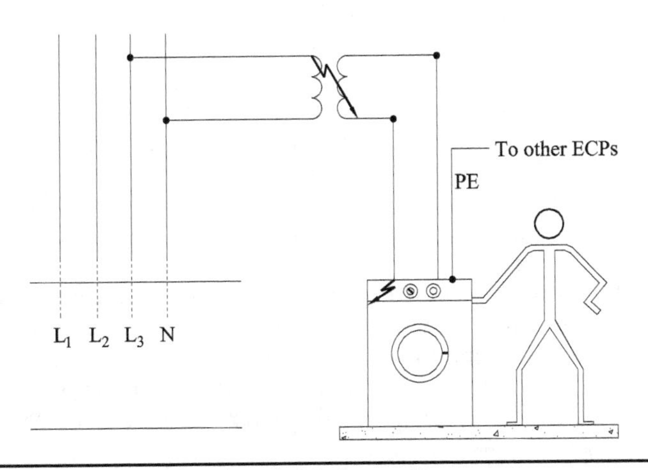

FIGURE 3.10 Risk as related to *electrical separation*.

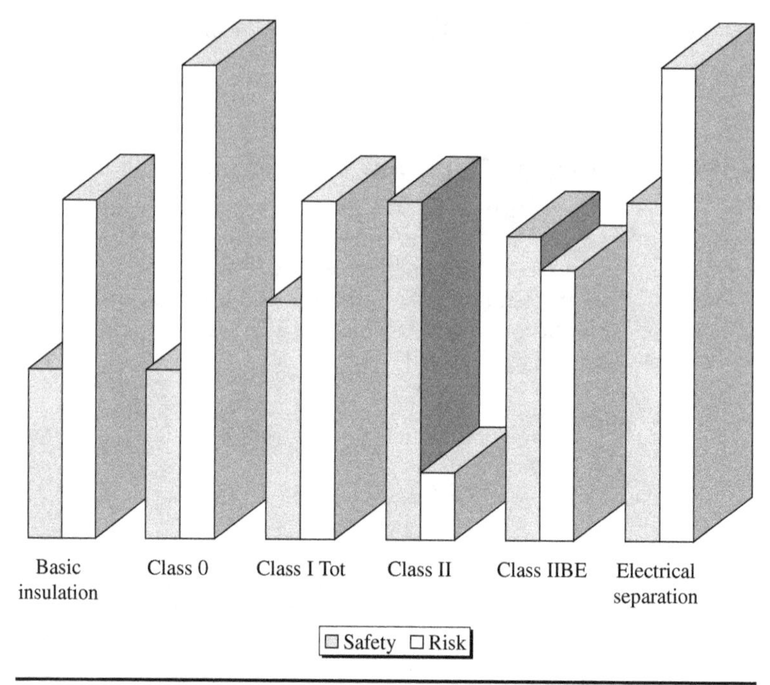

FIGURE 3.11 A qualitative comparison between safety and risk of protective measures.

3.10 A Qualitative Comparison Between Safety and Risk of Protective Measures

Figure 3.11 shows a qualitative comparison of the various PMs examined in this chapter.

As said in Sec. 3.6, we cannot establish if the risk $r_{BI}(t)$ of the protective measure basic insulation is greater, equal to, or less than the risk $r_{ITOT}(t)$ of the protective measure Class I. This is due to the increase in the probability that persons are subject to touch potentials caused by the metal enclosure. Because of this indetermination, $r_{BI}(t)$ and $r_{ITOT}(t)$ are shown of equal size in Fig. 3.11.

It should be noted that the above-calculated parameters are valid only if the manufacturer's instructions are followed during installation, the electrical equipment is maintained in relation to the environment, and we are in standard situations. Equipment in nonordinary locations (e.g., restrictive spaces, fire hazardous facility, etc.) may in fact offer greater risks.

FAQs

Q. If safety cannot ever be absolute (i.e., 100%), can we design safe electrical installations?

A. An electrical installation is deemed safe if the residual risk imposed to persons falls below the acceptable risk. The minimum acceptable risk is established, as a "legal" requirement, by local *Authority Having Jurisdiction (AHJ)*, which can enforce national electrical codes and/or "recommended practice" as provided by national (BS, CEI, DIN, etc.) and international (e.g., IEC, IEEE) standards. Thus, the answer is yes, we can design safe electrical installations within the above assumption.

Q. Can power systems be foolproof?

A. No, they cannot. Erratic human behaviors are difficult to predict and this makes it impossible to design for any possible scenario. In other words, should the electrical installations be safe with regard to untrained people? What amount of "awareness" of the danger of electricity should we assume for persons?

If this amount is assumed too high (e.g., each person is supposed to possess knowledge at electrical engineer level), we would have more permissible design criteria (e.g., live parts might even be exposed, as electrical engineers have a very high awareness of the danger caused by direct contacts). On the other hand, if the assumed level of awareness is set too low, we might not be able to design any installations, as the safety requirements would become too conservative.

In high-voltage substations, we can reasonably expect trained people, unlike in dwelling units, and therefore, we should assume two different kinds of "standard person" in the two realms.

The answer, once again, is to refer to code and standards to understand the amount of "electrical awareness," which is considered acceptable, and design accordingly.

Q. Does the failure of Class II equipment cause direct or indirect contact?

A. Indirect contact is defined as contact with metal parts not normally live. If double insulated items are in insulating enclosures, electric shock from Class II equipment can occur only by direct contact. Should Class II equipment be enclosed in a metal frame, the failure of the double insulation will cause, instead, indirect contact.

Q. If the failure rate of a circuit breaker is 0.0052 failure per year, how many devices, among a population of 100, will "survive" after 1 year?

A. In the discrete sense, the failure rate can be expressed as:

$$\lambda_i = \frac{1}{R_i(t)} \frac{R_i(t) - R_i(t + \Delta t)}{\Delta t} = \frac{1}{100} \frac{100 - R_i(1)}{1}$$

where $R_i(t)$ is the *reliability function*, which represents the number of devices that survived up to time t (100 in our case). Δt equals 1 year. We solve for $R_i(1)$:

$$R_i(1) = 99.48.$$

Q. Is traveling on an airplane more hazardous than riding a motorcycle?

A. The feared/expected damage to persons in case of failure of the airplane systems during a flight is very high, unlike the case of the motorcycle, whose failure will not necessarily cause a fatal outcome.

On the other hand, because the technology employed on an airplane is much more sophisticated than the one used on motorcycles, the probability of an aircraft failure is lower. Ergo, the airplane may be considered safer than the motorcycle.

Q. Would a piece of equipment in which, hypothetically, the basic insulation and the conductive enclosure have been inverted between each other be safer?

A. Hypothetically speaking, Class I equipment with the basic insulation covering the outside of the metal enclosure would not be safer! Eq. (3.10), in fact, does not take into account the mutual positions of the protective measures.

However, the risk offered by this hypothetical piece of equipment would decrease, as the probability of touching the energized enclosure would greatly reduce.

Endnotes

1. See Chap. 10.
2. IEEE Std. 493-2007, "*IEEE Recommended Practice for the Design of Reliable Industrial and Commercial Power Systems.*"
3. See Chap. 15 for further details.
4. Nuclear facilities, aircraft, and spacecraft may be the exception. Such installations typically use a radial distribution and the connecting wires are redundant (i.e., doubled).
5. In installations, a component with the sole basic insulation (e.g., cables in air) may become Class 0 equipment due to deposit of conductive dust or moisture eventually present in the environment, which may act as a metal enclosure.
6. We have already discussed in Chap. 2 that in supervised locations (e.g., nonconducting locations) Class 0 equipment can be used.
7. Bonding/grounding connections mechanically link protective conductors to metal enclosures, ground rods, steel structures, the neutral point of the source, etc.
8. The North American *National Electrical Code* states, instead, that double insulated pieces of equipment are *not required* to be bonded. Therefore, their bonding is permitted.

CHAPTER 4
The Earth

Is it a fact, or have I dreamt it, that by means of electricity, the world of matter has become a great nerve, vibrating thousands of miles in a breathless point of time?

NATHANIEL HAWTHORNE (1804–1864)

4.1 Introduction

Not all electric faults cause circulation of current through the earth. However, regardless of the grounding system adopted, if a fault energizes an *exposed-conductive-part*, a potential difference between its enclosure and the ground will appear. As a result, current will circulate through the body of the person touching the faulty item, as the actual earth becomes an available return path to the supply source.[1] This is because the ground is a conductor, with a definite resistivity, which varies according to its make up.

It is, then, important for the electrical safety of persons to understand the electrical behavior of the earth as a return means for fault currents.

4.2 The Earth Resistance

Currents can be impressed through the earth by means of ground electrodes.

Ground electrodes are conductive parts, which may be embedded in a specific conductive medium, in intimate electric contact with the earth. Their purpose is to guarantee safety by providing an effective access to the earth for fault currents.

When a potential difference is applied between two electrodes, a current field will be inevitably impressed into the soil. We now consider a homogeneous and isotropic[2] soil, and hemispherical

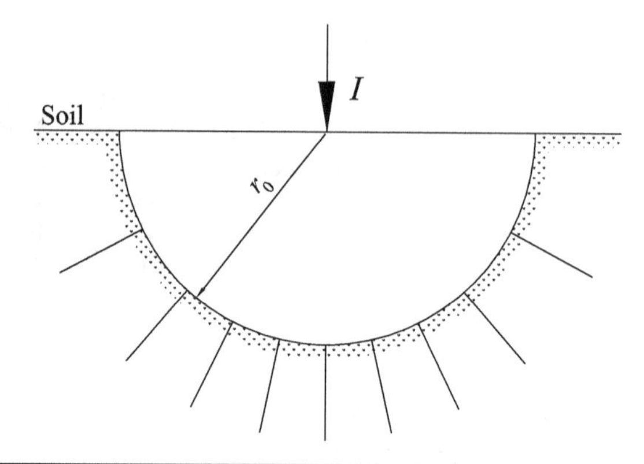

FIGURE 4.1 Current radially leaving a hemispherical electrode.

electrodes, which, due to their physical symmetry, uniformly leak current in any direction of the earth. Such electrode is, indeed, never used in practical applications, but its behavior can predict that of any other differently shaped electrode. In fact, at sufficient distance from any electrode and regardless of the shape, the current can be considered to radially flow, just as it were impressed by a hemispherical electrode.

Let us apply a potential difference between two identical hemispherical electrodes, of radius r_0, one transmitting and the other receiving current, displaced by a sufficiently large distance, in theory infinity. Because of its physical symmetry, the transmitting electrode will radially leak current I to ground, and in the same fashion the remote electrode will receive it.[3] The current leaving the electrode uniformly distributes in any available direction (Fig. 4.1).

The current is limited by the resistance of the soil surrounding the electrode, which can be modeled as the series of hemispherical "shells," each shell of increasingly large radius and infinitesimal thickness $\mathrm{d}x$ (Fig. 4.2).

We can consider the elementary resistance $\mathrm{d}R$ of a generic shell of radius r, and, then, once its expression is known, we can sum up the contribution of all the elementary resistances as they succeed from the electrode surface to the remote earth (in theory, the infinity).

Our assumptions of the uniformity of the soil allow us to apply the same formula as the resistance of a conductor[4]: $R = \rho\,(l/S)$, where ρ is the resistivity of the material ($\Omega \cdot \mathrm{m}$), l the length of the conductor, and S the cross-sectional area the current flows through. In our case, ρ is the earth resistivity, also referred to as specific earth resistance. The resistivity is defined as the resistance of 1 m^3 of earth.

By using this formula, we will obtain the resistance of the generic hemispherical shell[5] of Fig. 4.2, as follows:

$$\mathrm{d}R = \rho \cdot \frac{\mathrm{d}x}{2\pi r^2} \tag{4.1}$$

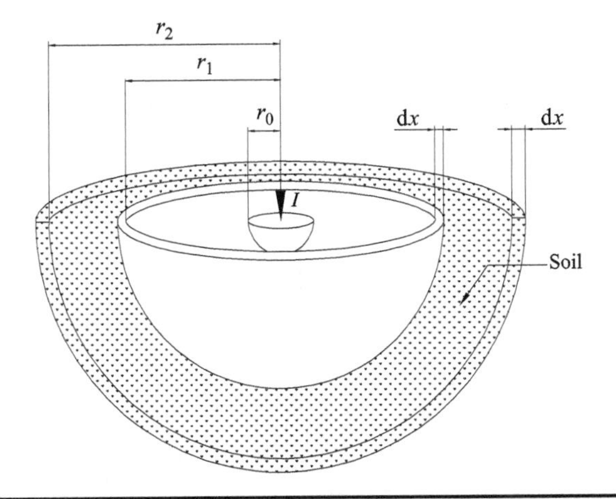

FIGURE 4.2 The earth can be modeled by the series of hemispherical shells.

It appears clear from Eq. (4.1) that the resistance of the generic hemispherical layer of the earth is smaller as its distance r from the electrode increases. This is due to the larger and larger section through which the current will circulate.

The assumption about the infinitesimal thickness dx of the shell is necessary, as it allows us to consider only the inner surface of the shell through which the current will flow. Therefore, the resistance of this ideal shell, with no thickness, is exclusively determined by its inner lateral area surface.

The total resistance R_G of the hemispherical electrode to the current, defined as *earth* (or *ground*) *resistance*, can be obtained by integrating Eq. (4.1) between the electrode's surface and the infinity:

$$R_G = \frac{\rho}{2\pi}\int_{r_0}^{\infty}\frac{dr}{r^2} = \frac{\rho}{2\pi}\left[-\frac{1}{r}\right]_{r_0}^{\infty} = \frac{\rho}{2\pi r_0} \tag{4.2}$$

We have already mentioned that the resistance of the earth becomes smaller and smaller as its distance from the electrode increases. For a better understanding of this concept, let us consider replacing each hemispherical shell of soil, as represented in Fig. 4.2, with a circular layer of equivalent area, and, therefore, equivalent resistance and same thickness dx. The surfaces of the layers increase with the distance r_i from the electrode, just as it does for the lateral area of the hemispherical shells, hence $R_i < R_2 < R_1 < R_0$ (Fig. 4.3).

For example, in correspondence of the distance r_i, the equivalent circumference has area $2\pi r_i^2$. We can determine the radius of these equivalent circumferences from the following expression, as applied

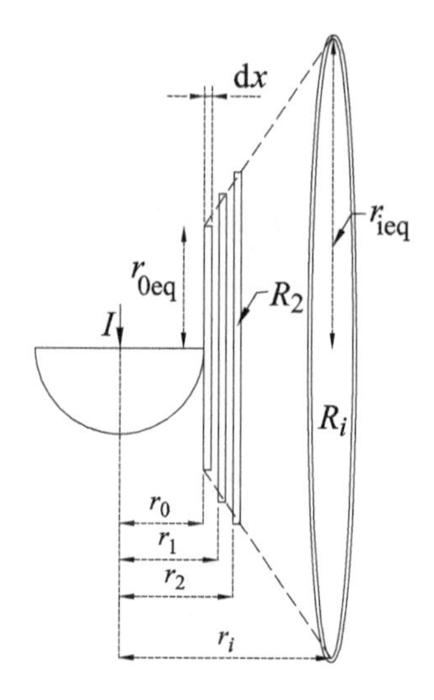

FIGURE 4.3 The hemispherical electrode's earth resistance equivalent conductor.

to the generic ith hemisphere:

$$2\pi r_i^2 = \pi r_{ieq}^2 \Rightarrow r_{ieq} = \sqrt{2} r_i \tag{4.3}$$

where the left-hand side of Eq. (4.3) is the lateral area of the hemisphere and the right-hand side is the area of the equivalent circumference.

If we assemble together the circular layers as in Fig. 4.3, we obtain a frustum of a cone, whose initial and final radii are, respectively, $\sqrt{2}r_0$ and $\sqrt{2}r_i$. We can, thus, consider the resistance of the soil comprised between the electrode and the distance r_i as the resistance of an equivalent conductor, of similar resistivity, shaped as a frustum of a cone (Fig. 4.3).

It can be noted that the earth resistance occurs in higher concentration around the electrode itself. In fact, due to the increase of the cross section of the frustum of a cone, the contribution of the subsequent layers is smaller and smaller.

To demonstrate mathematically, let us calculate, for instance, the resistance of the soil between the hemispherical electrode's surface and the distance $2r_0$:

$$\begin{aligned} R|_{r_0}^{2r_0} &= \frac{\rho}{2\pi}\int_{r_0}^{2r_0}\frac{dr}{r^2} = \frac{\rho}{2\pi}\left[-\frac{1}{r}\right]_{r_0}^{2r_0} = \frac{\rho}{2\pi}\left(-\frac{1}{2r_0}+\frac{1}{r_0}\right) \\ &= \frac{\rho}{2\pi}\frac{1}{2r_0} = \frac{1}{2}R_T \end{aligned} \tag{4.4}$$

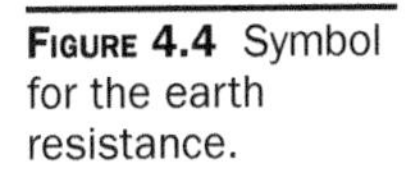
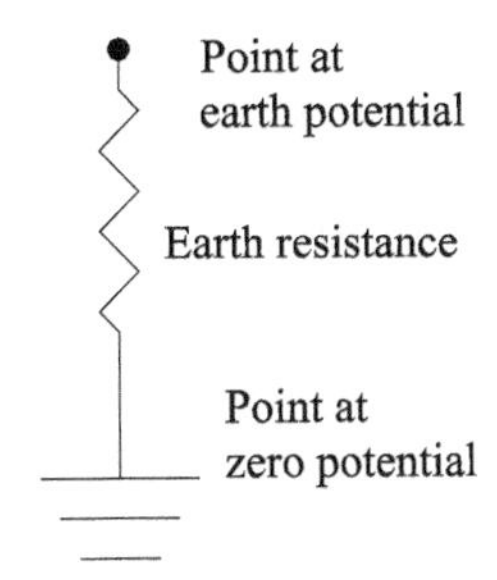

FIGURE 4.4 Symbol for the earth resistance.

It is apparent that 50% of the total earth resistance is concentrated in a hemispherical volume of soil of radius $2r_0$. This result has a general validity, regardless of the shape of the electrode.

A good connection to ground may be successfully achieved by replacing the aforementioned volume of dirt of radius $2r_0$ with earth enhanced with special substances with low resistivity (i.e., not exceeding $0.12\ \Omega \cdot \mathrm{m}$). The same result can be achieved by using, for example, the concrete-encased rods of a building's foundation footings. The concrete, in fact, absorbs and retains moisture better than the actual earth.

The symbol for the earth resistance is given in Fig. 4.4.

4.3 The Earth Potential

The ground current will raise the electric potential of each point of the earth, with respect to a remote point (i.e., infinity) conventionally assumed as zero potential. In this assumption, the potential of a generic point P located at the distance r from the electrode, as caused by the ground current I, is

$$V_{r-\infty} = R_G I = \frac{\rho I}{2\pi}\int_r^{\infty}\frac{1}{r^2}\,\mathrm{d}r = \frac{\rho I}{2\pi}\left[-\frac{1}{r}\right]_r^{\infty} = \frac{\rho I}{2\pi}\left(-\frac{1}{\infty}+\frac{1}{r}\right)$$

$$= \frac{\rho}{2\pi}\frac{I}{r}, \quad \text{for } r \geq r_0 \tag{4.5}$$

$$V_{r-\infty} = V_G = \frac{\rho}{2\pi}\frac{I}{r_0}, \quad \text{for } 0 \leq r \leq r_0 \tag{4.6}$$

Thus, the electric earth potential, as a function of the distance x in any direction from the electrode, is a rectangular hyperbola,[6] which asymptotically approaches zero as r approaches to infinity (Figs. 4.5 and 4.6).

At $r = r_0$, we obtain the total earth potential V_G, also referred to as *ground potential rise* (GPR), that is, the potential difference between any point on the electrode's surface and infinity. If we evaluate the

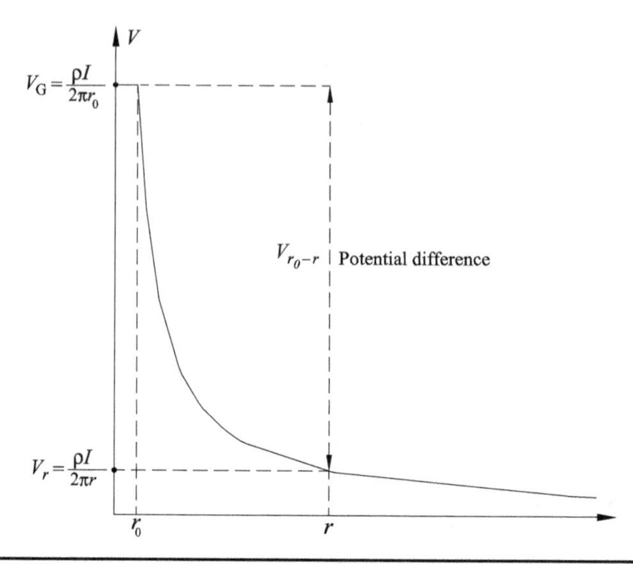

FIGURE 4.5 The earth potential as a function of the distance from the electrode.

earth potential at $r = 5r_0$, we will obtain from Eq. (4.5)

$$V_{5r_{0-\infty}} = \frac{V_G}{5} \tag{4.7}$$

Equation (4.7) shows that 80% of V_G falls between the electrode's surface and the equipotential surface of radius $5r_0$. This is the reason

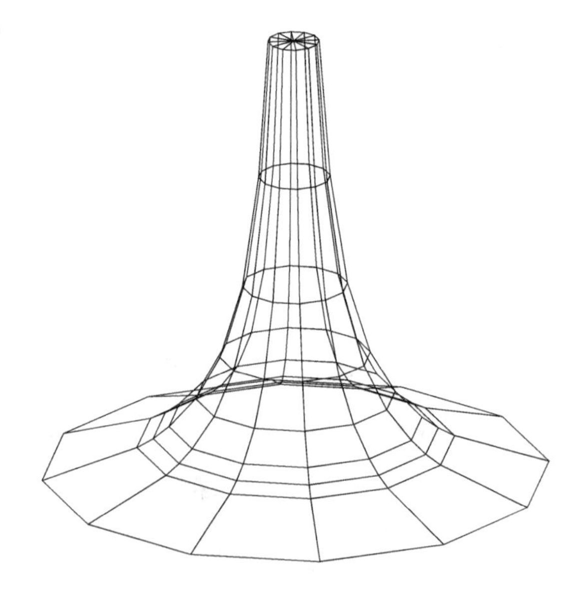

FIGURE 4.6 The earth potential in its three-dimensional development as a function of the distance from the electrode.

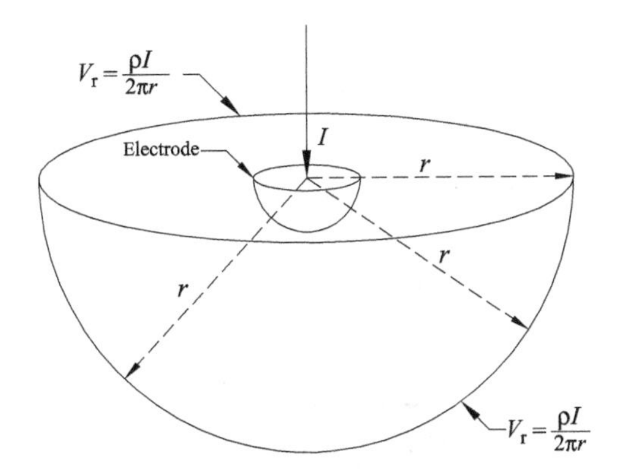

FIGURE 4.7 Hemispherical equipotential surface at distance r from the electrode.

why we can consider virtually zero the earth potential impressed by any electrode at five times the length of their radii.

If the electrode has no circular symmetry (i.e., there is no actual radius r_0), an equivalent radius r_e can be calculated. In this way, any electrode of earth resistance R_G, regardless of its shape, can be considered as a hemispherical one, as long as the hemisphere's radius equals

$$r_e = \frac{\rho}{2\pi R_G} \tag{4.8}$$

In Fig. 4.5, V_G represents the potential of the electrode with respect to infinity, while V_{r_0-r} is the potential difference between the electrode's surface and point r, referred to as *perspective touch voltage*.

The electric potential curve, as shown in Fig. 4.5, allows us to determine the equipotential surfaces surrounding the electrode. These are defined as the loci of points at the same constant electric potential[7] (Fig. 4.7).

It is important to note that hemispherical equipotential surfaces produce radial electric fields and vice versa, in the presence of radial electric fields, we will find hemispherical equipotential surfaces.

4.4 Independent and Interacting Earth Electrodes

In many cases, earth electrodes are connected together in order to lower both the earth resistance and the earth potential. We will, now, calculate the total ground resistance due to the parallel connection of two identical hemispherical electrodes, A and B, of radius r_0 displaced by distance d from center to center (Fig. 4.8). Let I be the leakage

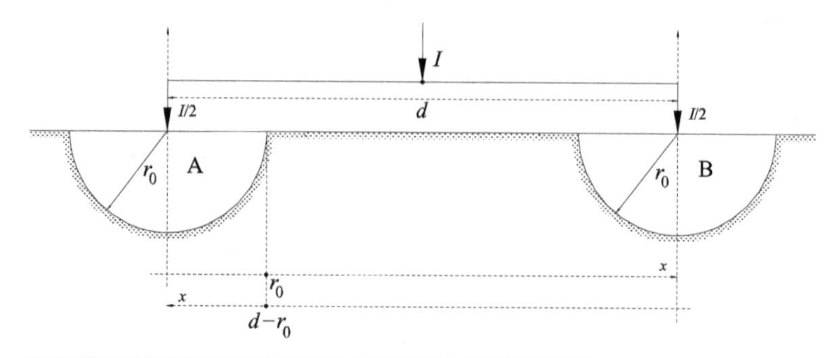

Figure 4.8 Identical hemispherical electrodes connected in parallel and displaced by distance *d*.

current and ρ the uniform resistivity of the soil. Each electrode will leak current $\frac{1}{2}I$, as per the symmetry of the system.

If the electrodes are too close to each other, they interfere, causing a change in the shape of the total potential with respect to their own single potential curve. The curve of the total potential can be determined by means of the superposition principle, considering the contributions of each electrode separately, and, then, adding them up. In Fig. 4.9, a possible result of this procedure is shown.

Because of their metal connection, whose resistance we consider negligible, the two electrodes will attain the same earth potential V_{GTOT}, which we can calculate as follows. If we consider two different systems of reference, originating at the electrodes A and B (Fig. 4.8 or 4.9), and define R_{GTOT} as the total earth resistance of the two-electrode

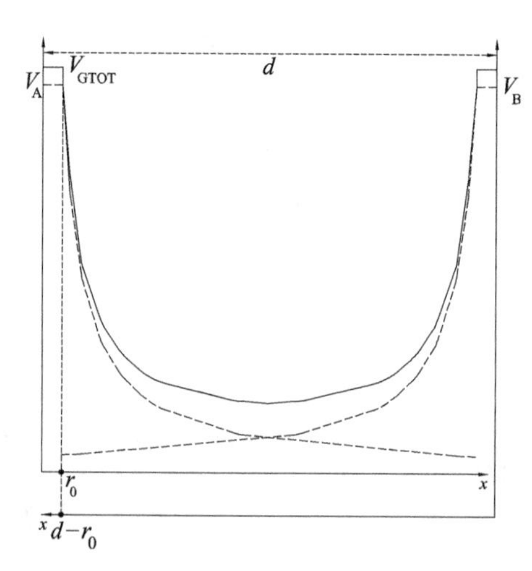

Figure 4.9 Equivalent potential due to two interfering electrodes.

system, we obtain

$$
\begin{aligned}
V_{GTOT} &= R_{GTOT} I = V_A(r_0) + V_B(d - r_0) \\
&= \frac{I}{2}\frac{\rho}{2\pi r_0} + \frac{I}{2}\frac{\rho}{2\pi(d - r_0)} = \frac{I\rho}{4\pi}\left(\frac{1}{r_0} + \frac{1}{(d - r_0)}\right) \\
&= \frac{I\rho}{4\pi}\left(\frac{d - r_0 + r_0}{r_0(d - r_0)}\right) = \frac{I\rho}{4\pi r_0}\left(\frac{d}{d - r_0}\right) \\
&= \frac{\rho}{4\pi r_0}\left(\frac{1}{1 - (r_0/d)}\right) I
\end{aligned} \tag{4.9}
$$

Thus, R_{GTOT} equals:

$$
R_{GTOT} = \frac{\rho}{4\pi r_0}\left(\frac{1}{1 - (r_0/d)}\right) \tag{4.10}
$$

The multiplicand in Eq. (4.10) corresponds to the parallel of the earth resistances of two identical hemispherical electrodes, obtained by dividing by 2 the result of Eq. (4.2). The equivalent ground resistance of the two electrodes of Fig. 4.8, then, is not merely the parallel between their corresponding earth resistances, as shown by the presence of the multiplier in parenthesis in Eq. (4.10). Such multiplier may be ≥ 1 and be considered 1 only if the mutual distance $d \gg r_0$, in which case the two electrodes will result connected in a "true" parallel.

As the multiplier increases the earth resistance of the parallel, it can be thought as an additional resistance R_a in series with the aforementioned parallel. In formulas:

$$
R_{GTOT} = \frac{\rho}{4\pi r_0}\left(\frac{1}{1 - (r_0/d)}\right) = \frac{\rho}{4\pi r_0} + R_a \tag{4.11}
$$

$$
\begin{aligned}
R_a &= \frac{\rho}{4\pi r_0}\left(\frac{1}{1 - (r_0/d)}\right) - \frac{\rho}{4\pi r_0} = \frac{\rho}{4\pi r_0}\left(\frac{1}{1 - (r_0/d)} - 1\right) \\
&= \frac{\rho}{4\pi r_0}\left(\frac{(r_0/d)}{1 - (r_0/d)}\right) = \frac{\rho}{4\pi}\frac{1}{(d - r_0)}
\end{aligned} \tag{4.12}
$$

The earth resistance R_{GTOT}, calculated in Eq. (4.10), is shown in the circuit of Fig. 4.10, where R_{GA} (or R_{GB}) is the earth resistance of the electrode A (or B) when the other is not present.

In some cases, real estate constraints may impose a very close placement of earth electrodes. This arrangement limits the grounding system "interface" capability with the earth and lowers the effectiveness of its performance.

In sum, two or more ground electrodes can be considered independent when, due to their own geometries and relative positions,

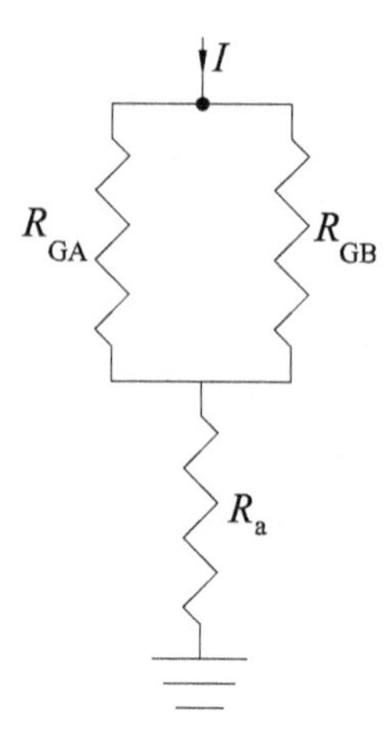

FIGURE 4.10 Earth resistance's equivalent circuit.

their electric potentials are not significantly affected by their reciprocal influence (Fig. 4.11).

Equation (4.9) can be generalized to the case of two dissimilar electrodes, buried at different depths and leaking different currents I_1 and I_2, by writing the following system of equations [Eq. (4.13)]:

$$\begin{cases} V_{GTOT} = R_{11}I_1 + R_{12}I_2 \\ V_{GTOT} = R_{12}I_1 + R_{22}I_2 \\ I_{GTOT} = I_1 + I_2 \end{cases} \tag{4.13}$$

R_{11} (R_{22}) is the earth resistance of the first (second) electrode, obtained by disconnecting the second (first) one; R_{12} (R_{21}) is the ratio of the ground potential attained by the first (second) electrode, not connected to the grounding system, to the current flowing through the second (first) one. R_{12} and R_{21} are referred to as mutual resistances.

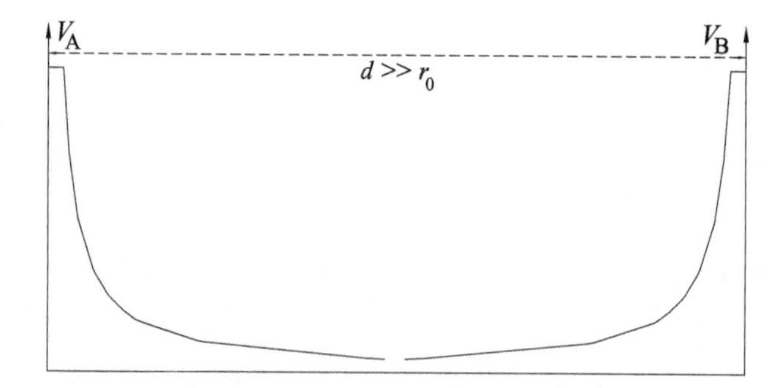

FIGURE 4.11 Potentials due to two independent electrodes.

The previous system yields the following solutions:

$$I_1 = \frac{V_{GTOT}(R_{22} - R_{12})}{R_{11}R_{22} - R_{12}R_{21}} \tag{4.14}$$

$$I_2 = \frac{V_{GTOT}(R_{11} - R_{21})}{R_{11}R_{22} - R_{12}R_{21}} \tag{4.15}$$

Thus, R_{GTOT} from the two electrodes can be so calculated [Eq. (4.16)]:

$$R_{GTOT} = \frac{V_{GTOT}}{I_{GTOT}} = \frac{R_{11}R_{22} - R_{12}R_{21}}{R_{11} + R_{22} - (R_{12} + R_{21})} \tag{4.16}$$

If the two electrodes are independent,[8] the mutual resistances are negligible and the total resistance R_{GTOT} coincides with the parallel of the electrodes' ground resistances.

It is important to note that the optimization of the grounding system requires as little interactions as possible between electrodes. This allows the minimization of the total earth resistance, which benefits the safety of the installation. In practice, we consider as in parallel electrodes separated by at least five times their (equivalent) radii, as the earth potential has greatly decayed at that distance [Eq. (4.7)].

It may not be always economically convenient to connect more electrodes to a given grounding system, when this addition reduces their reciprocal distances. The value of the grounding resistance, in fact, may "saturate," and no longer linearly decreases with the number of electrodes and therefore with the additional cost.

4.5 Spherical Electrodes

Another electrode never used in practical applications, but with a very interesting behavior, is the spherical electrode. A spherical electrode embedded at infinite depth within homogeneous soil would produce spherical equipotential surfaces solely developing within the earth. This does not happen if the spherical electrode, of radius r_0 and leaking current I, is buried at a finite depth D. In this case, in fact, the equipotential surfaces will develop through two different media of different resistivity: soil (resistivity ρ_2) and air (resistivity ρ_1 equal to infinity) (Fig. 4.12).

The resulting nonhomogenous medium can be studied by using the theory of images, which yields the equivalent configuration of Fig. 4.13.

The two semi-infinite media (i.e., air and soil) are replaced by a single medium coinciding with the soil. An additional "virtual" electrode, a symmetrical image with respect to the ground of the real

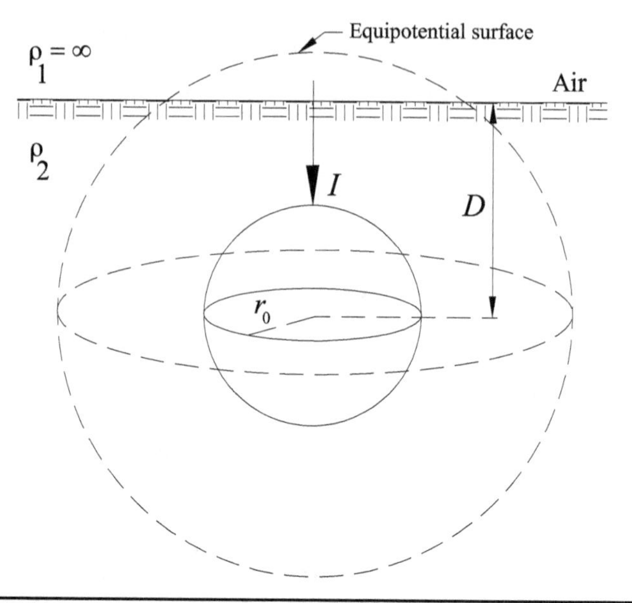

FIGURE 4.12 Spherical electrode buried in earth at depth D.

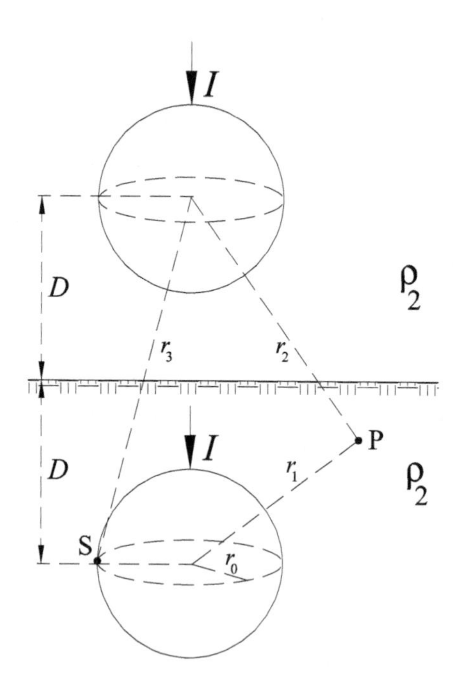

FIGURE 4.13 Spherical electrode buried in earth at depth D: equivalent configuration as per the theory of images.

one, is added. The image electrode leaks the same current as the real one.

This configuration, which is electrically equivalent to the actual arrangement given in Fig. 4.12, allows a simpler determination of the electrode's earth potential and earth resistance, as the interface air–soil is eliminated and the electrode medium is rendered homogeneous again. We also assume that the depth D is much greater than the radius r_0 of the sphere.

The earth potential in correspondence of a generic point P from the spherical electrode can be analyzed by superimposing the contributions of actual and virtual spheres. Thus, by using Eq. (4.5) and the lateral area of the sphere (i.e., $4\pi r_0^2$), we obtain

$$V_{\mathrm{P}\infty} = \frac{\rho_2}{4\pi}\frac{I}{r_1} + \frac{\rho_2}{4\pi}\frac{I}{r_2} \tag{4.17}$$

The total earth potential V_{G}, that is, the potential difference between any point on the actual electrode's surface, the point S, and infinity, can be calculated as follows:

$$V_{\mathrm{G}} = \frac{\rho_2}{4\pi}\frac{I}{r_0} + \frac{\rho_2}{4\pi}\frac{I}{r_3} = \frac{\rho_2}{4\pi}\frac{I}{r_0} + \frac{\rho_2}{4\pi}\frac{I}{\sqrt{r_0^2 + 4D^2}} \tag{4.18}$$

Thus, dividing Eq. (4.15) by current I, the earth resistance R_{G} of the spherical electrode is

$$R_{\mathrm{G}} = \frac{\rho_2}{4\pi}\left(\frac{1}{r_0} + \frac{1}{\sqrt{r_0^2 + 4D^2}}\right) \tag{4.19}$$

The values of R_{G} (and V_{G}) decrease as D increases, but the rate of change with high values of D is very modest, as the values tend to saturate (Fig. 4.14). Therefore, large, and therefore expensive, depths are not necessary.

The earth potential in correspondence of a generic point Q, located at distance x over the soil surface (Fig. 4.15), can be calculated by superposing the effect of the actual and the virtual spheres:

$$V_{x\infty}^{\mathrm{Q}} = \frac{\rho_2}{4\pi}\frac{I}{r_1} + \frac{\rho_2}{4\pi}\frac{I}{r_1} = \frac{\rho_2}{2\pi}\frac{I}{\sqrt{x^2 + D^2}} \tag{4.20}$$

An example of the variation of the surface potential as a function of distance x from the electrode's center buried at 0.1 m is shown in Fig. 4.16.

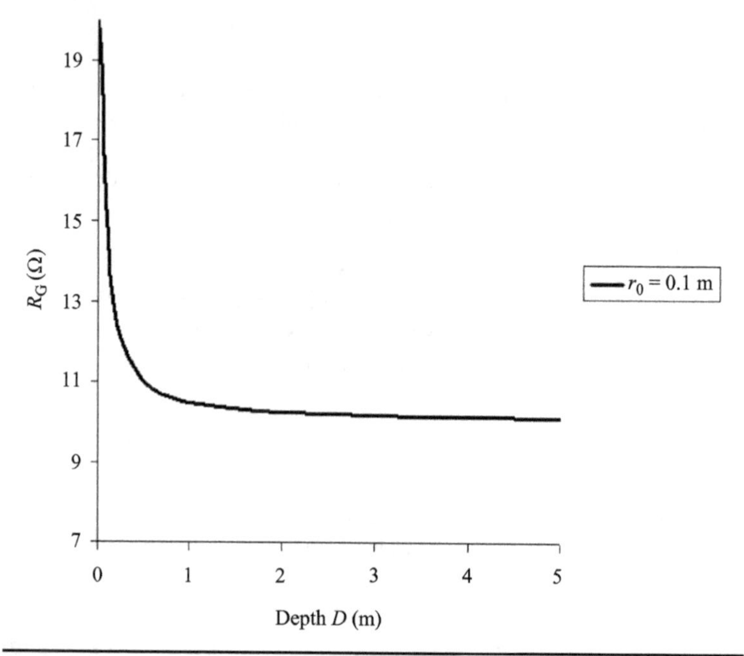

FIGURE 4.14 R_G of a spherical electrode as a function of depth D in the soil.

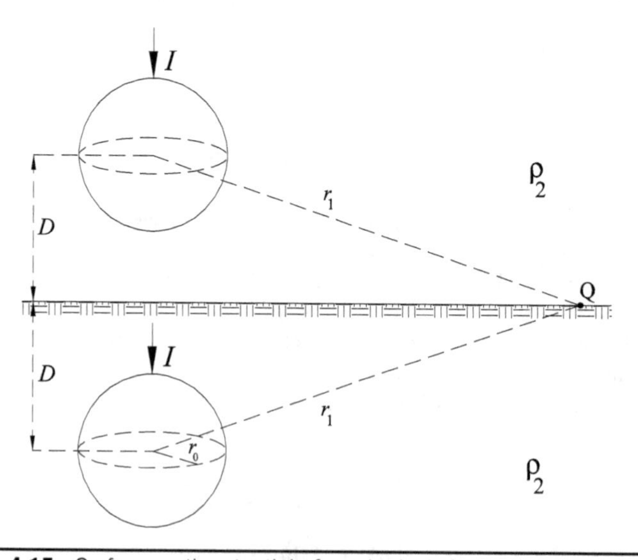

FIGURE 4.15 Surface earth potential of a spherical electrode.

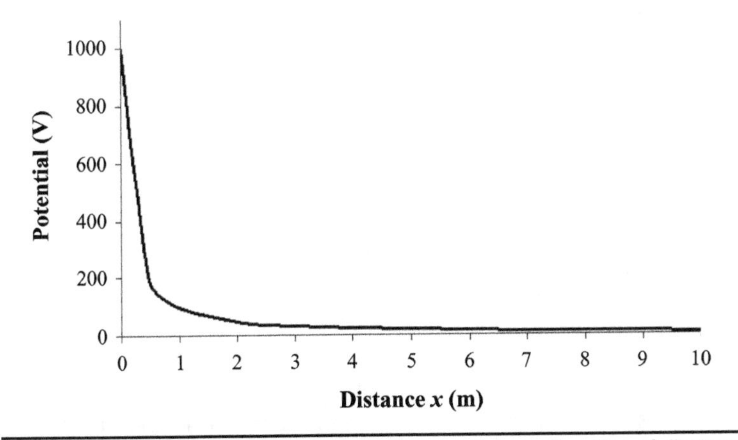

FIGURE 4.16 Variation of the surface earth potential as a function of distance x from the spherical electrode's center.

In correspondence with the vertical straight line passing through the center of the spherical electrode (i.e., $x = 0$), the ground potential assumes finite[9] value:

$$V_{x=0} = \frac{\rho_2 I}{2\pi D} \tag{4.21}$$

4.6 Voltage Exposure Upon Ground Faults

If the basic insulation of a grounded ECP fails, the metal enclosure may be energized and persons are exposed to the risk of electrocution.[10] Persons touching the enclosure, in fact, close the loop between the faulty metal frame and the earth, which, due to its conductive nature, will carry the fault current toward the source.

If the current impressed to ground is I and the earth resistance is R_G, the potential assumed by the ECP is $V_G = R_G I$ (Fig. 4.17).

Once again, V_G is the potential difference between any point on the electrode's surface and infinity; its magnitude is generally less than the phase voltage.

If the faulty ECP is disconnected from the grounding system (e.g., because of the accidental interruption of the protective conductor, PE), in case of fault-to-ground, the enclosure will attain phase-to-ground potential. This is a particularly hazardous situation, as in the absence of a clear path to ground the fault cannot be cleared.

4.6.1 Touch Voltage

The amount of current possibly flowing through a person depends on, and is limited by, the series of the body resistance[11] R_B and the resistance of person-to-ground R_{BG}.

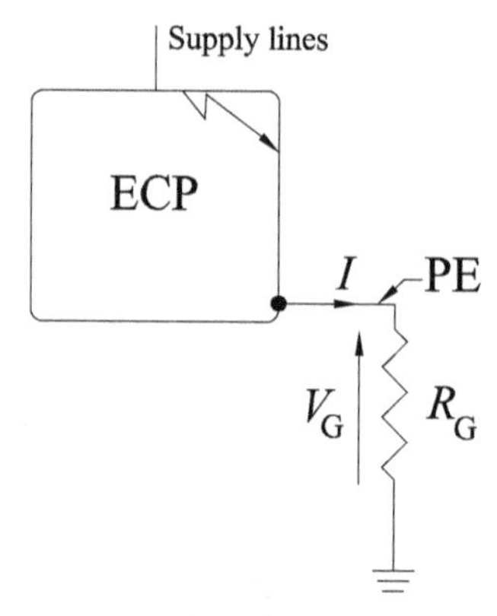

Figure 4.17 The earth potential V_G assumed by the faulty ECP.

R_{BG} is due to the presence of the floor, whose resistance in dry conditions is at least 1 kΩ. In the absence of floor (i.e., outdoor locations), R_{BG} can be calculated by considering the person's feet as two circular plates, of radius $r = 0.1$ m, in parallel on the soil. Each foot/plate has a ground resistance approximately equal to

$$R_{\text{Foot}} \cong \frac{2\rho}{5r} = 4\rho \tag{4.22}$$

where ρ is the superficial soil resistivity in ohm meters (Ω · m). Thus, the human body resistance-to-ground R_{BG} equals 2ρ. The tendency of international standards for low-voltage installations is not to consider the resistance of footwear, by assuming the person, conservatively, shoeless. The presence of shoe resistance in series to the body, in fact, would limit the body current's circulation, benefiting the person's safety.

Let us assume a person standing in an area sufficiently far from the electrode to be considered at zero potential. In this worst-case scenario, be the person exposed to indirect contact by touching a faulted ECP. Assume the ECP leaks the ground current I, so that the person's hand is subject to the potential $V_{ST} = V_G$ (Fig. 4.18).

The prospective (or source) touch voltage V_{ST} is defined as the potential difference between the faulted ECP (i.e., the dryer of Fig. 4.18) and the earth occupied by the person, at a distance of 1 m from the ECP,[12] when the ECP is not being touched by the person.

In reality, the person touching the enclosure is not subject to the potential difference V_{ST} but to the touch voltage V_T, which is $\leq V_{ST}$. A voltage divider, in fact, takes place between the body resistance R_B and the person resistance-to-ground R_{BG} (Fig. 4.19).

In low-voltage systems, V_T is defined as the voltage differential which a person,[13] standing at 1 m from the grounded ECP, may be subject to, between both hands and both feet.[14]

Upon the human touch of the faulted ECP, the natural presence of R_{BG} causes an "elevation" of the electric potential in correspondence

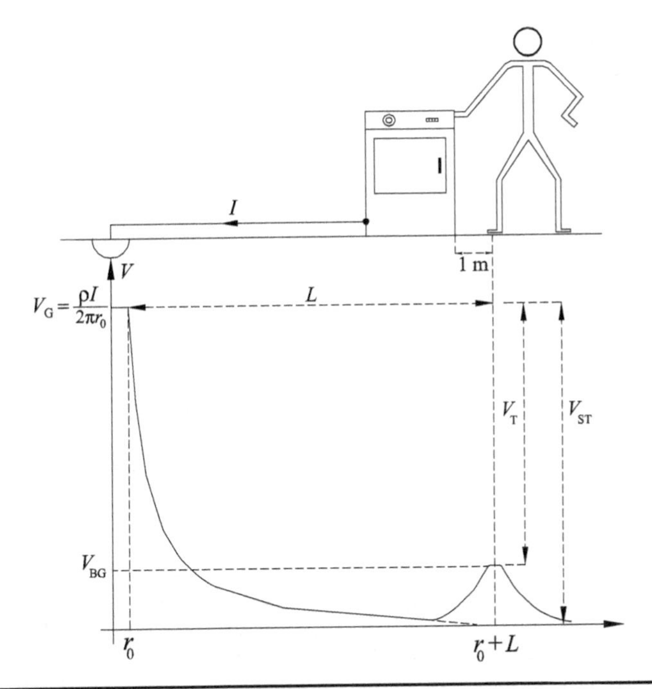

FIGURE 4.18 Person standing in an area at zero potential, while touching a faulted enclosure (worst-case scenario).

of the area occupied by the person's feet (i.e., at distance L from the electrode's surface). The ground potential rises from zero to $V_{BG} = R_{BG} I_B$, as shown in Fig. 4.18.

A better-case scenario for safety occurs when the person is standing in an area at nonzero potential at a distance $l < L$ from the electrode (Fig. 4.20).

The perspective touch voltage V_{ST} is less than the earth potential V_G and, in addition, the presence of the person rises the potential under his/her feet.

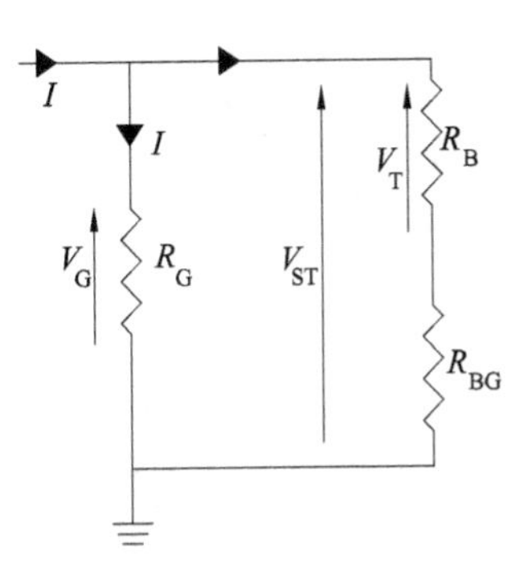

FIGURE 4.19 Equivalent fault-loop for a person standing in an area at zero potential.

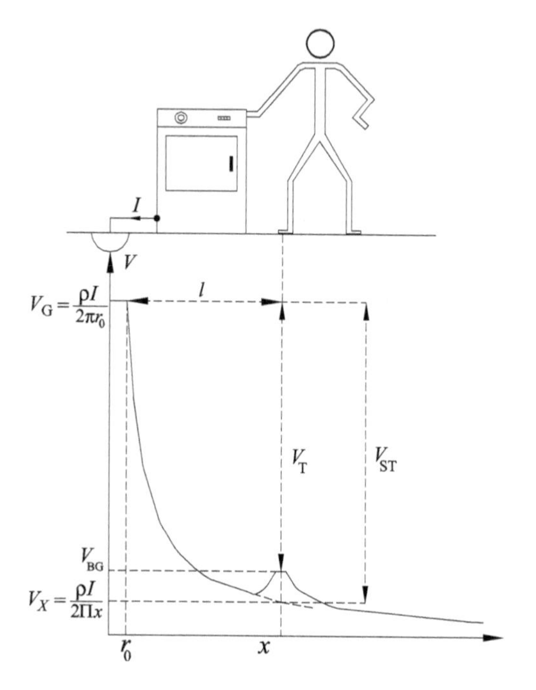

FIGURE 4.20 Person standing in an area at nonzero potential, while touching a faulted enclosure.

The person is subject to the potential difference V_T as imposed by the voltage divider (Fig. 4.21). In general, we can say that $V_T \leq V_{ST} \leq V_G$. In low-voltage systems, international standards conservatively impose permissible limits to the prospective touch voltage, instead of the touch voltage. As a result, protective devices must automatically disconnect the supply to faulty circuits, so that the prospective touch voltage does not persist long enough to cause harm to people. Given the truth of the previous inequality, by limiting the source touch voltage in case of a fault, we also limit the touch voltage, which is the true voltage exposure to people.

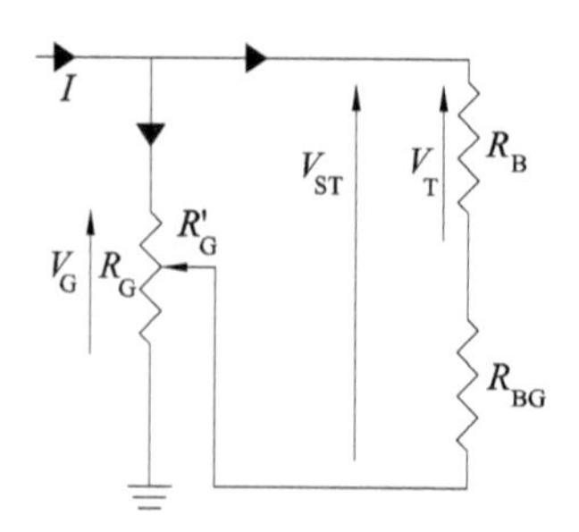

FIGURE 4.21 Equivalent fault-loop for a person standing in an area at nonzero potential.

4.6.2 Extraneous-Conductive-Part (EXCP)

EXCP is defined as a conductive part not belonging to the electrical system, which can be touched and is liable to introduce a dangerous potential into the premises, for example, the earth potential.

The danger introduced by EXCPs is caused by the possibility that a person may be in simultaneous contact with these and with faulty ECPs, whose enclosure is energized.

In this case, the feet potential of a person will be immediately lowered to zero, even if he/she initially stands over an area at nonzero potential as shown in Fig. 4.20. The EXCP, in fact, will "short circuit" R_{BG} (Fig. 4.22), and the advantage of having an additional resistance between the person's body and the earth will be lost.

The person will face an electric shock driven by V_{ST}, instead of V_T, and as a result his/her safety will be lowered.

The solution to this problem is the equipotentialization between ECPs and EXCPs via an equipotential bonding conductor (EBC), which, by linking together ECPs and EXCPs, keeps them at the same potential in fault conditions.[15] The person, not insulated from earth, is still at risk of electrocution, but will at least have "recuperated" his/her R_{BG} in series to his/her body resistance R_B, as shown in Fig. 4.23.

Properly identifying EXCPs is, then, crucial in order to implement the safe equipotentialization within the installation. Any metalwork, even if entering the premises not from the earth, should be considered as an EXCP "candidate." Metal conduits, metal sheath, or armor of cables entering the premises from another building may, in fact, have separate earthing connections and, therefore, introduce the earth potential into the premises, or even nonzero potentials caused by faults occurring in their building of origin.

In ordinary conditions, we can conventionally assume a R_{BG} of at least 1000 Ω in series to the body resistance, even in the absence of a floor. In particular locations (e.g., hospitals, construction sites, and agricultural buildings), where human beings' resistance to ground may be lower, the value of 200 Ω for R_{BG} is used because additional safety requirements must be met (e.g., the limit for the permissible touch voltage is reduced to 25 V from 50 V).

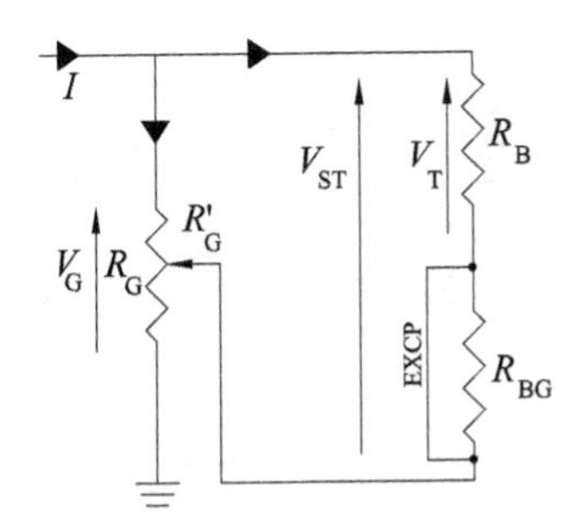

FIGURE 4.22 Equivalent fault-loop for a person standing in an area at nonzero potential and in contact with an EXCP.

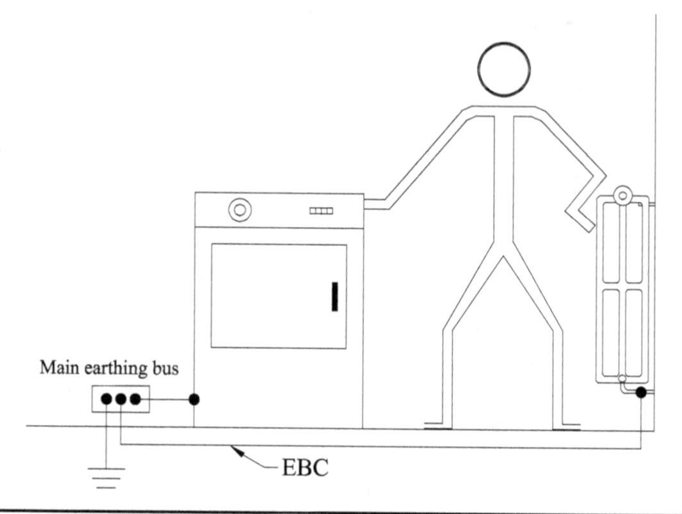

Figure 4.23 Equipotentialization between ECPs and EXCPs.

In light of the above considerations, any metal part, with a resistance to ground less than 1000 Ω (or 200 Ω) and in contact with the person's body, lowers the conventional value of R_{BG}, which we rely on in the electrical design. Thus, 1000 Ω (or 200 Ω) can be considered the limit value of ground resistance of metal parts so to recognize them as EXCPs for bonding purposes.

4.6.2.1 Should We Bond Each and Every EXCPs?

Figure 4.24 proposes a dilemma regarding the necessity to bond each and every EXCPs.

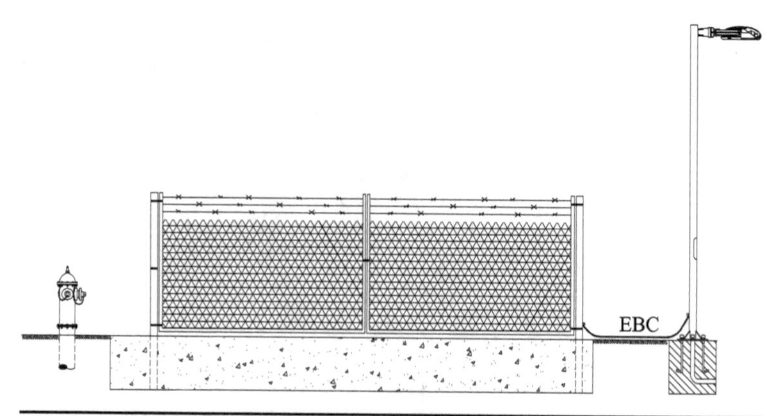

Figure 4.24 Equipotentialization between ECP (light pole) and EXCP (fence).

Now consider the metal fence an EXCP, as its ground resistance is less than 1000 Ω, and the light pole an ECP. Upon failure of the pole's basic insulation, persons might be in simultaneous contact with the pole and the fence, and, therefore, exposed to the total earth potential. This hazardous condition can generally be avoided by means of the EBC.

This bonding connection, though, transfers the touch voltage along the entire length of the fence, which could be miles long. Persons in simultaneous contact with it and another EXCP (e.g., a fire hydrant) are once again exposed to the whole earth potential. Should we bond to the fence the fire hydrant too and, thereby, transferring the fault potential even farther?

There is no general answer to this question: the electrical engineer must decide on an individual case basis whether to bond each and every EXCPs in order to minimize the risk of electric shock.

Another case is presented in Fig. 4.25. The fence is connected to the grounding system, while the fire hydrant is not. In the case where the grounding system becomes energized, the fence does too, and the total earth potential establishes between it and the hydrant, which acts as an EXCP. This hazard can be avoided by means of an EBC between the two elements.

4.7 Voltage or Current?

The human body is sensitive to, and endangered by, current, not voltage. In addition, it has been proved that the human body's impedance Z_B, as offered by the same person, is not a constant value, but depends on the voltage of the energized object to which he/she is exposed.[16] There is a nonlinear relationship between voltage and body impedance: the greater the potential difference applied to the body, the lower its resistance and the greater the hazard.

Consequently, two different touch potentials may, in fact, "provoke" two different body resistances, but cause the circulation of the same current. This makes the touch voltage a not very effective

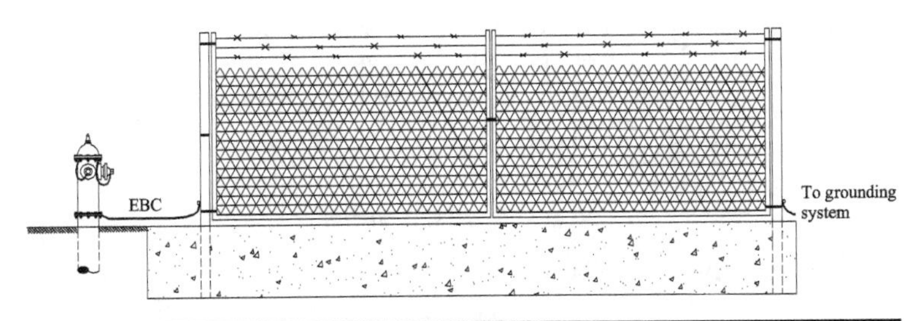

Figure 4.25 Equipotentialization between fence and fire hydrant.

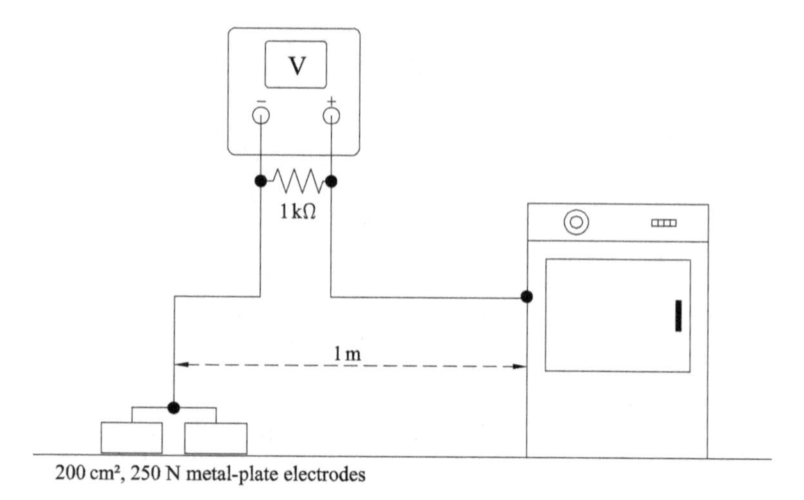

FIGURE 4.26 Standard touch voltage measurement.

parameter to establish the hazard in electrical systems, even though it is the easiest one to measure. The capability of the voltage source to impress a current greater than the body value deemed dangerous is, instead, the right quantity to consider.

To solve this dilemma, and normalize the measurement of touch voltages, international standards have established, as a reference, a conventional human body resistance of 1 kΩ.

A standard touch voltage measurement is shown in Fig. 4.26.

The person's resistance to ground R_{BG} is "simulated" by a pair of 200 cm^2 metal plates as electrodes. Each plate should weigh at least 250 N and be 1 m apart from the faulted ECP being tested. The standardized body resistance is modeled through a 1-kΩ resistance connected in parallel to the voltmeter leads. The person is conservatively supposed to be shoeless.

If we measured the voltage across an open circuit, instead of the 1-kΩ resistance, we would be measuring the perspective touch voltage V_{ST}.

Example 4.1 A hemispherical electrode is embedded in a soil with resistivity of 200 Ω · m (e.g., poorly graded gravel). Calculate the size of the electrode's radius in order to achieve an earth resistance not exceeding 10 Ω.

Solution Equation (4.2) yields:

$$R_G = \frac{\rho}{2\pi r_0} = \frac{200}{2\pi r_0} \leq 10\,\Omega$$

By solving for r_0, we obtain

$$r_0 \geq \frac{200}{20\pi} = 3.18\,\text{m}$$

Example 4.2 A hemispherical electrode, embedded into a soil of resistivity 300 $\Omega \cdot$ m (e.g., sand–clay mixtures), has a radius of 2 m. The maximum fault current I allowed by the protective device is 100 A. Determine:

1. The earth resistance of the electrode
2. The potential difference between the electrode and a point on the earth's surface at distance $r = 10$ m

Solution The earth resistance of the electrode is

$$R_G = \frac{\rho}{2\pi r_0} = \frac{300}{4\pi} = 23.8\ \Omega$$

Reference is made to Fig. 4.5.

$$V_{r_0-r} = \frac{\rho I}{2\pi r_0} - \frac{\rho I}{2\pi r} = \frac{\rho I}{2\pi}\left(\frac{1}{r_0} - \frac{1}{r}\right) = \frac{300 \times 100}{2\pi}\left(\frac{1}{2} - \frac{1}{10}\right) = 1911\ \text{V}$$

Example 4.3 A hemispherical electrode of radius 1 m is embedded into a soil of resistivity 600 $\Omega \cdot$ m (e.g., well-graded gravel). The center of the electrode is 14 m away from an ECP. Determine the perspective touch voltage V_{ST} and the touch voltage V_T a person may be subject to, upon the circulation of the ground-fault current $I = 25$ A.

Solution

$$V_{ST} = \frac{\rho I}{2\pi r_0} - \frac{\rho I}{2\pi r} = \frac{\rho I}{2\pi}\left(\frac{1}{r_0} - \frac{1}{r}\right) = \frac{600 \times 25}{2\pi}\left(\frac{1}{1} - \frac{1}{(14+1)}\right) = 2299\ \text{V}$$

$$R_{BG} = 2\rho = 1200\ \Omega$$

$$V_T = V_{ST} \times \frac{R_B}{R_B + R_{BG}} = 2299 \times \frac{1000}{1000 + 1200} = 1045\ \text{V}.$$

FAQs

Q. What is the relationship between the ground potential, as shown in the curve in Fig. 4.5, and the equipotential surface in Fig. 4.7 for a hemispherical electrode?

A. As shown in Fig. 4.5, by "entering" with a generic distance r from the electrode, we can read the relative potential V_r. Once this pair of coordinates is known, we can determine the equipotential surface by drawing a hemisphere of radius r to which we associate the ground potential V_r. By doing this for any distance r, we will obtain all the possible equipotential surfaces.

Q. Are a person's feet two electrodes in parallel?

A. Human feet are conventionally modeled as two circular plates and can be considered in parallel only if they are at least five times their radii, of length 10 cm, apart. We conventionally assume feet separated by 60 cm. In reality, this might not always happen during faults. However, it is assumed that the mutual ground resistance between the two feet, as calculated in Sec. 4.4, is negligible.[17] Thus, we can consider a person's feet as electrodes in parallel.

Q. Why it may not be economically convenient to add more electrodes to a given grounding system, if this reduces their interdistance?

A. Connecting electrodes to a grounding system always lower its total earth resistance. This reduction, though, may not follow the rule of the parallel of resistances, if the electrodes mutually interfere. In layman terms: if by adding an electrode to a grounding system we lower its earth resistance from 19 to 10 Ω, by adding a second one we may not reduce the total resistance to 5 Ω (i.e., the result of the parallel) due to their mutual couplings. The total value may be 7 Ω. Therefore, the second element is less economically effective than the first with respect to the same cost.

Q. What is the difference between V_{ST} and V_T?

A. V_T is the potential difference actually experienced by a person standing at point A on the earth surface, while in hand-contact with a faulty grounded part B. V_{ST} is the potential difference between the same two points A and B, but in the absence of the person.

Endnotes

1. As further discussed, this also applies to ungrounded systems (IT systems).
2. An isotropic soil has physical properties that do not vary with direction.
3. To fix ideas, imagine that the hemispherical electrode acts as a colander to water.
4. The symmetry of the electrode yields uniform current density in the ground; therefore, the formula for the resistance of conductors is still applicable.
5. The lateral area of a sphere equals $4\pi r^2$, and therefore the lateral area of a hemisphere is half of that.
6. The general equation of a rectangular hyperbola is $y = m/x$.
7. The electric potential at a point $P(x,y,z)$ is a scalar quantity, which is solely a function of the coordinates (x,y,z) being considered, once a reference (e.g., infinity) has been established.
8. They are independent if, upon their alternate disconnection, a current impressed through the first one does not appreciably change the earth potential of the other one.
9. In the case of the hemispherical electrode, the ground potential for $r = 0$ equals infinity.
10. As already discussed in previous chapters, the voltage exposure is present during the time the protective device employs to disconnect the supply.
11. As further discussed in Chap. 5, at 50/60 Hz, the body can be considered a resistance instead of impedance; R_B depends, among other things, on the current path through the body, and, in our calculations, we conventionally consider the most critical one consisting of the pathway both hands-to-both feet.

12. Conventionally, 1 m is considered the distance equal to a man's normal maximum horizontal reach.
13. As persons are not necessarily electrically similar, we refer to a "standard" person, whose resistance is conventionally defined as it is explained in Chap. 5.
14. In electrical systems exceeding 1 kV (i.e., defined as high-voltage systems), international standards conventionally define touch voltage as a contact with one hand and both feet.
15. An exception to this rule can be made for EXCPs of such small dimensions that cannot be effectively contacted by persons.
16. See Chap. 5 for further details.
17. IEEE Std. 80–2000, "*IEEE Guide for Safety in AC Substation Grounding.*"

CHAPTER 5

Effects of Electric Currents Passing Through the Human Body, and Safety Requirements

When reason and unreason come into contact,
an electrical shock occurs.
FRIEDRICH VON SCHLEGEL (1772–1829)

5.1 Introduction

It has been discussed in Chap. 4 that the earth, due to its conductive nature, can be an available return path to ground currents and, therefore, allows circulation of current through the person in contact with an energized object. Current passing through a human body, upon touch of an energized object, cause electric shock, also referred to as *macroshock*.

Electric currents are harmful to human beings primarily in view of the fact that they interfere with the biological electrical activity of the body. This interference disrupts the proper natural electrical pattern in the organism and can cause lethal consequences, even in the absence of other effects, such as burns. Currents, in fact, also increase the temperature of body tissues due to the Joule effect. In some cases, temperatures may rise up to 3000°C.

The physiologic effects of currents must be comprehended as they dictate the minimum safety requirements in electrical installations.

5.2 The Human Body as an Electrical System

On a cellular level, the human body is an electrical system, as the flux of information necessary to its proper functioning is propagated by the means of electric charges, positive and negative, constituted by ions. A clear example is the cardiac muscle, whose contractions are the result of the biological electrical system.

5.2.1 On the Electrical Nature of the Cells

Biological tissues are formed by cells, in contact with each other, immersed in the extracellular (or interstitial) fluid. The cell is enclosed in a membrane, which contains the intracellular fluid. Both intracellular and interstitial fluids contain electrically charged ions, whose relative concentration is indicated by the size of the circles in Fig. 5.1.

Measurements indicate that the charges of the undisturbed cell, uniformly distributed on each side of the membrane, create a permanent potential difference across the cell, called *resting* (or *membrane*) *potential*. This voltage can assume a value as high as −70 mV, where the negative sign indicates that there is an excess of negative charge inside of the cell, with respect to the interstitial liquid.

The resting potential is the result of the equilibrium between two different forces. Ions on both sides of the membrane are subject to the electric field reciprocally exerted by their charges. Ions are also exposed to the forces of diffusion due to their chemical gradient. Ions,

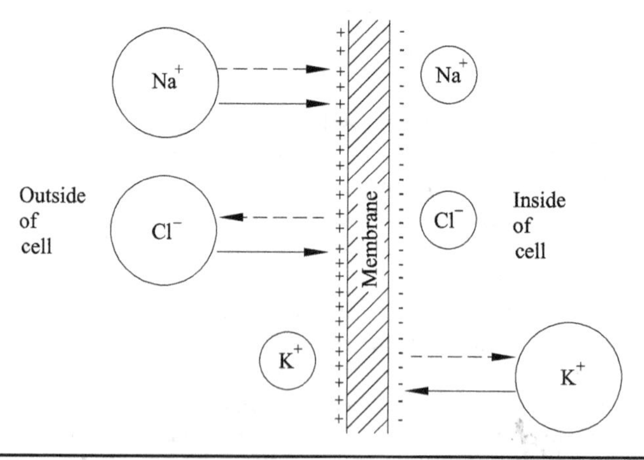

Figure 5.1 Relative concentrations and forces of ions in equilibrium across the cell's membrane.

in fact, diffuse from areas of high concentration to areas of low concentration. In Fig. 5.1, the dotted line symbolizes the direction of the electric field, while the solid one shows the force of diffusion.

The concentration of potassium K^+ is larger inside the cell; therefore, the diffusion force tries to force these ions out. On the other hand, the extracellular fluid is positively charged, with respect to the intracellular fluid, and the resulting electric field will oppose the diffusion of the positive ions. The two actions, then, balance each other. The above explanation can be similarly applied to the Cl^- ions.

The process just described would not seem to explain, though, the behavior of the Na^+ ion. Sodium ions, in fact, are subject to both forces, electrical and diffusion, which act in the same direction toward the inside of the cell. The forces do not cancel each other. In equilibrium, then, the largest concentration of Na^+ ions should be inside the cell and not in the interstitial fluid. Studies have shown that another active process, called the *sodium–potassium pump*, explains the lower concentration of sodium in the intracellular fluid. Due to this process, protein molecules, at the expense of the body metabolism, continuously transport Na^+ ions out of the cell and replace them with K^+ ions.

In stable conditions, that is, in the absence of applied stimuli, the concentrations of ions inside and outside the cell allow the establishment and the holding of the membrane potential. The cell, therefore, can be thought as a capacitor C. The two fluids, intracellular and extracellular, are good electrolytic conductors[1] and act as the armatures of a capacitor, whose dielectric is the membrane itself. The membrane, in fact, has a high resistivity of approximately $10^7\ \Omega \cdot \mathrm{m}$ and dielectric constant of $7\varepsilon_0$.

The above-mentioned capacitor, though, is not an ideal component, as leak currents can flow through the dielectric. As a consequence, the model of the cell must include a leakage resistor R_L in parallel to the capacitor (Fig. 5.2).

The resting potential is represented by the electromotive force V_{RP}.

5.2.2 Action Potential

Excitable cells have the property to remarkably increase the permeability of their membrane to sodium ions upon application of a depolarizing stimulus,[2] whose intensity and duration exceed the cell's threshold of excitation (Fig. 5.3).

The stimulus causes the voltage-sensitive sodium ions channels in the membrane to open, allowing the "inrush" of Na^+ ions into the cell, driven by both electric and diffusion forces. This dramatically changes the cell potential, which becomes positive from its resting negative value. When the depolarization is complete, that is, the inside of the cell is positive with respect to the outside, the sodium ion channels

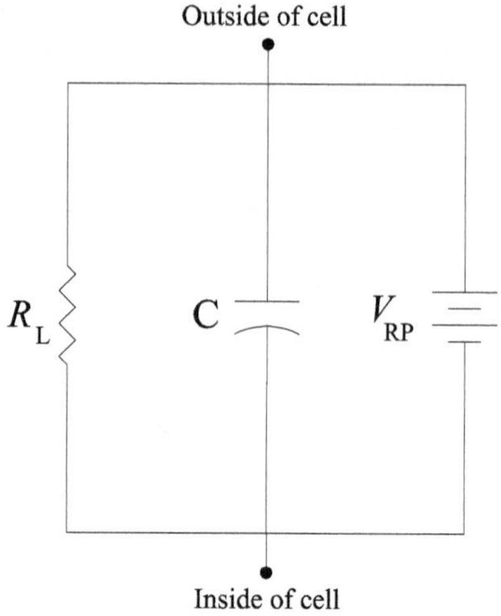

FIGURE 5.2 Electrical model of a cell.

close and the potassium ions channels open. These channels are gates to the K^+ ions, which, by leaving the cell, cause it to repolarize and the potential to reach negative values again. Eventually, the membrane potassium conductance drops and the cell returns to its original resting potential. This total process takes about 2 ms.

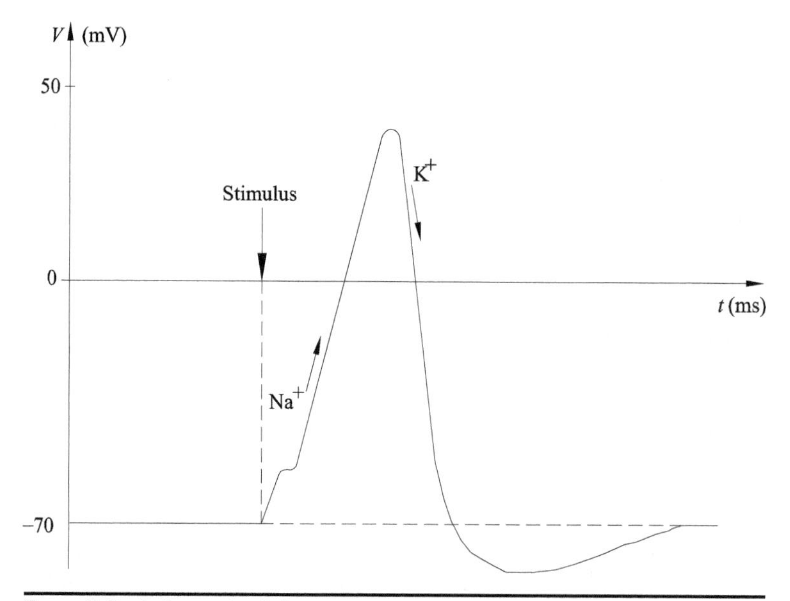

FIGURE 5.3 Action potential.

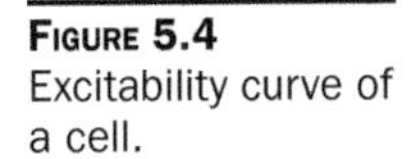

FIGURE 5.4 Excitability curve of a cell.

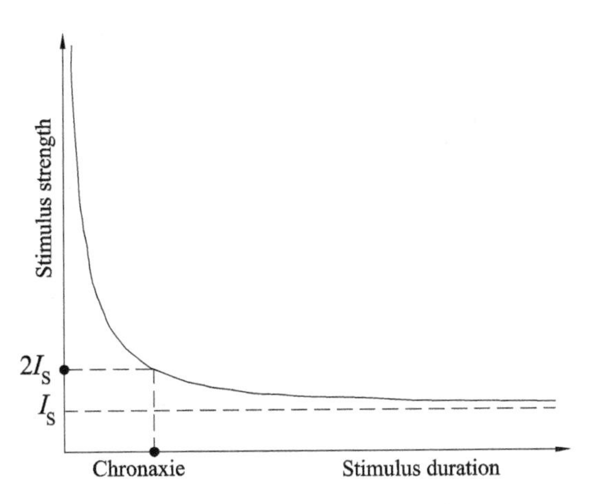

This variation of the resting potential is called an *action potential,* and represents the way information is carried within, and between, tissues in the human body. The amplitude of the membrane potential of the cell is not proportional to the stimulus, but obeys the *all-or-none* law, that is, the action potential either occurs or does not occur.[3]

The combination of stimulus duration and stimulus strength must lie above the *excitability* curve of a cell, as qualitatively shown in Fig. 5.4, to elicit depolarization of the cell membrane and cause the action potential.

I_s represents the strength of the stimulus required to activate the action potential if applied for infinite time (defined as *Rheobase*). *Chronaxie* is the duration of a stimulus of intensity $2I_s$ that needs to be applied to trigger the action potential.

The above excitability curve describes the behavior of the cell only when following stimuli are sufficiently spaced in time. The cell does not respond to close subsequent stimuli, showing periods of refractoriness. The *absolute refractory* period T_{AR} is the time interval following the inception of the cell excitation, during which no action potential can be triggered, regardless of the strength of the stimulus (Fig. 5.5).

The *relative refractory* period T_{RR} immediately follows T_{AR} and is the time frame in which the membrane can be activated, but only by a greater stimulus (Fig. 5.5). The cell will return to its standard response after a time given by the summation of the two aforementioned periods, called *refractory* period T_R.

If an effective stimulus persists for a time exceeding T_R, a natural phenomenon known as *accommodation* takes place. The cell "adapts" by elevating its excitation threshold necessary to trigger subsequent action potentials after the refractory period has expired.

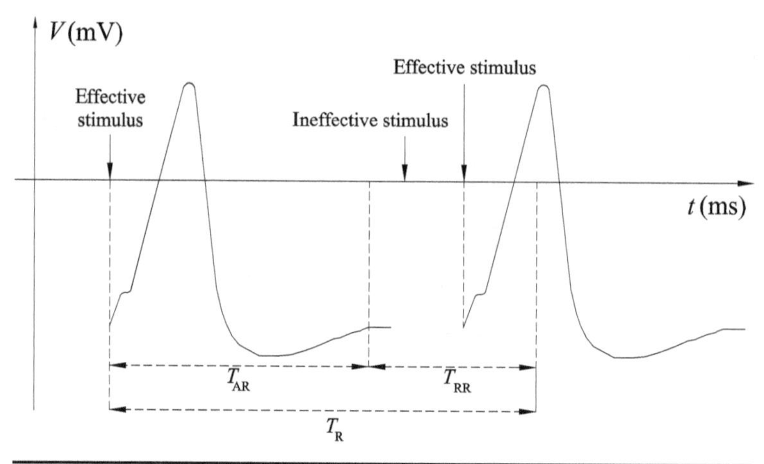

FIGURE 5.5 Refractory periods.

5.3 Influence of Frequency on the Effects of Current

A sine wave of frequency f can be thought as a collection of impulses, regardless of the sign, constituting electrical stimuli of duration $T/2$ (Fig. 5.6).

An equal intensity current, of frequency $f_1 > f$, applies equal strength stimuli, but of a shorter duration $T_1/2 < T/2$. In this case, the stimuli may not excite the cell, because of their lower application time, which put them below the excitability curve of Fig. 5.4. High-frequency (>100 Hz) currents are, therefore, less dangerous than low-frequency ones of same intensity, and their threshold of perception is higher than those at low frequency.

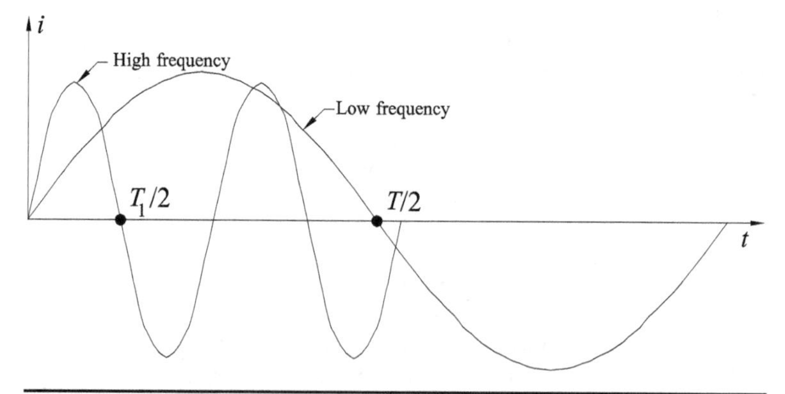

FIGURE 5.6 Sine waves as collection of stimuli.

Another important result that we can infer from the previous paragraph is that the direct current (i.e., frequency equals zero) is generally less dangerous than the alternating current. The accommodation phenomenon earlier described as occurring due to the prolonged stimulus caused by the d.c. current, causes the cell excitability threshold to increase.[4]

5.4 Physiological Response to Electrical Currents

The physiological response of the body to electric current depends on its magnitude and duration.

The *threshold of reaction* is the minimum value that causes involuntary muscle contraction. Normally, at 50/60 Hz, the most common power frequencies, a conventional value for threshold of reaction is assumed to be 0.5 mA (r.m.s), independently of the contact time (2 mA for direct currents). This value, in reality, varies with the conditions of contact (i.e., dry, wet, contact pressure, etc.), the area of the body in touch with the live part, and the physiology of the individual.

The threshold of *let-go* is the maximum value of touch current that allows the subject to voluntarily be able to release his/her grasp on the energized part. This threshold depends on the contact area, the shape of the live part, as well as on the physiology of the person. The conventional threshold of let-go is assumed to be 10 mA (r.m.s) for adult males.

Current can cause momentary or permanent pathologies to the body, such as paralysis (*tetanization*), extensive burns, inability to breathe, unconsciousness, ventricular fibrillation, and cardiac arrest. In the following sections, we will examine the most important ones.

5.4.1 Tetanization

As previously stated, the right combination of strength and duration of a stimulus can produce an action potential, which, by propagating along nerves, can "order" muscle fibers to mechanically contract.[5] After the contraction, the muscle slowly returns to the resting state, unless, before the end of the contraction cycle, another effective stimulus has elicited a subsequent action potential.[6] In this case, the original mechanical contraction, not expired yet, is re-initiated. If subsequent action potentials occur, like a "burst," the resulting contraction of the muscle is maintained as a constant global spasm (Fig. 5.7). This phenomenon is called tetanization.[7] The injured "can't let go" of the energized part an individual is in contact with, and the paralysis of the respiratory muscles, can induce asphyxiation, with subsequent oxygen deprivation resulting in death or irreparable brain damages. When the burst stops the muscle slowly expands towards the resting state.

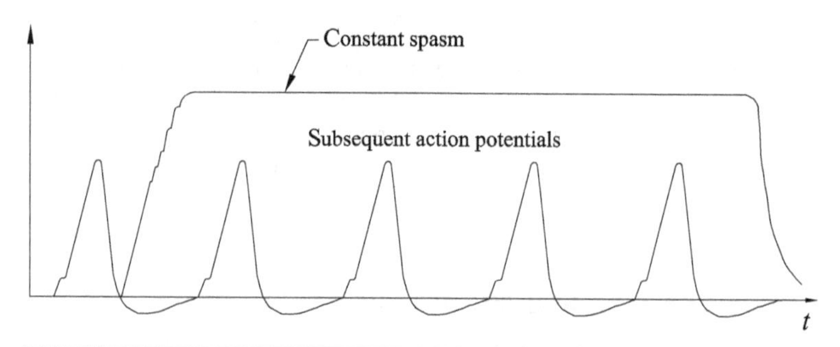

Figure 5.7 Burst of action potentials and resulting tetanization.

Tetanization occurs for currents ranging between 21 and 50 mA. Higher values of current do not cause tetanization, but, as we see in the next section, can cause other serious pathologies.

5.4.2 Ventricular Fibrillation

The ventricular fibrillation is a nonspontaneous reversible condition of the heart, during which the cardiac muscle disorderly contracts, denying the proper blood circulation, which is crucial to supply oxygen to the body. This is considered to be the main cause of death in the case of electrocution.

The cardiac muscle (also known as *myocardium*) is an involuntary muscle found in the heart, whose function is to "pump" blood throughout the circulatory system. It has the capability to contract, like any other muscles, but, in addition, also has the ability to generate and conduct electricity. The *sinoatrial node* (SA), located in the right atrium of the heart, acts as an impulse generator (i.e., a biological pacemaker) and generates action potentials that drive the heart contractions. The action potentials propagate through the whole cardiac muscle and reach the *atrioventricular node* (AV). As atria and ventricles are insulated by nonconducting tissues, the AV node will receive and transmit the action potentials to the ventricles, after applying a functional delay to this transmission. The myocardium can, then, contract and perform its important and continuous duties. After contracting, the heart relaxes and fills up with blood again.

The propagation of the action potential through the heart during the cardiac cycle generates potential differences $V(t)$ between different points of the entire body, which vary with time. By monitoring such potentials, by means of electrocardiograms (EKGs), it is possible to study the electrical activity of the heart over time (Fig. 5.8).

It is in this period of time dt, at the beginning of the "T" wave, that the heart relaxes and awaits for a new stimulus to contract itself again. In this time interval, which is approximately 150-ms long and

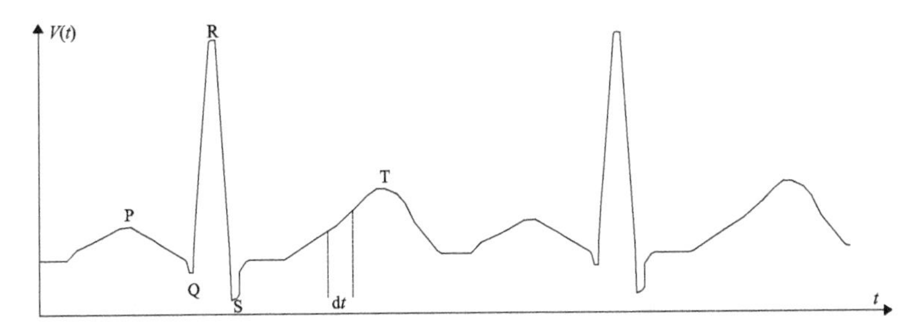

FIGURE 5.8 Normal heart potentials as shown by an electrocardiogram.

corresponds to nearly 10% of the whole cardiac cycle, the myocardium, no longer driven by the SA node, is very vulnerable to foreign stimuli. Any external current of sufficient magnitude applied to the heart in the vulnerable interval will trigger the ventricular fibrillation.

The determination of the component of the actual current flowing through the heart upon contact with a live part, with respect to the total current flowing through the body, is extremely difficult. This component is the true culprit of the ventricular fibrillation and depends on both the individual and the actual current pathway through the human body. For this reason, we conventionally use the total body current to identify the *threshold of ventricular fibrillation*, defined as the minimum value of body current that causes ventricular fibrillation. This is a conservative definition as the current through the heart is generally less than 10% of the body current (Fig. 5.9).

For shock durations below 0.1 s, fibrillation may be triggered by current in excess of 500 mA. For longer exposure, lower current intensities will elicit fibrillation.

5.4.3 Thermal Shock

The circulation of electric current I through tissues, for a time Δt, generates heat due to the Joule effect, and, thus, possible burns.[8] Let us consider a sample of tissue of length l, cross-sectional area S, and resistivity ρ. In addition, let us conservatively assume that there is no thermal exchange between the body and the surrounding environment (i.e., adiabatic process). In this case, all the heat developed by the current is absorbed in the tissues, whose temperature rises.

This adiabatic process is described by the thermal balance of Eq. (5.1):

$$\rho \frac{l}{S} I^2 \Delta t = S l c \Delta \theta \tag{5.1}$$

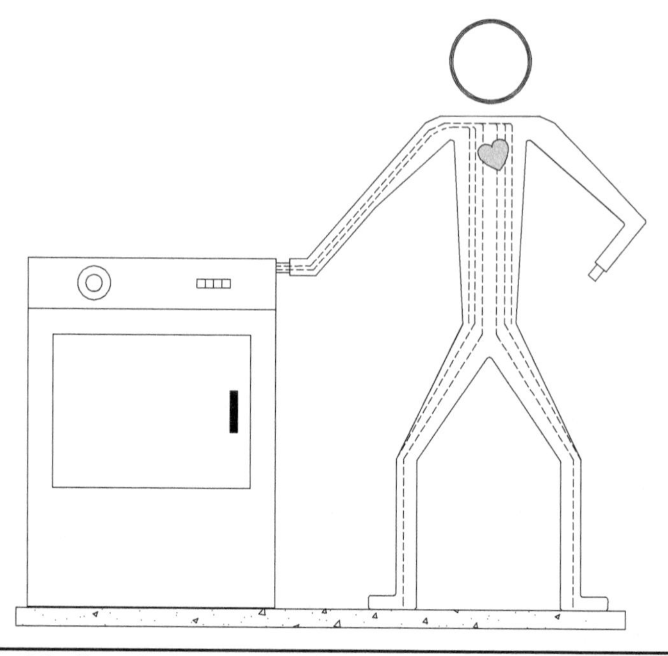

FIGURE 5.9 The actual current through the heart is a small percentage of the total body current.

The left-hand side represents the heat developed by the current due to the Joule effect during the time Δt, while the right-hand side represents the heat accumulated in the tissues. c is the volumetric specific heat capacity[9] of the tissue, a mean value assumed constant with the temperature. $\Delta\theta$ represents the difference between the initial temperature θ_0, at the inception of the current circulation, and the final temperature θ_f, after the time Δt.

Equation (5.1) can be solved for $\Delta\theta$:

$$\Delta\theta = \frac{\rho}{c}\left(\frac{I}{S}\right)^2 \Delta t = \frac{\rho}{c} J^2 \Delta t \qquad (5.2)$$

Equation (5.2) shows that the temperature rise $\Delta\theta$ to which the human tissue is subject depends on the square of the current density J and on the duration of current circulation. If the area of contact S is sufficiently large, J may be low enough that there would be no damage to the tissues, despite the eventual high value for the current. In ordinary conditions, current densities less than 10 mA/mm^2 do not cause any visible mark on the skin.

The above rationale confirms that the primary hazard caused by electric currents is not burns, but the result of the currents interaction with the body's own electricity.

5.5 Permissible Body Current and Person's Body Mass

The electric current through the human body, as the result of an accident, should not exceed the threshold of ventricular fibrillation. This is the fundamental criterion that electrical engineers must adopt in their design so that fault currents passing through human body will not have lethal magnitudes and durations.

Over the years, researchers have investigated the threshold of fibrillation to determine the permissible body current limit, in relation to its duration. It has been statistically extrapolated that 99.5% of people can be exposed, without undergoing ventricular fibrillation, to circulation of currents I_B (r.m.s. mA) for durations t (s) given in Eq. (5.3)[10]:

$$I_B(t) = \frac{k}{\sqrt{t}} \tag{5.3}$$

In Eq. (5.3), k is a factor that, basically, depends on body weight. I_B is the current that has the probability of triggering ventricular fibrillation of 0.5%. For a standard person of body mass 50 kg, k equals 116, whereas for a person of body mass 70 kg, k equals 157. Equation (5.3) is the result of tests carried out for touch times ranging between 0.03 and 3 s, and therefore cannot be used for times not in this interval.

To confirm the study that led to Eq. (5.3), we also note that the graph of $I_B(t)$ (Fig. 5.10) has the same shape as the excitability curve in Fig. 5.4.

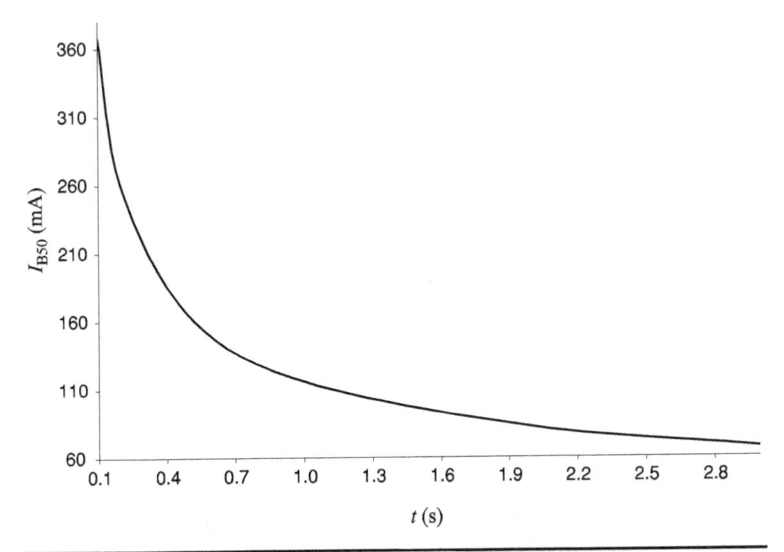

FIGURE 5.10 Nonfibrillating body current as a function of the duration of the contact (standard person of 50 kg of mass).

5.6 Permissible Body Current Independent of Human Size

Conventional current–time curves, which describe the effects of a.c. currents (15–100 Hz) on human beings, when a left-hand-to-feet contact occurs, have also been elaborated.[11] These curves do not take into consideration the individual's size and identify four current–time zones of increasing hazard for people (Fig. 5.11).

In *Zone 1* (0 up to 0.5 mA, curve a), the perception of current is possible, but no reactions will be induced.

In *Zone 2* (0.5 mA up to curve b), involuntary muscular contractions are likely, but there are no harmful effects.

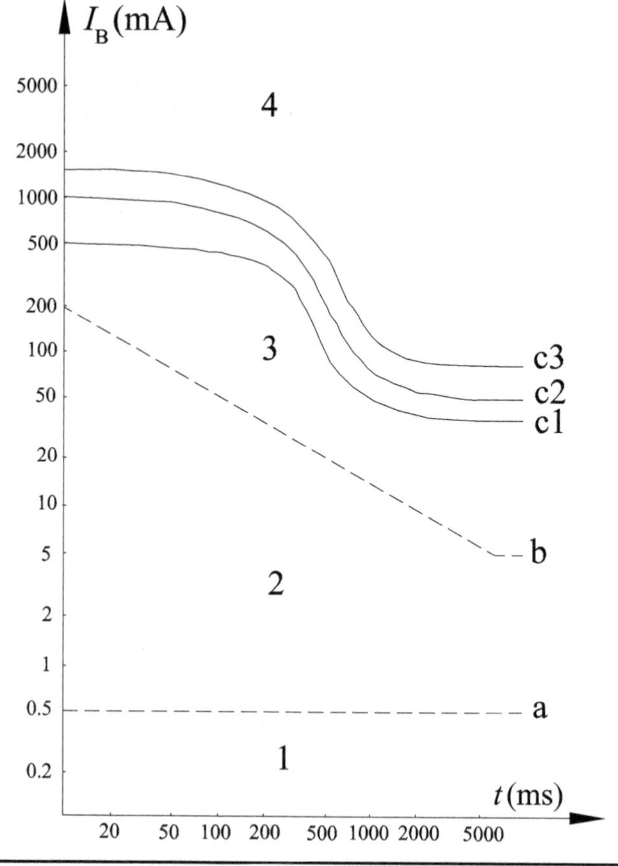

FIGURE 5.11 Conventional a.c. current–time curves and hazardous zones (15–100 Hz).

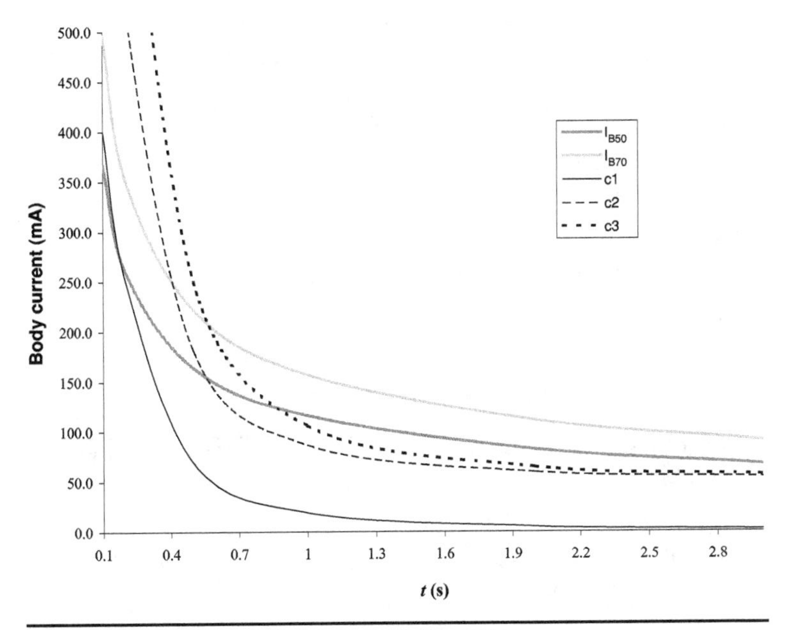

FIGURE 5.12 Comparison between fibrillation curves.

In *Zone 3* (curve b and above), physiological effects, usually reversible, will occur, such as strong muscular contractions, respiratory difficulties, atrial fibrillation, and temporary cardiac arrest, but no ventricular fibrillation.

In *Zone 4* (above curve c1), ventricular fibrillation is likely to occur and its probability increases with magnitude and duration of the current; curves c1–c2 enclose the area characterized by the probability of 5% of ventricular fibrillation, while in between curves c2 and c3 this probability increases up to 50%; beyond curve c3 this probability exceeds 50%.

A comparison between IEC and IEEE curves shows that both methodologies roughly provide the same permissible values of body currents (Fig. 5.12).

5.6.1 Heart Current Factor

Figure 5.11 allows the determination of fibrillating currents only for a left-hand-to-feet pathway. Accidental contacts, though, may involve paths other than the aforementioned one. The probability of fibrillation depends on the course of the current passing through the body, as this changes the direction of the electric field through to the heart.

F	Left Hand	Right Hand	Both Hands	Right Foot
Left hand	n/a	0.4	n/a	1
Right hand	0.4	n/a	n/a	0.8
Both hands	1	n/a	n/a	1
Left foot	1	0.8	1	0.04
Right foot	1	0.8	1	n/a
Both feet	1 (reference)	0.8	1	n/a
Chest	1.5	1.3	n/a	n/a
Back	0.7	0.3	n/a	n/a
Glutei	0.7	0.7	0.7	n/a

TABLE 5.1 Heart-Current Factor *F* Applicable to Current Not Flowing Through the Reference Path

For other paths, then, the following corrective heart-current factor F must be considered:

$$F = \frac{I_{\text{LH-2F}}}{I_{\text{F}}} \tag{5.4}$$

where $I_{\text{LH-2F}}$ is the fibrillating current for a left-hand-to-feet pathway, assumed as reference, while I_{F} is the current flowing through a different route that has the same probability to cause ventricular fibrillation as $I_{\text{LH-2F}}$, but not necessarily the same intensity.

Values for F for various paths as the result of possible combinations are reported in Table 5.1.

Example 5.1 A chest-to-left-hand current ($F = 1.5$) equal to 50 mA has the same probability to cause ventricular fibrillation as a 75-mA left-hand-to-feet current (i.e., reference current); a left-foot-to-right-foot current ($F = 0.04$) must be 25 times larger than a left-hand-to-feet current in order to cause ventricular fibrillation with equal probability; a left-foot-to-right-foot current $I_{\text{F}}^{\text{LF-RF}}$ ($F = 0.04$) must be 20 times larger than a right-hand-to-feet current ($F = 0.8$) in order to cause ventricular fibrillation with the same probability, in fact:

$$I_{\text{LH-2F}} = 0.8 I_{\text{F}}^{\text{RH-2F}},$$

$$0.04 = \frac{I_{\text{LH-2F}}}{I_{\text{F}}^{\text{LF-RF}}} = \frac{0.8 I_{\text{F}}^{\text{RH-2F}}}{I_{\text{F}}^{\text{LF-RF}}} \Rightarrow I_{\text{F}}^{\text{LF-RF}} = \frac{0.8 I_{\text{F}}^{\text{RH-2F}}}{0.04} = 20 I_{\text{F}}^{\text{RH-2F}}$$

5.7 Human Body Impedance

The human body impedance Z_{B} consists of resistive and capacitive elements. The skin, by acting as an insulating dielectric between the conductive tissue underneath it and live parts, can be thought as a

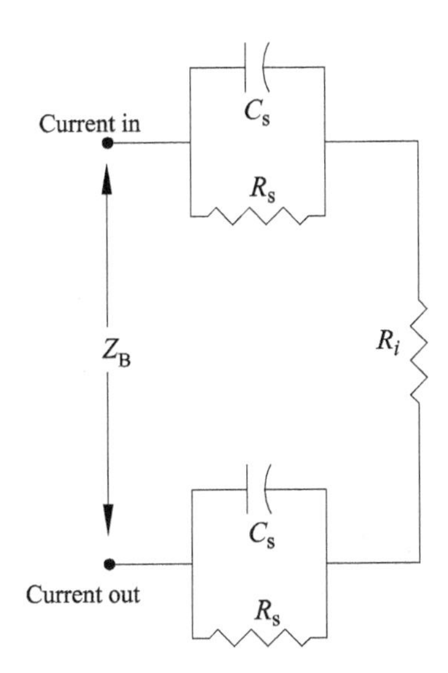

FIGURE 5.13 Impedance of the human body.

capacitance C_S. In parallel to C_S, we consider the resistance R_s offered by the skin pores, which are small conductive elements. In series to the skin impedance Z_S, there is the internal resistance R_i of the body (Fig. 5.13).

At 50/60 Hz, the capacitive reactance is extremely high (i.e., the capacitors of Fig. 5.13 are open circuits) and we can consider the human body as a purely resistive element (i.e., $\underline{Z}_B \approx R_B$). In addition, for touch voltages increasingly greater than 200 V, the skin ruptures and no longer plays any insulating role from live parts. In this case, $R_B \approx R_i$.

R_i essentially depends on the current path and is mainly concentrated in the lower and upper limbs, as they have a small cross section with respect to the rest of the body. The trunk resistance, in fact, is much lower due to its larger section and the presence of conductive fluids in it. If we consider the trunk as a "short circuit" and suppose arms and legs of equal resistance, we obtain a "Π" quadripole for modeling the resistance of the human body (Fig. 5.14).

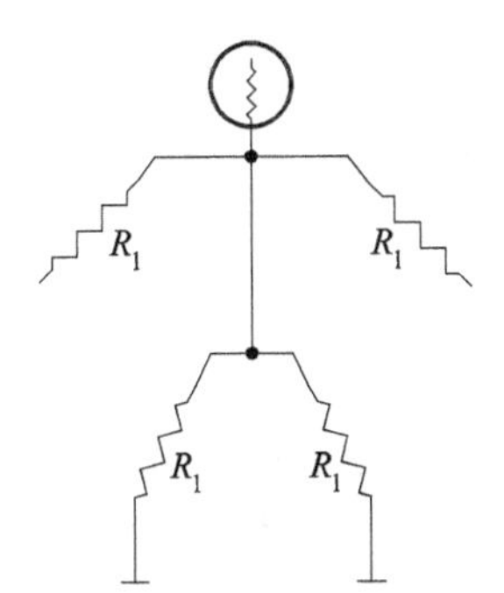

FIGURE 5.14 Equivalent "Π" quadripole for the body resistance R_B (= R_i) of the human body.

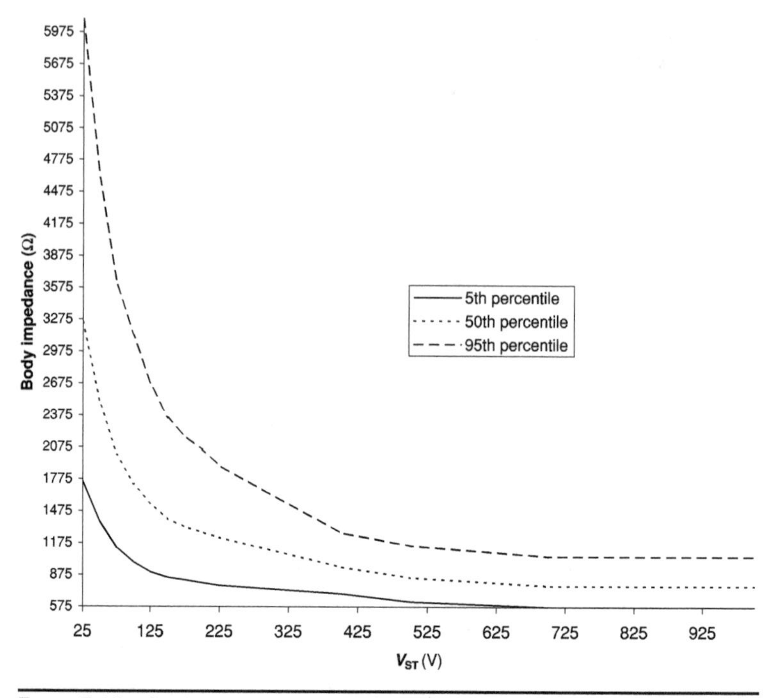

FIGURE 5.15 Statistical values of the human body impedance for hand-to-hand current path.

As anticipated in Chap. 4, IEC standards consider the human body impedance as a nonlinear function of the prospective touch voltage V_{ST}.[12] IEEE standard,[13] instead, considers the human body resistance as a constant value of 1000 Ω for hand-to-hand, hand-to-feet, and foot-to-foot paths, independent of the voltage.

In Fig. 5.15, statistical values of hand-to-hand impedance are reported, in the case of large surface areas of contact (i.e., order of magnitude 10^4 mm^2) and in dry conditions; 5%, 50%, and 95% of the population does not exceed these statistical values indicated by the curves in Fig. 5.15,[14] respectively.

In Fig. 5.15, note the conservative use of the source voltage V_{ST} instead of the touch voltage V_T. If safety is achieved for values of V_{ST}, it will surely be achieved for the lesser values of V_T, since $V_T < V_{ST}$.

5.8 Current Paths

Different current paths through the body cause different hazards to persons. The hand-to-hand path in Fig. 5.16 is less dangerous than the hand-to-feet path in Fig. 5.17.

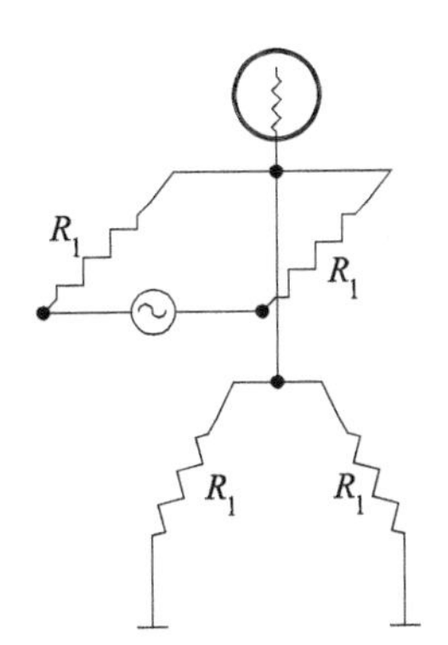

FIGURE 5.16 Hand-to-hand current path.

In the first case, in fact, the current is limited by the series of the two arms' resistances $R_{B}^{Hand\text{-}Hand} = 2R_{l}$, while in the latter case, the two legs are in parallel; therefore, $R_{B}^{Hand\text{-}Feet} = 1.5R_{l}$, and hence, $R_{B}^{Hand\text{-}Feet}$ results to be 75% of $R_{B}^{Hand\text{-}Hand}$.

Some measurements would indicate that the impedance hand-to-foot $R_{B}^{H\text{-}F}$, for large areas of contact, is 10% to 30% lower than $R_{B}^{Hand\text{-}Hand}$. Hence, an average reductive factor of 0.8 may be applied to $R_{B}^{Hand\text{-}Hand}$ to calculate the hand-to-foot body resistance.

The worst case, assumed in the electrical design, is a person touching a live part with both hands (Fig. 5.18). In this case, $R_{B}^{Hands\text{-}Feet} = R_{l}$, as all the limbs are in parallel. $R_{B}^{Hands\text{-}Feet}$ is 50% of $R_{B}^{Hand\text{-}Hand}$.

In relation to the above worst-case scenario, the values for the total body impedance for large areas of contact are detailed in Table 5.2 as a function of the prospective touch voltage and the percentages of population.

5.9 Permissible Prospective Touch Voltage V_{ST}^{p}

If we assume in series to the human body a resistance-to-ground R_{BG} of 1000 Ω in standard conditions (see Chap. 4), Table 5.2 allows

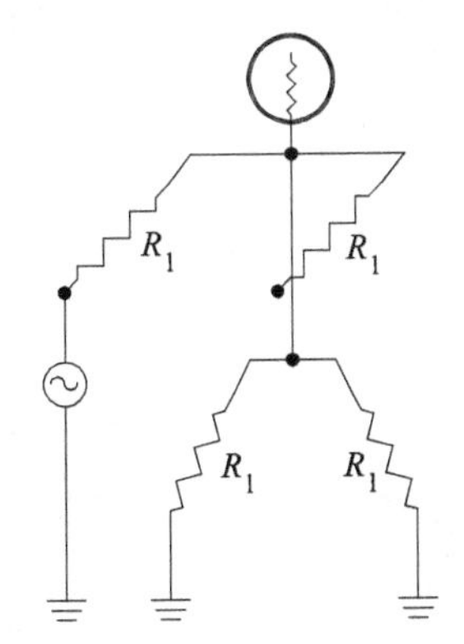

FIGURE 5.17 Hand-to-feet current path.

FIGURE 5.18 Both hands-to-both feet current path.

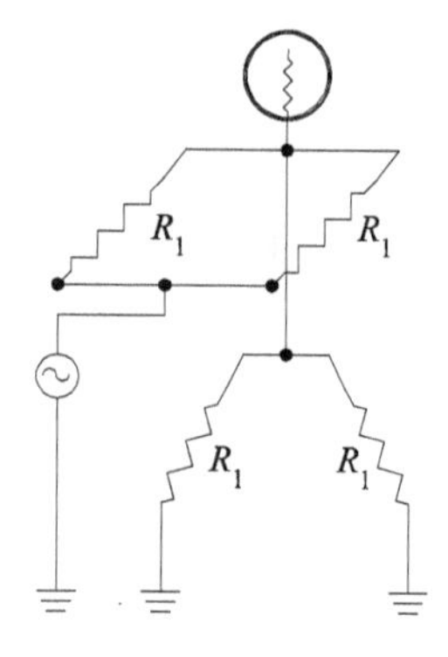

the definition of the *permissible prospective touch voltage* V_{ST}^{P} through Eq. (5.5):

$$V_{ST} = I_B\left(Z_B^{\text{Hands-Feet}} + R_{BG}\right) \tag{5.5}$$

We can solve Eq. (5.5) for the body current I_B, assuming for $Z_B^{\text{Hands-Feet}}$ the values of body resistance not exceeded by 95% of the population at any chosen value of V_{ST}. In correspondence to the obtained I_B, we can read on curve c1 in Fig. 5.11 the maximum time t that does not cause ventricular fibrillation. As a result, the chosen value

	$Z_B^{2H\text{-}2F}$ (Ω)		
	Percentage of Population		
V_{ST}	**5%**	**50%**	**95%**
25	875	1625	3050
50	687	1250	2300
75	562	1000	1800
100	495	862	1562
125	450	775	1337
150	425	700	1175
175	412	662	1087
200	400	637	1025
225	387	612	950
400	350	475	637
500	312	425	575
700	287	387	525
1000	287	387	525

TABLE 5.2 Statistical Values of the Human Body Impedance $Z_B^{2H\text{-}2F}$ for Both Hands-to-Both Feet Current Path

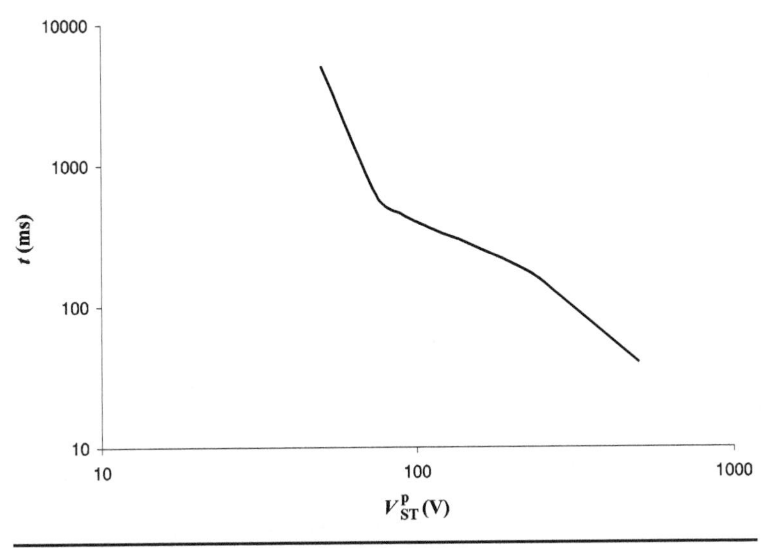

FIGURE 5.19 Time–voltage safety curve in ordinary locations.

of V_{ST} becomes the permissible prospective touch voltage V_{ST}^{P} human beings can withstand for the maximum amount of time t, without suffering ventricular fibrillation.

By solving Eq. (5.5) for all the values of V_{ST} and $Z_{B}^{2H\text{-}2F}$ in Table 5.2, we can build the *time–voltage safety curve* in standard conditions, as indicatively shown in Fig. 5.19.

It is apparent that as the touch voltage increases, the maximum permissible contact time decreases. The *time–voltage safety curve* is the "damage" curve for human beings: any combination of time and voltage above this curve is dangerous.

Example 5.2 Calculate the body current I_B due to a touch with a metal part energized at $V_{ST} = 200$ V, in the case of dry conditions; current path hands-to-feet with medium surface area of contact for hands (order of magnitude 10^3 mm^2, $R_B^{\text{Hand-Hand medium area}} = 2.2\,\text{k}\Omega$); large surface area of contact for feet; $R_{BG} = 1000\ \Omega$. Figure 5.20 schematically represents the data of the example.

Solution $R_B^{\text{Hand-Hand large area}} (= 2R_l) = 1.275\,\text{k}\Omega$ (from Fig. 5.15 in correspondence to 200 V).

$R_B^{\text{Hand-Foot large area}} = 0.8 R_B^{\text{Hand-Hand large area}} = 1.02\,\text{k}\Omega$ (the hand-to-foot body resistance is obtained by reducing by 20% the value $R_B^{\text{Hand-Hand large area}}$).

$R_B^{\text{Hand-Trunk large area}} = 0.5 R_B^{\text{Hand-Hand large area}} = 0.637\,\text{k}\Omega$ (as said, the trunk has negligible resistance).

$$\begin{aligned} R_B^{\text{Trunk-Foot large area}} &= R_B^{\text{Hand-Foot large area}} - R_B^{\text{Hand-Trunk large area}} \\ &= 0.8 R_B^{\text{Hand-Hand large area}} - 0.5 R_B^{\text{Hand-Hand large area}} \\ &= 0.3 R_B^{\text{Hand-Hand large area}} = 0.382\,\text{k}\Omega. \end{aligned}$$

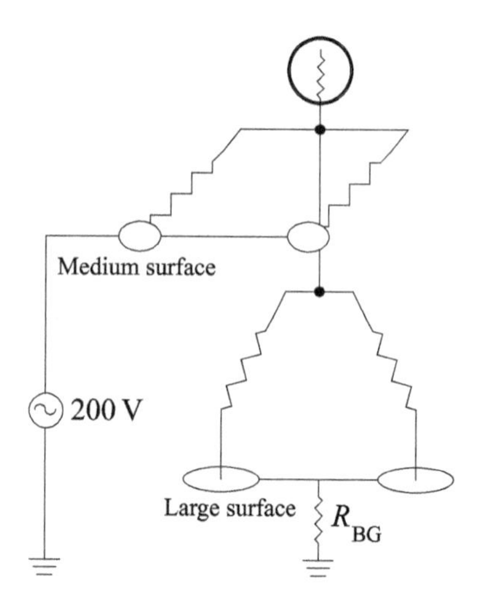

FIGURE 5.20 Schematic representation of the data of the example.

$R_B^{\text{Hand-Hand medium area}} = 2.2\,\text{k}\Omega$ (the hand-to-hand body resistance for medium contact area is given in the example).

$$R_B^{\text{Hand-Trunk medium area}} = 0.5 R_B^{\text{Hand-Hand medium area}} = 1.1\,\text{k}\Omega$$

The total hand-to-foot body impedance, in the conditions required by the problem, is

$$\begin{aligned} R_B^{\text{Hand-Foot}} &= R_B^{\text{Hand-Trunk medium area}} + R_B^{\text{Trunk-Foot large area}} \\ &= 1.1\,\text{k}\Omega + 0.382\,\text{k}\Omega = 1.482\,\text{k}\Omega. \end{aligned}$$

The total hands-to-feet body impedance is

$$R_B^{\text{Hands-Feet}} = 0.5 R_B^{\text{Hand-Foot}} = 0.741\,\text{k}\Omega.$$

The body current I_B is

$$I_B = \frac{V_{ST}}{R_B^{\text{Hands-Feet}} + R_{BG}} = \frac{200}{1741} = 114\,\text{mA}.$$

5.10 Effects of Direct Currents

Direct currents have come into the spotlight in the recent years due to the increased development of renewable sources of energy (e.g., photovoltaic cells, wind turbines, etc.), which generate d.c. currents. By the term direct current, international standards intend a constant current to which may be superimposed a sinusoidal ripple, whose r.m.s value does not exceeds 10% of the d.c. current itself. In this case, the expression ripple-free current is used.

Direct current is generally less dangerous, since the thresholds of let-go, and of ventricular fibrillation, for contacts longer than the cardiac cycle, are significantly higher than for a.c. current.

Data from electrical accidents and experiments on animals corroborate the fact that the risk of fibrillation is generally present only for longitudinal currents, that is, for currents flowing upward/downward through the person, for which the feet represent the positive/negative polarity. The most dangerous current path is upward (i.e., feet at positive polarity), which is characterized by a threshold of fibrillation about one-half of the downward current. For transverse currents circulating, for example, from hand to hand, ventricular fibrillation may only occur with high current intensities.

Conventional body current–time curves, which describe the effects of d.c. currents on persons for a longitudinal upward current path, have been elaborated[15] (Fig. 5.21).

In *Zone 1* (0 up to 2 mA, curve a), there is generally no reaction.

In *Zone 2* (2 mA up to curve b), no harmful physiological effects will usually occur.

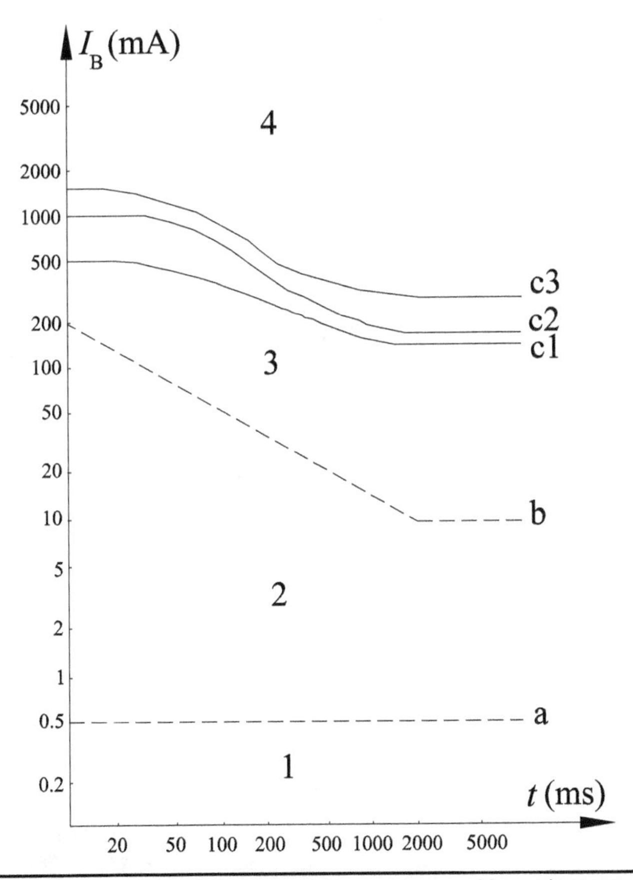

FIGURE 5.21 Conventional d.c. current–time curves and hazardous zones (longitudinal upward current path).

In *Zone 3* (curve *b* and above), no organic damage will usually occur; by increasing current magnitude and exposure time reversible disturbances of formation and conduction of impulses in the heart may occur.

In *Zone 4* (above curve c1), ventricular fibrillation is likely to occur with a probability that increases with magnitude and duration of the current; pathophysiological effects, such as burns, may occur in addition to those of *Zone 3*. Curves c1–c2 enclose the area characterized by the probability of 5% of ventricular fibrillation, while in the area c2–c3, this probability increases up to 50%; beyond curve c3 this probability exceeds 50%.

In order to compare the fibrillation curves of alternating and direct currents, we can place them in a single chart (Fig. 5.22).

We note that for contact times below certain values, the direct current is more dangerous than the alternating one. For instance, if

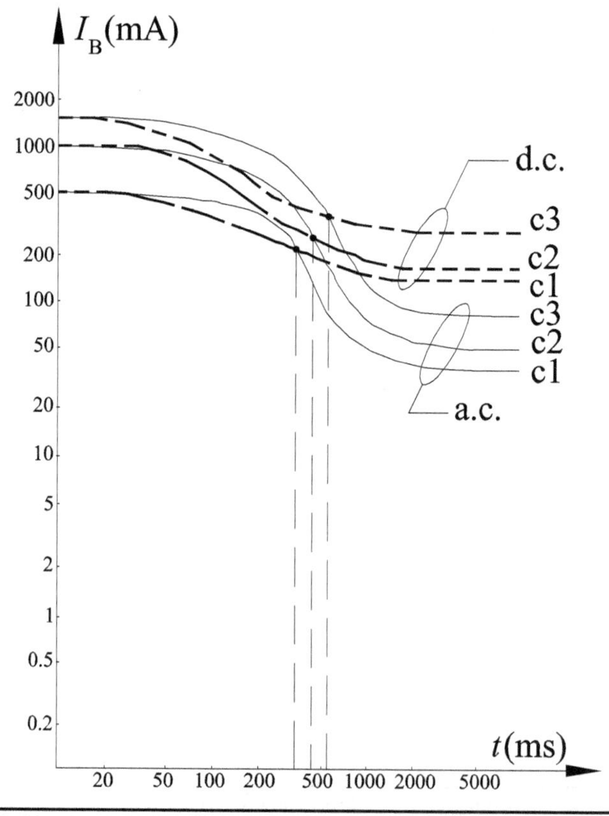

FIGURE 5.22 Comparison between a.c. and d.c. fibrillation curves.

we consider curves c3 in d.c. and c3 in a.c. for any time below 600 ms, the threshold value that triggers the ventricular fibrillation is lower in d.c. than in a.c.

FAQs

Q. Are the body current values expressed in Eq. (5.3) safe?

A. Equation (5.3), based on statistical extrapolations, gives values that provoke ventricular fibrillation "only" in 0.5% of the population examined. As said in Chap. 3, this is the acceptable risk as set by IEEE standard 80, well knowing that we might not belong to the lucky 99.5% of the sample!

Q. What is the difference between the human body impedance Z_B and the internal resistance R_i?

A. Z_B is the complex of resistive and capacitive components, which includes the resistance and the capacitive reactance of the skin. R_i is the resistive core component of the human body impedance, which coincides with Z_B at low frequencies (i.e., 50/60 Hz) and for voltages exceeding 200 V. R_i is basically due to the resistances of the limbs, as the trunk is highly conductive.

Q. If the body current is alternating, why do we mark the directions "in" and "out" in Fig. 5.13?

A. We are here taking into consideration alternating body current (50/60 Hz), which changes direction 100/120 times in a second. Therefore, in Fig. 5.13, the directions "in" and "out" alternate themselves at the same rate. Conventionally, though, we deem the direction "in" as the point of touch with live parts.

Endnotes

1. Their resistivity approximately equals 0.6 and 2 $\Omega \cdot$ m, respectively.
2. A depolarizing stimulus, that is an electric impulse, must have the opposite potential of the cell.
3. The *all-or-none* law is no longer true if applied to functional units, such as the heart. In fact, a unit's global response is proportional to the stimulus.
4. As a curiosity, the discussion on the presumed higher danger of the a.c. current versus d.c. initiated at the end of 1800 between Thomas Alva Edison, inventor of the d.c. utilization and transmission, and George Westinghouse, supporter of the a.c. system. The debate was animated by the New York State officials' efforts in finding a more humane means to carry out capital punishments than hanging. Eventually, New York's decision for the electric chair in alternating current, implicitly suggested its greater hazard to humans (W. Long, S. Nilsson, "HVDC Transmission: Yesterday and Today," *IEEE Power and Energy Magazine*, Vol. 5, No. 2, March/April 2007).
5. A muscle cell contracts with a force that nearly equals 9.8 mN.
6. As previously said, the action potential can be elicited by an equal strength stimulus only if this occurs after the refractory period.

7. The term has been coined after *tetanus,* a disease of the nervous system that shows identical symptoms.
8. IEEE 1584–2002 *"Guide for Performing Arc-Flash Hazard Calculations"* assumes that persons will suffer second-degree burns if their skin is exposed to incident energies of at least 5 J/cm^2.
9. Defined as the heat necessary to increase the temperature of a unit volume of a substance by 1°C.
10. Based on IEEE Std. 80–2000, *"IEEE Guide for Safety in AC Substation Grounding."*
11. IEC TS 604479–1: 2005, *"Effects of Current on Human Beings and Livestock."*
12. IEC TS 604479–1: 2005, *"Effects of Current on Human Beings and Livestock,"* Table 1.
13. IEEE Std. 80–2000, *"IEEE Guide for Safety in AC Substation Grounding,"* paragraph 7.1.
14. Besides dry conditions, body impedance data are also available for water-wet and saltwater-wet conditions, as well as for medium and small surface areas of contact. IEC 604479–1 deems the values described by the 50th percentile curve the most statistically reliable.
15. IEC TS 604479–1: 2005, *"Effects of Current on Human Beings and Livestock."*

CHAPTER 6

TT Grounding System

Electricity, water, gas, and steam course through the walls of my building, keeping it alive.

MASON COOLEY (1991)

6.1 Introduction

The TT system (Terre-Terre, or earth-earth) is the grounding method for low-voltage public supply employed in several countries in the world; for example, Algeria, United Arab Emirates, Belgium, Denmark, Egypt, France, Greece, Italy, Japan, Kenya, Luxemburg, Morocco, Tunisia, Spain, Portugal, Turkey, etc.

The supply system is solidly grounded, and the neutral is usually carried in order to provide power to single-phase loads, as is typical of dwelling units.

The consumer's ECPs (exposed-conductive-parts) are connected to a house ground electrode, independent of the earthing of the utility. In these conditions, the ground-fault current will return to the supply through the soil (Fig. 6.1), flowing through both the earth electrode of the installation (R_G) and the earth electrode of the source (R_N).

As a result, the ground-fault current is limited in its magnitude by the two above-mentioned grounds, as the impedances of phase and PE conductors are negligible if compared to R_G and R_N. Therefore, $\underline{I}_G$[1] is independent of where the ground fault occurs within the system.

TT grounding systems are used when electrical utilities cannot make available safe means of earthing for their users. The owner of the installation, therefore, must provide its own connection to ground, employing suitable earth electrodes [i.e., conductive element(s) in intimate contact with earth].

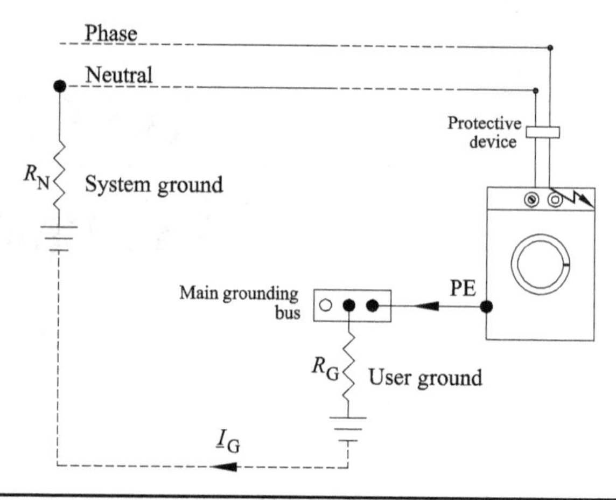

FIGURE 6.1 Fault-loop in TT systems.

6.2 Voltage Exposure in TT Systems

Upon ground faults, persons, conservatively assumed as standing in areas at zero potential, are subject to the touch voltage caused by the ground current flowing through the grounding electrode of resistance R_G. The equivalent circuit of the fault-loop is represented in Fig. 6.2, where $R_{GT} = R_G + R_{PE} + R_{GEC}$, R_{PE} is the resistance of the protective conductor of the ECP, and R_{GEC} is the grounding electrode conductor's resistance. In most cases, it is not always necessary to add up the last two components, because they are very small if compared to R_G and, therefore, may be neglected. There may be, though, cases when the ground resistance R_G is very low, that is, of the same magnitude as R_{PE} and/or R_{GEC}, and these two terms can no longer be neglected.[2] In addition, we can reasonably assume that the internal impedance of the source $\underline{Z}_i$ and the impedance of the phase conductor $\underline{Z}_{ph}$ are negligible.

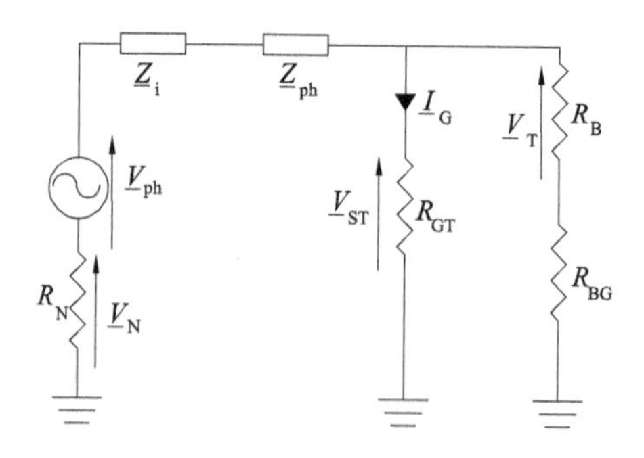

FIGURE 6.2 Equivalent circuit of the fault-loop in TT systems.

As explained in Chap. 4, in low-voltage systems, we conservatively consider persons exposed to the prospective touch voltage V_{ST}, instead of the touch voltage V_T ($V_{ST} > V_T$). Consequently, we disconnect the branch containing ($R_B + R_{BG}$) in Fig. 6.2, as we calculate the touch voltage prior to the person's contact.

By applying Millmann's theorem[3] to the circuit in Fig. 6.1, we obtain the magnitude of the prospective touch voltage V_{ST}:

$$V_{ST} = \frac{(V_{ph}/R_N)}{(1/R_{GT}) + (1/R_N)} = V_{ph} \times \frac{R_{GT}}{R_{GT} + R_N} = V_{ph} \times \frac{1}{1 + (R_N/R_{GT})} \tag{6.1}$$

The negligibility of the phase and protective conductors' impedances in Eq. (6.1) show that the location of the fault, directly related to the aforementioned impedances, has no influence on the magnitude of the prospective touch voltage. V_{ST}, therefore, is a constant value regardless of where the ground fault occurs, but does depend on the system ground resistance of the distributor R_N, which is generally unknown to the designer and is out of his/her control.

In order for V_{ST} to be harmless to persons, R_N should ideally be very large and R_{GT} very low. Figure 6.3 shows the prospective touch voltage V_{ST} as a function of R_{GT} for three increasing values of R_N (1, 10, and 100 Ω) in correspondence with a phase-to-ground voltage V_{ph} of 230 V.

It is clear that a larger R_N improves safety for any fixed value of the user ground. In practice, in urban areas, the utility grounding system at the supply substation may be interconnected in parallel to the earths of other substations via the metal sheath/armor of cables and/or the overhead lightning protection wire. Therefore, the value of R_N is typically very low (i.e., fraction of ohms). This may not be true in rural environments, where pole transformers may only be grounded locally and have an earth resistance R_N of tens of ohms.

It is important to note that in urban areas, the protection offered by the sole user's grounding system does not guarantee the safety of persons, because, as shown in Fig. 6.3, for $R_N = 1$, the prospective touch voltage is not sufficiently low to be harmless and nearly coincides with the phase-to-ground potential as R_{GT} increases. To clarify this concept, let us calculate by applying the voltage divider, the value that R_{GT} should reach in order to limit V_{ST} to the nondangerous value of 25 V (see Fig. 5.19), when V_{ph} equals 230 V and R_N equals 1 Ω:

$$25 = 230 \times \frac{R_{GT}}{R_{GT} + 1} \Rightarrow R_{GT} = 0.12\,\Omega \tag{6.2}$$

The above value is rather difficult to achieve for the users, whose earth resistance ranges in the order of tens of ohms, depending on

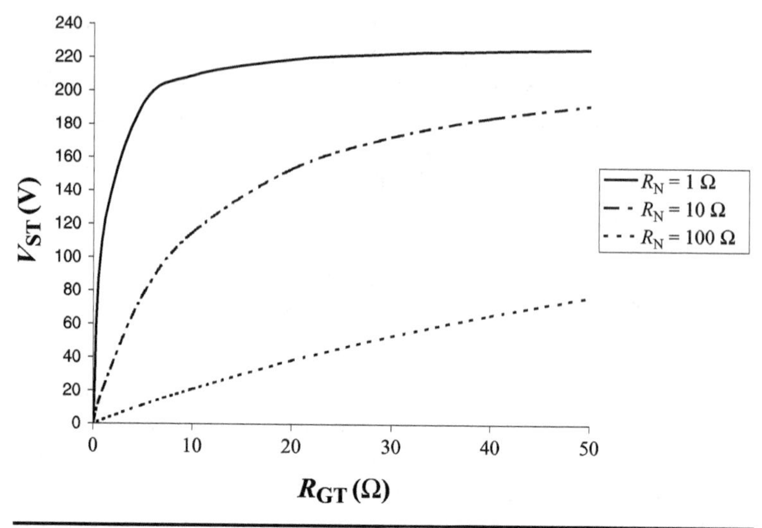

FIGURE 6.3 V_{ST} as a function of R_{GT} for three increasing values of R_N (1, 10, and 100 Ω).

the resistivity of the soil and the size and shape of the electrode. For example, a theoretical hemispherical electrode, buried in a fine sandy soil of resistivity $\rho = 300\,\Omega \cdot \text{m}$, should have the following radius to contain the prospective touch voltage to 25 V:

$$R_{GT} = \frac{\rho}{2\pi r_0} = 0.1\,\Omega \Rightarrow r_0 = 477\,\text{m} \qquad (6.3)$$

The above value, not feasible for use in practice, shows that in TT systems the limitation of V_{ST} to harmless values by the sole means of the house grounding system is not easily achieved due to the extremely low value of its earth resistance that is required.

Safety must be provided by employing protective devices, overcurrent and/or residual, which can promptly disconnect the supply upon ground faults. Each protective device has a built-in an inverse time–current operating curve, which allows the disconnection of the circuit in a time inversely proportional to the magnitude of the fault current. Safety is assured if the protective device operates within the permissible time as per the *time–voltage safety curve* (Fig. 5.19). The protective device's inverse time–current curve must always be below the safety curve (Fig. 6.4).

During a ground fault, the protective device senses the current I_G and initiates the opening action, which will part its contacts. Let us assume that its fault clearing time is t_G. During this period, the enclosure will be energized at the potential $V_G = R_{GT} I_G$. As per the *safety curve* in Fig. 6.4, this voltage can be withstood for the permissible

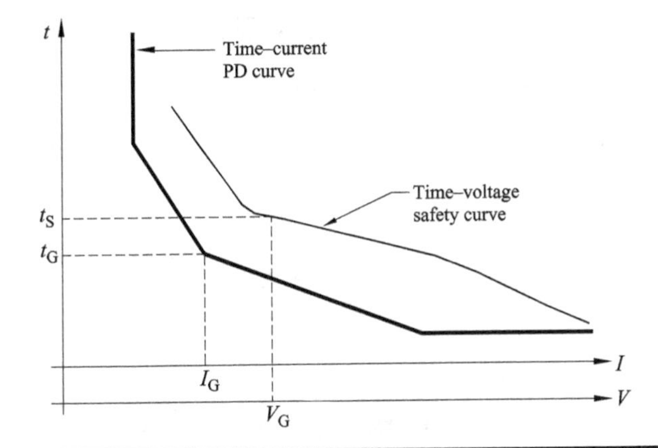

FIGURE 6.4 *Time–voltage safety curve* and inverse time–current curve of the protective device.

time t_P. Safety is achieved if $t_G < t_P$ for any given value of ground-fault current I_G.

6.3 Protection Against Indirect Contact in TT Systems by Using Overcurrent Devices

Overcurrent devices [i.e., electrical circuit breakers (CBs) and fuses] are present in any installation to protect cables and equipment against overloads and short circuits. Standard values for CBs for household and similar applications[4] are (in amperes): 6, 10, 13, 16, 20, 25, 32, 40, 50, 63, 80, 100, 125.[5] The North American *National Electrical Code* lists, among other sizes, for the fixed-trip inverse time CB the following (in amperes): 15, 20, 25, 30, 35, 40, 45, 50, 60, 70, 80, 90, 100, 110, 125.

Circuit breakers (or fuses) can be employed as a protection against indirect contact if the following condition is satisfied:

$$I_G = \frac{V_{ph}}{Z_{Loop}} \geq I_a \tag{6.4}$$

where Z_{Loop} is the series of the impedances of the components that form the ground-fault loop, and specifically the source, the line conductor up to the fault point, R_{GT} and R_N. V_{ph} is the nominal voltage to ground and I_a is the operating current causing the automatic operation of the overcurrent protective device within the time specified in IEC Table 6.1 as a function of the nominal voltage of the system.

Equation (6.4) requires that the ground current be so high as to allow a prompt disconnection of the supply within a time not exceeding

Voltage Range (V)	Maximum Disconnection Times t_a (s)
$50 < V_{ph} \leq 120$	0.3
$120 < V_{ph} \leq 230$	0.2
$230 < V_{ph} \leq 400$	0.07
$V_{ph} > 400$	0.04

TABLE 6.1 Maximum Disconnection Times as a Function of the Nominal Voltage of the System

t_a. If Eq. (6.4) is fulfilled, the overcurrent device will trip within a time that will prevent harmful effects to persons touching live parts.

Table 6.1 applies to final circuits not exceeding 32 A. Final circuits directly supply loads or receptacles. Distribution circuits, instead, supply more than one final circuit, for example, an electric panel where final circuits originate (Fig. 6.5).

As regarding distribution circuits, a maximum disconnection time of 1 s is allowed, as they are conventionally deemed less susceptible to faults than are final circuits.

To clarify the actual applicability of Eq. (6.4) in TT systems, let us consider a typical low-voltage thermal magnetic molded case circuit breaker rated 16 A. Its time–current characteristic is composed of two trip regions: *overload* (also referred to as *thermal*) and *instantaneous* (also referred to as *magnetic*) (Fig. 6.6).

Circuit breakers trip with constant time for fault currents above their instantaneous trip setting I_i, which is a multiple of their ratings.

For instance, in Table 6.1, the maximum permissible clearing time for a system voltage of 230 V is 0.2 s. In correspondence with this safe time, the breaker will trip for a current I_a ranging between 200

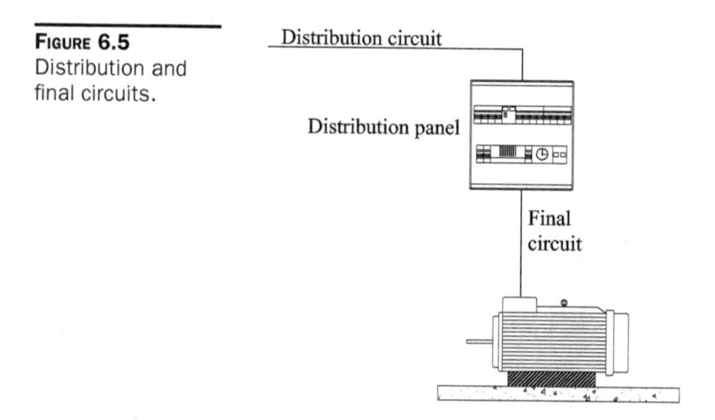

FIGURE 6.5 Distribution and final circuits.

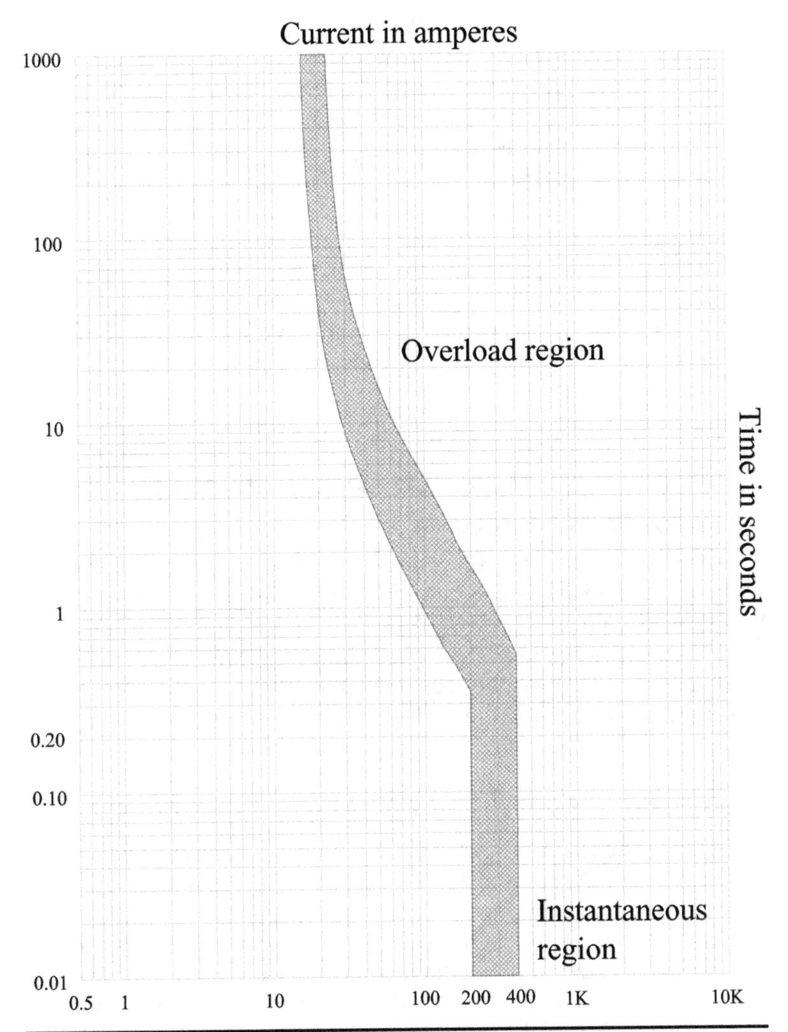

FIGURE 6.6 Time–current characteristic of a thermal magnetic molded case circuit breaker.

and 400 A, depending on the temperature of its contacts. Such currents correspond to 12 to 25 times the breaker's rating (i.e., 16 A in our case). Hence, I_a may be too high to satisfy the condition imposed in Eq. (6.4). The earth loop impedance, in fact, may limit I_G to values below I_a and, therefore, the CB may not defend persons against indirect contact in TT systems.

The general goal of overcurrent devices is to protect installations against abnormal currents that would compromise the integrity of cables and equipment. On the other hand, they must allow the circulation of the normal continuous currents required by the loads. Therefore, in correspondence with large loads, the CB's rating will also be large. This implies that Eq. (6.4) is more difficult to fulfill for large loads than for small ones, as I_a for large loads is, in fact, a multiple of a larger nominal current.

Thus, one might think that small electrical loads are safer than large loads only because they are protected by smaller CBs, and the fulfilling of Eq. (6.4) is, therefore, facilitated. This is, of course, not true, as the risk of indirect contact does not change with the power of the load. The complication, if not the impossibility, of fulfilling Eq. (6.4) can be resolved by using *residual current devices* (RCDs; already introduced in Chap. 2).

6.4 Protection Against Indirect Contact by Using Residual Current Devices

The presence of RCDs in TT systems does not exclude, of course, the overcurrent devices, which must still be employed against overloads and short circuits.

When RCDs are used, the following safety condition, which ties together residual operating current, permissible touch voltage, and the electrode's earth resistance R_{GT}, must be fulfilled:

$$R_{GT} \leq \frac{50}{I_{dn}} \tag{6.5}$$

where I_{dn} is the residual operating current of the RCD, whose standard values are (in mA) 10, 30, 100, 300, 500, and 1000. Equation (6.5) fixes at 50 V the maximum permissible touch voltage upon circulation of the earth current I_{dn} through R_{GT}. If the ground-fault current exceeds I_{dn}, the prospective touch voltage will be greater than 50 V, but it is assumed that the RCD will open the circuit in a shorter time, following the *time–voltage safety curve*, thereby, "compensating" for a larger touch potential. On the other hand, if the earth current is less than I_{dn}, the RCD may not trip at all, but the prospective voltage appearing over the enclosure would be less than 50 V, which is assumed a safe value.

It appears clear that Eq. (6.5) allows higher values of R_{GT} than Eq. (6.4) and therefore is easier to fulfill. Permissible maximum values of R_{GT}, calculated per Eq. (6.5), are shown in Fig. 6.7 as a function of the residual operating currents of the RCD.

In the presence of a RCD rated 30 mA, the ground resistance must not exceed 1667 Ω, which is a fairly easy condition to meet. RCDs, therefore, are the most effective way to protect against indirect contact in TT systems.

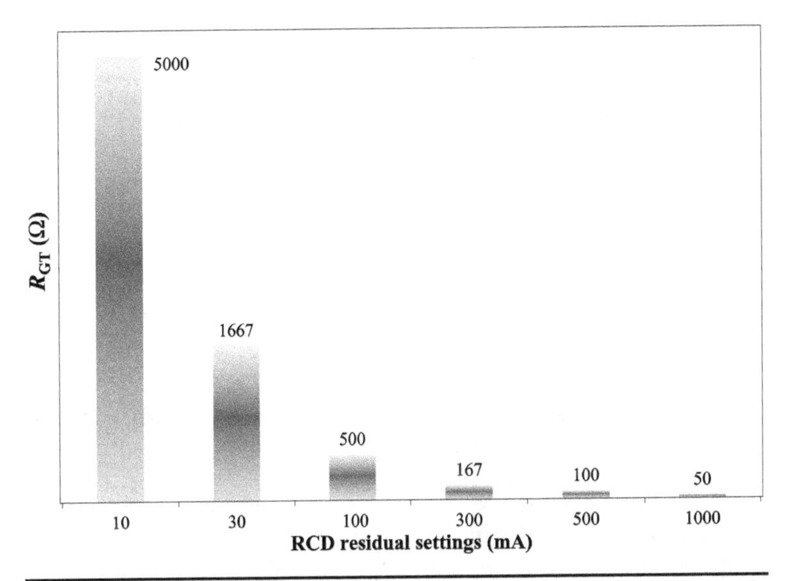

FIGURE 6.7 Values of R_{GT} as a function of the residual operating currents of the RCD.

6.5 Neutral-to-Ground Fault in TT Systems

If the neutral conductor comes into contact with a grounded enclosure, or the protective conductor, or if neutral and ground connections are inverted at the load, a neutral-to-ground first fault occurs (Fig. 6.8).

When the load is off, the RCD will trip only if the neutral is energized and able to impress a sufficient earth current through the user's ground. In this abnormal condition of the neutral,[6] the RCD will

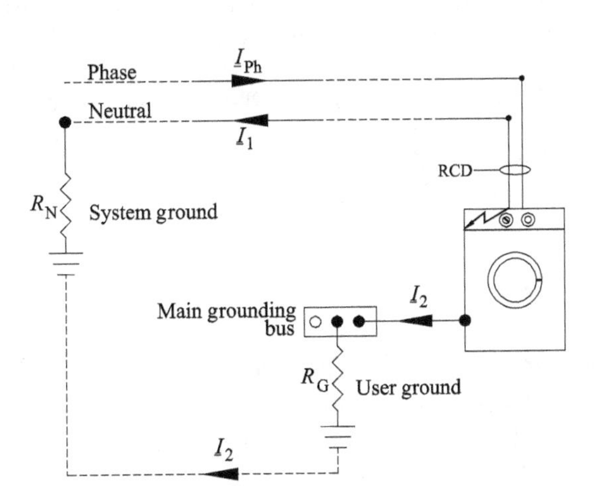

FIGURE 6.8 Neutral-to-ground fault in TT systems.

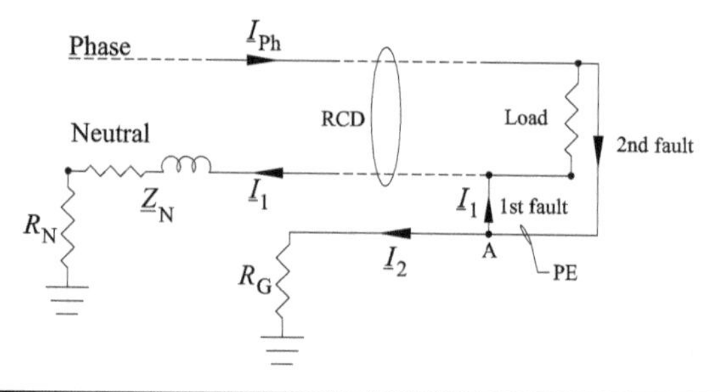

FIGURE 6.9 Second fault phase-to-ground in TT systems.

disconnect the supply to the load before any attempt to turn it on. This may cause nuisance tripping and disruption of the continuity of service.

If the load is on, the RCD may sense an unbalance, as the earth will return part of the neutral current to the source, and a nuisance tripping may occur, if the operating threshold is exceeded (i.e., R_G must be low enough for this to happen).

Upon a phase-to-ground second fault, a current divider takes place at node A (Fig. 6.9) between R_G and the impedance of the neutral conductor $\underline{Z}_N$, which is much lower than R_G.

Only if the current through R_G exceeds the RCD residual operating current, the protection against indirect contact offered by the RCD is effective.

In addition, the presence of a neutral-to-ground first fault may void the additional protection offered by RCDs against direct contact, by creating a latent hazardous situation for persons, even when the load is off.

If a person comes in direct contact with a live part and, simultaneously with the ECP, which the neutral is faulting to, the neutral conductor establishes an alternative, and additional, path to the fault current, which circulates through the person (Fig. 6.10).

Such a path will carry the component $\underline{I}_1$ of the ground current to the source through the RCD's toroid and, therefore, desensitizes it. The RCD will sense only the earth current $\underline{I}_4 = \underline{I}_2 + \underline{I}_3$. $\underline{I}_4$ may not be large enough to cause the RCD to intervene and persons are exposed to the risk of electric shock despite a perfectly functioning protective device.

6.6 Independently Grounded ECPs in TT Systems

ECPs protected by the same RCD must not be connected to independent grounds, because a neutral-to-ground fault can lead to hazardous situations (Fig. 6.11).

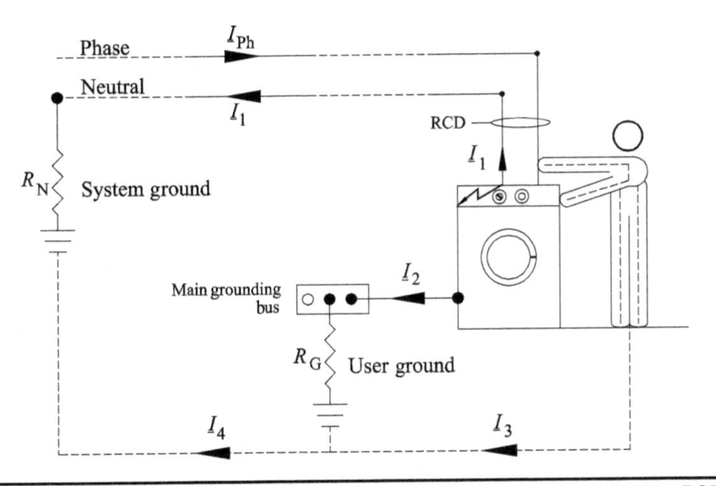

Figure 6.10 Neutral-to-ground first fault in TT systems desensitizing the RCD.

At the occurrence of a ground fault on ECP 1, the faulting-to-ground neutral on ECP 2 conduces current $\underline{I}_2$, which desensitizes the RCD. The RCD will not sense, in fact, the total ground current $\underline{I}_1$, but only the difference $\underline{I}_1 - \underline{I}_2$ and, therefore, might not trip. In addition, $\underline{I}_2$, by flowing through R_{G2}, energizes all the ECPs connected to it, even if healthy, with risk for persons.

It is important to stress that the presence of two independent grounds in the same facility is extremely dangerous, since the ECPs

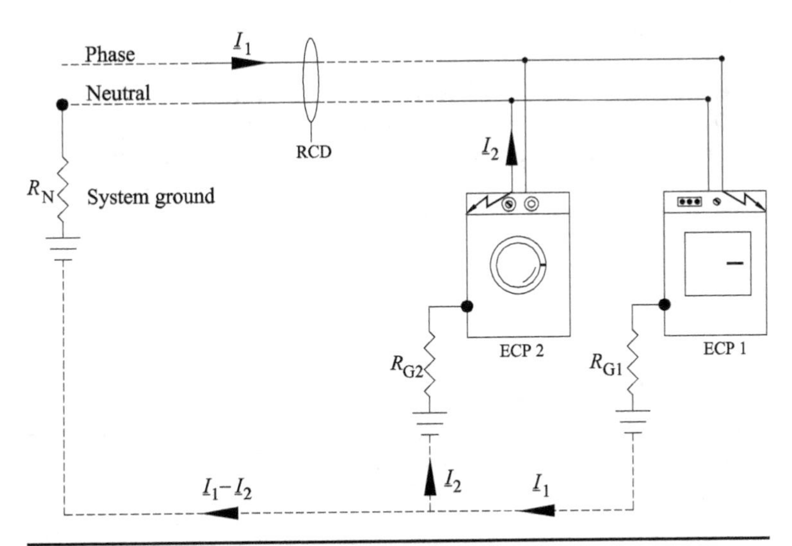

Figure 6.11 Independently grounded ECPs in TT systems.

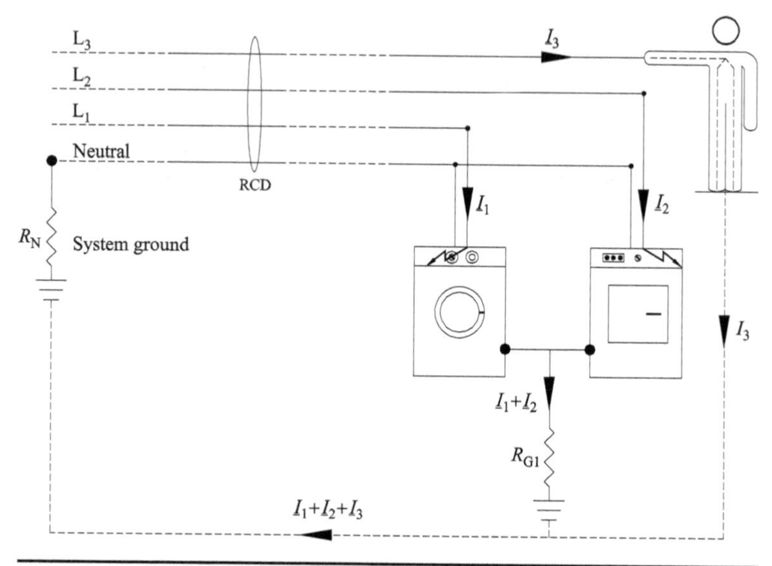

FIGURE 6.12 ECPs that leak to ground in three-phase TT systems.

linked to one ground become EXCPs (extraneous-conductive-parts) to the ECPs connected to the other ground and vice versa.

6.7 Leaking-to-Ground ECPs in Three-Phase TT Systems

As previously explained in Chap. 2, RCDs execute the vectorial summation of all alternating currents flowing through a circuit's wires and compare the result to their operating threshold.

Let us assume that two ECPs, supplied by different phases, are leaking to ground currents,[7] whose vectorial summation is below the RCD operating setting. If direct contact occurs with the healthy phase L_3 (Fig. 6.12), the ground current $\underline{I}_3$ will flow through the person.

The RCD will sense the vectorial summation of these currents, whose magnitude may be less than the magnitude of $\underline{I}_3$ and below its operating threshold (Fig. 6.13). This may cause the RCD not to trip, despite a potentially dangerous shock current $\underline{I}_3$ circulating through the person.

6.8 Electrical Interferences in TT Systems

Let us consider the case of dwelling units, as a part of a building or a complex, sharing a common grounding system. At the occurrence of a ground fault in one unit, the common grounding system becomes live and reaches the voltage rise V_G. Such fault potential is transferred to other dwelling units' healthy ECPs (Fig. 6.14).

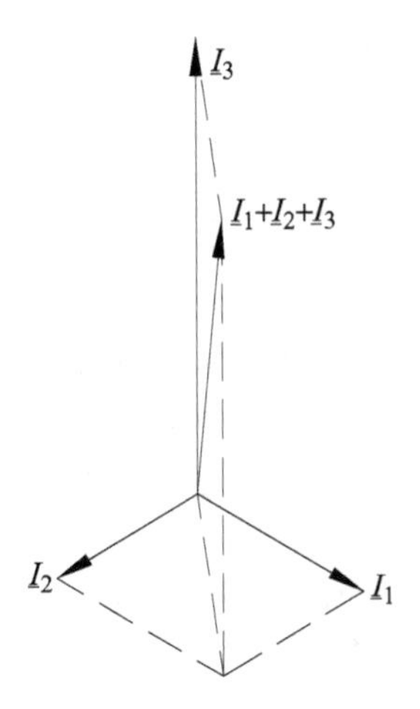

FIGURE 6.13 Vectorial current summation sensed by the RCD.

If the faulty Unit B has no, or malfunctioning, protective device, the voltage rise will persist in the system, as the RCDs of other dwelling units will not be able to sense and clear the ground fault.

Interferences can occur even in the absence of faults. Dwelling units might impress a current to ground below their RCDs residual settings. Some equipment (e.g., computers, high-frequency luminaries, UPS, etc.), in fact, incorporate radio frequency filters, with capacitors connected between the "hot" conductors and the ground. These filters cause functional leakage current (e.g., in excess of 3.5 mA) during the regular functioning of the equipment in order to limit electromagnetic effects.[8] The earth currents from all the units flow through protective conductors and converge into the common R_{GT}. The leakage currents may not cancel each other if they are supplied by the same phase conductor (worst-case scenario). Therefore, they can induce a permanent ground potential on all the grounded metal parts.

In addition, the continuous circulation of the ground currents interferes with the sensitivity of the RCD, by increasing it. If, for instance, 10 mA constantly flow through a 30-mA-rated RCD, it will

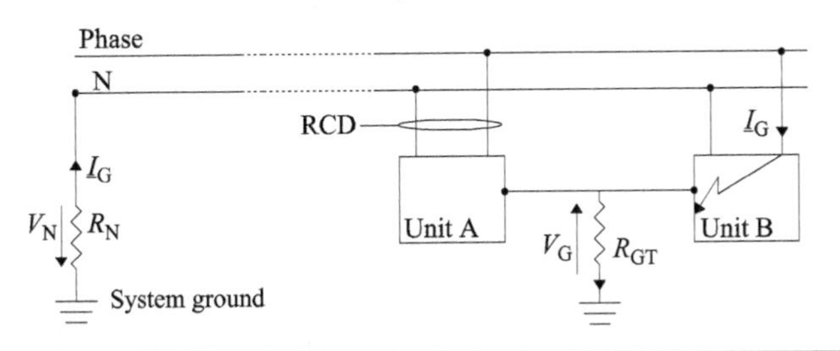

FIGURE 6.14 Electrical interferences upon a fault between dwelling units in a condominium.

take only an extra residual current comprising between 5 and 20 mA to unnecessarily operate the device and cause nuisance trippings. To preserve the continuity of the service, in the presence of known leakage currents, sometimes physiological to electrical systems, it is important that $I_{dNO} > I_d$ (see Chap. 2).

6.9 The Neutral Conductor in TT Systems

In TT systems, the neutral wire is connected to the common point of the three secondary windings in a typical star-connected utility power transformer and is locally earthed. The neutral wire is shipped to the customer together with the line conductors. In normal conditions, this conductor is at zero potential, but due to faults, and also in other nonfault conditions, can assume a nonzero voltage with respect to ground.

Any current circulating in the utility's ground R_N, due to ground faults on both the high and the low-voltage side of the utility's transformer, at the customer's or along the distribution line, causes a ground potential rise V_N on the neutral. If R_N is not low enough, V_N may reach dangerous values.

If the neutral conductor is accidentally interrupted, the neutral wire downstream of the interruption becomes live because phase and neutral conductors result at the same potential (Fig. 6.15).

For the above reason, the neutral wire in TT systems must be considered a "live" conductor and needs to be switched off at the same time as the line conductors.

Also in three-phase systems, the accidental interruption of the neutral wire causes hazardous situations (Fig. 6.16).

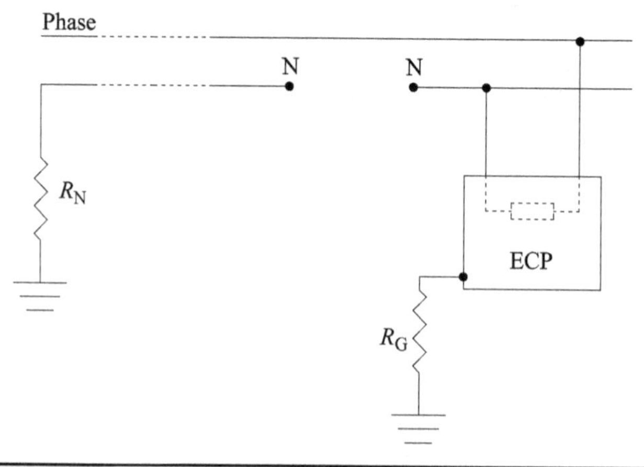

FIGURE 6.15 Accidental interruption of the neutral conductor in single-phase TT systems.

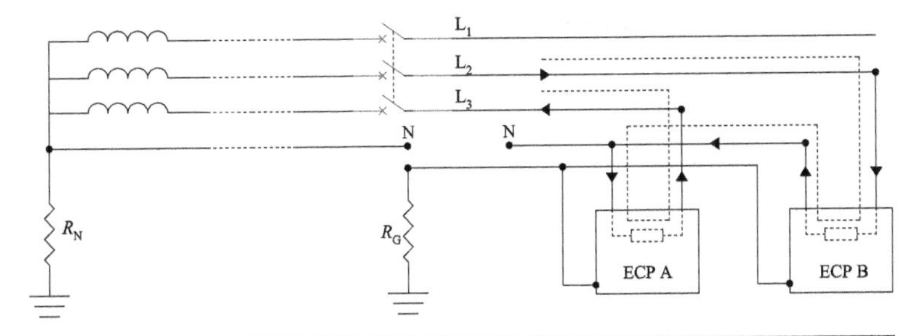

Figure 6.16 Accidental interruption of the neutral conductor in three-phase TT systems.

The absence of the neutral as the return path to the sources causes the phase-to-phase voltage to supply the two ECPs, which will divide across them according to their own impedances. This condition may permanently damage the insulation of the ECPs and provoke ground currents (i.e., order of 500 mA). These ground currents, if not promptly interrupted, may trigger fires.

6.9.1 Resistance of the Utility Neutral in TT Systems

As previously substantiated, safety against indirect contact in TT systems is achieved through prompt disconnection of supply upon the first ground fault. The RCD needs to sense the ground-fault current in order to operate within the limits established by the *time–voltage safety curve*. The amount of earth current depends on both the ground resistances R_G[9] and R_N.

The certain disconnection of the faulty circuit will occur only if

$$\frac{V_{ph}}{R_N + R_G} \geq I_{dn} \tag{6.6}$$

Let us replace in Eq. (6.6) the expression of R_G as per Eq. (6.5), and solve for R_N. We obtain

$$\frac{V_{ph} - 50}{I_{dn}} \geq R_N \tag{6.7}$$

For example, if we conservatively assume I_{dn} equal to 1 A at the user and V_{ph} equal to 230 V, the maximum value that R_N should assume to guarantee the positive tripping of the RCD is 180 Ω. Thus, utilities must keep the resistance of their grounding electrode systems, which earths the neutral point of the supply, below the above-calculated threshold. If the condition expressed in Eq. (6.7) is not fulfilled, the ground fault cannot be cleared by the customer, and the grounding system from TT turns into IT, without having the safety requirements characteristic of this system (see Chap. 9).

6.10 Main Equipotential Bonding

As anticipated in Chap. 4, the equipotentialization between ECPs and EXCPs, which can be simultaneously touched, reduces to safe values, or eliminates, potential differences arisen between them, in earth faults conditions.

Equipotentialization is practically achieved by connecting all the EXCPs to the grounding system. In particular, the following items must be linked together at the main grounding bus of the building to realize the *main equipotential bonding* (MEB) (Fig. 6.17):

- Pipes supplying services within the building (e.g., gas, cold water, etc.)
- Central heating and air-conditioning systems (if present)
- Structural metallic parts of building
- Reinforcing bars embedded in concrete
- Circuit protective conductors
- Main grounding conductor

The EXCPs originating outside of the building must be bonded as close as is practical to their point of entry within the building.

The remarkable "by-product" of the main equipotentialization is the reduction of the resistance of the house grounding system because the EXCPs act as electrodes in parallel to the made-electrode(s)

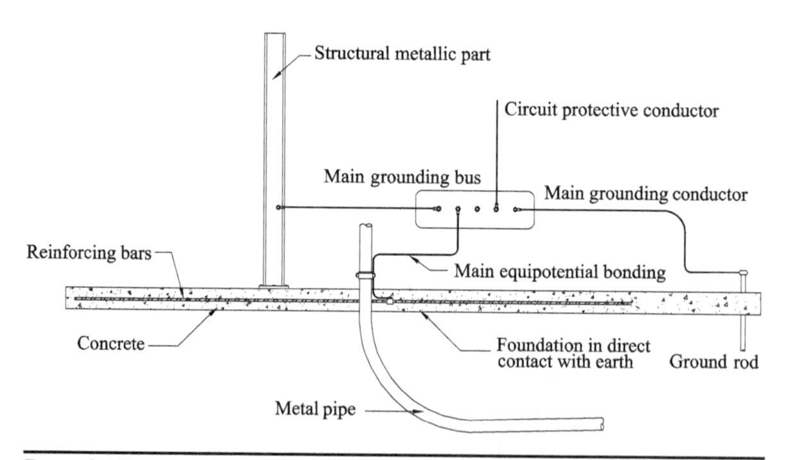

FIGURE 6.17 Main equipotential bonding.

(e.g., the ground rod). This fortunate result attenuates the touch voltages eventually occurring in the building.

6.10.1 Should We Bond Incoming Pipes Made of Plastic?

In some cases water pipes entering the building are made of plastic. There is no need to say that it is not possible, nor required, to bond such pipe. Formally speaking, the plastic pipe is neither an EXCP, as its resistance to ground is much greater than 1000 Ω, nor an ECP. For this reason there is no requirement to link it to the equipotential system, even if it originates outside of the building.

There should be no concern either about the possibility that the tap water present in the plastic pipe can conduct electricity. It has been demonstrated via measurements, in fact, that tap water is a poor conductor of electricity, even in the presence of impurities.[10]

If the incoming pipe is made of plastic, but within the building is made of metal, the main bonding is still necessary (Fig. 6.18).

If, in fact, ECP A, accidentally not bonded, energized the metal water pipe, the fault potential would be transferred within the building. Persons simultaneously touching pipe and ECP B (bonded equipment) would be exposed to the earth potential in the absence of the main equipotential connection. The bonding connection between the metal pipe and the grounding system, realized downstream of the water meter, would virtually cancel this potential difference, by equalizing the pipe and ECP B.

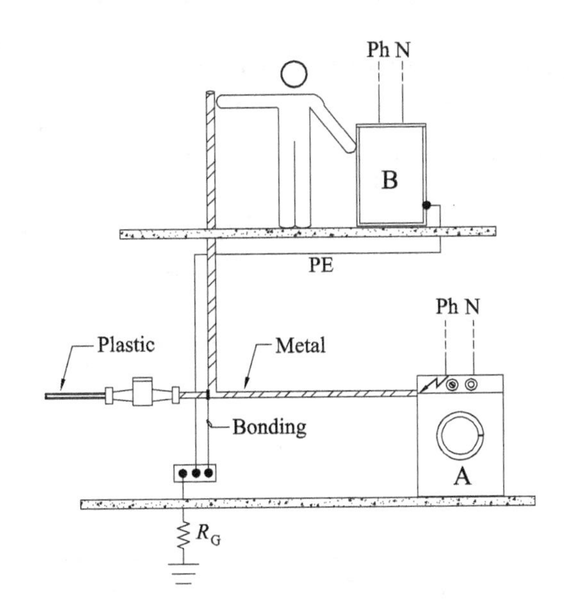

FIGURE 6.18 Main equipotential bonding of the metal pipe.

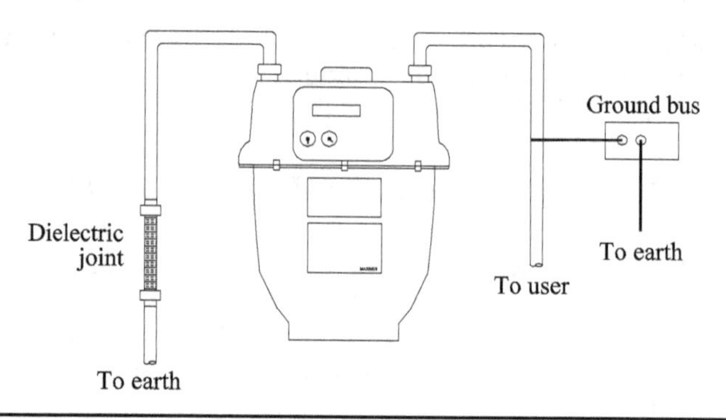

Figure 6.19 Dielectric joint across a natural gas metal pipe.

6.10.2 Should We Bond Incoming Pipes Electrically Separated by a Dielectric Joint?

Natural gas utilities may interpose a dielectric joint, which is highly resistant to currents to electrically separate their metal pipes from the user's in a permanent fashion in order to protect their metal work (Fig. 6.19).

The electrical connection of unlike conductive materials (i.e., steel pipe and copper ground rod) embedded in electrolytes such as the earth, by forming a galvanic cell (i.e., a battery) might trigger corrosion to the detriment of the pipe. Corrosion occurs when direct currents, by leaving metal parts, "drag" out their constituent materials. As copper has a negative potential (i.e., −0.34 V), while the steel is positive (i.e., 0.04 V), the latter will suffer corrosion (Fig. 6.20).

The dielectric joint prevents the above-described corrosion of the pipe by interrupting the circuit of the galvanic cell.

Because of the dielectric joint, the metal pipe is insulated from ground; however, for the same reasons earlier explained, it needs to be connected to the main equipotential system for safety reasons. The

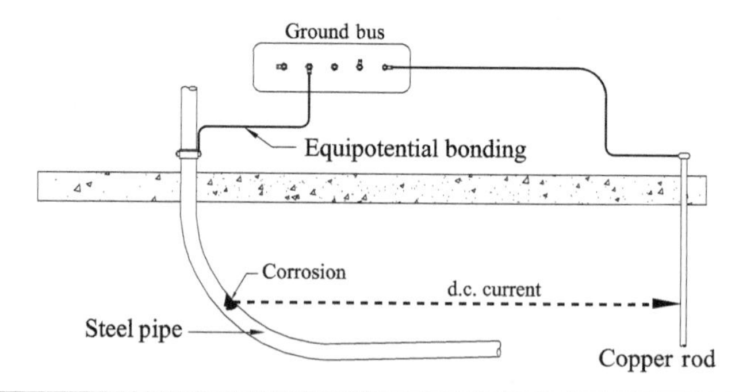

Figure 6.20 Galvanic corrosion.

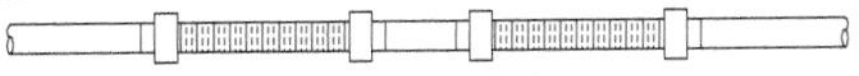

FIGURE 6.21 Two dielectric joints configuration.

bonding should take place downstream of the joint so as to not compromise the cathodic protection of the pipe. The dielectric joint is not under the user's control, even if installed on his/her premises. A utility might, therefore, remove it without warning the client of the hazard of having an ECP not equipotentialized in the building.

The dielectric joint prevents stray currents, impressed by foreign systems and flowing through any underground metal work, from causing the energization of the pipe in the customer's premises. Such energization is potentially dangerous if it involves a piece of equipment fuelled by natural gas.

There is a further safety issue introduced by the insulating joint caused by the separation of the metal pipe in two parts. The two extremities, in fact, may be at different potentials upon faults. A potential difference between the two segments is a hazard for persons simultaneously touching them. For this reason, the joint should be at least 2 m long, or two joints should be used instead of one (Fig. 6.21).

6.11 Supplementary Equipotential Bonding

Areas containing baths, showers, or pools provide a further shock risk due to the presence of water and humidity. Moisture, in fact, by decreasing the human body resistance to values below the ones shown in Table 5.2, may cause protective devices, designed to operate in standard situations, not to be effective. In bathrooms, both hot and cold taps are EXCPs and belong to different systems; therefore, they may be subject to potential differences under fault conditions.

To decrease the hazard in these areas at increased electrical risk, all EXCPs, within the reach of ECPs, must be bonded to a local supplementary ground bus (Fig. 6.22).

The supplementary bonding (SB), consisting of extra connections between ECPs and EXCPs at a more local level, does not substitute for the MEB. The MEB is made by the PEs originating from the main grounding bus and linking all the ECPs and EXCPs. The SB "reinforces," and does not replace, the equipotentiality already created by the MEB.

6.12 Potential Differences Among Metal Parts in Fault Conditions in TT Systems

The previous sections allow an important consideration on the potentials attained by metal parts in fault conditions. As exemplified in

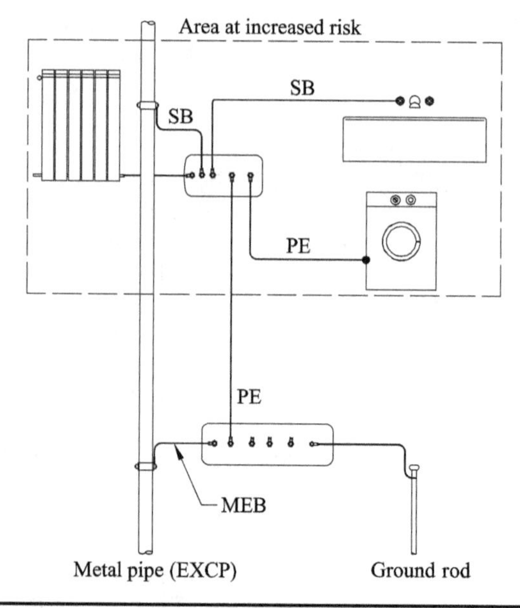

FIGURE 6.22 Supplementary equipotential bonding (SB) in area at increased risk.

Fig. 6.23, upon failure of an ECP, any other earthed metal parts (i.e., ECPs and EXCPs) simultaneously attain the same value of fault potential, even if healthy.

This "signature" feature of the TT systems prevents the appearance of dangerous potential differences among grounded metal parts.

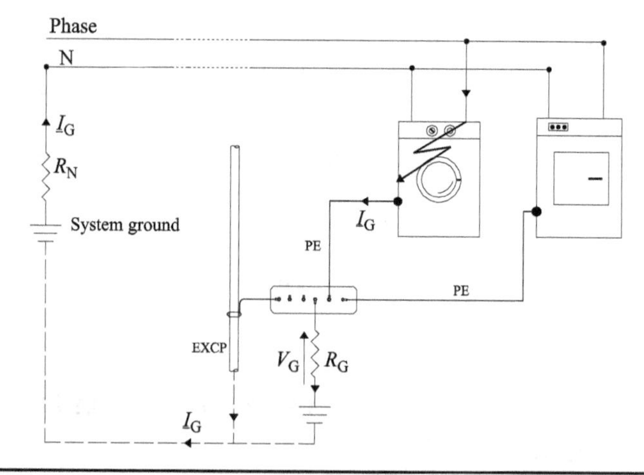

FIGURE 6.23 In TT systems, all the ECPs and EXCPs attain the same potential upon ground faults.

This important characteristic compensates the unavoidable energization of all the healthy ECPs and EXCPs, thereby benefiting the persons' safety.

FAQs

Q. What is the difference between underlined (i.e., $\underline{V}_{ST}$) and nonunderlined (i.e., V_{ST}) symbols?

A. The underlined symbols represent the phasor quantities, that is, vectors or complex numbers, characterized by magnitude and angular displacement. The symbol V_{ST}, as well as the symbol enclosed between bars $|\underline{V}_{ST}|$, indicate the magnitude of the phasor (see Appendix A).

Q. Equipotential bonding connections cancel potential differences between EXCPs and ECPs within reach. What about the touch voltage between faulty ECPs and the floor?

A. The main equipotential bonding links together all the ECPs and EXCPs to the grounding system, including structural metallic parts of building and reinforcing bars embedded in concrete. Ergo, if the floor is conductive, and in contact with these components, the main equipotential bonding also eliminates ECP-to-floor potential differences. If the floor is made of insulating materials, no ECP-to-floor potential differences arise.

Q. Do we need to supplementarily bond metal window frames in areas at increased risk?

A. Metal frames of windows are generally not EXCPs, that is, they are not likely to introduce remote or dangerous potentials. In addition, they are not the part most likely to be touched in a window. Therefore, there is no need for their supplementary bonding. On the other hand, their connection to the equipotential system may result in their undue energization, in the case of a fault occurring somewhere else in the system.

Q. To improve the performance of grounding systems in TT systems, should we connect it to the utility neutral wire that is earthed?

A. The TT grounding system is used when the electrical utility cannot guarantee a safe means of earthing for their users. The neutral wire, in fact, although earthed, might assume dangerous potentials with respect to ground, because of high values of the utility's ground R_N. The connection between the user grounding system and the neutral wire would transfer the neutral potential over the user's enclosures. For these reasons, this bond is not permitted.

Endnotes

1. The underlined quantities indicate complex numbers (or phasors) as representative of sine waves in the circuit. Symbols not underlined designate magnitudes of complex numbers (see App. A).

2. For example, the earthing systems in Venice, Italy, where foundations of building are immersed in the salty water of the lagoon.
3. See App. B.
4. As per IEC 60898-1, "*Electrical Accessories—Circuit-Breakers for Overcurrent Protection for Household and Similar Installations—Part 1: Circuit-Breakers for A.C. Operation*," Consolidated Edition 1.2 (2003–07).
5. The above standard CB values were internationally adopted before World War II and are based on the Renard geometric progression of common ratio $10^{1/10}$. The Renard progression is named after its inventor, a French army officer, who created it in the 1870s. Each term of the progression is generated by multiplying the previous one by the common ratio.
6. Causes of the energization of the neutral will be explained in Sec. 6.9.
7. The leakage may be, for example, due to the aging of equipment. "Spilling" current to ground causes additional consumption of electric energy.
8. See Chap. 15 for further details.
9. For simplicity, we can assume that $R_{GT} = R_G$.
10. The case of the water circulating in heating systems may be different, as it may contain additives necessary to prevent corrosion. The presence of such additives lowers the resistivity of the water.

CHAPTER 7

TN Grounding System

Meglio agitarsi nel dubbio che riposare nell'errore.
Better to seethe with doubt than to rest with mistake.

ALESSANDRO MANZONI (1785–1873)

7.1 Introduction

Users in industrial facilities may receive their power supply from the local utility in medium or high tension, and therefore, install, and own, front-end substations. In this case, within the facility, customers employ TN grounding system (e.g., Terre Neutral), even if the outside low-voltage earthing system is TT.

The user's substation may contain one or more transformers, whose windings are typically wound as a delta at the primary side and as a wye at the secondary side. The transformers are necessary in order to step down the medium/high tension to low-voltage levels suitable for the customer. The user must solidly ground the center of the transformer's wye and directly connect all the exposed-conductive-parts (ECPs) to it via protective conductors. The neutral conductor may be carried in order to provide power to single-phase loads.

If separate neutral and protective conductors are used throughout the facility, the system is defined as TN-S (Sec. 1.2.22). If in the electrical system, or a part of it, neutral and protective functions are combined in a single conductor, referred to as PEN conductor, and the system is defined TN-C or TN-C-S (Secs. 1.2.20 and 1.2.21).

In TN systems, the ground-fault current will return to the transformer through the protective conductor and, unlike in TT systems, will not circulate through the earth (Fig. 7.1).

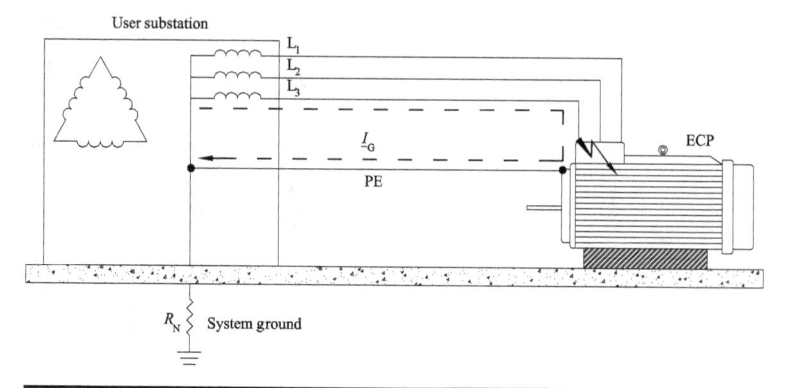

FIGURE 7.1 Ground fault in TN-S systems (three-phase load with no neutral).

In case of a zero-impedance ground fault (i.e., bolted fault), the current in the fault-loop is limited by the series of the following impedances: transformer (i.e., $\underline{Z}_{\mathrm{i}}$), phase wire (i.e., $\underline{Z}_{\mathrm{ph}}$), and protective conductor (i.e., $\underline{Z}_{\mathrm{PE}}$). All as seen at the point of fault. Of course, the farther the fault occurs from the transformer, the larger is the loop impedance.

If the fault is not too far from the source, the loop impedance is low because metal conductors offer high conductivity to currents. As a result, the ground-fault current is of the same magnitude as a short-circuit current and can be easily detected by overcurrent devices. Consequently, in these conditions, the presence of RCDs in TN systems is not strictly necessary for safety.

In only one circumstance can the fault current circulate through the earth in TN systems. This is the case when the ground fault occurs toward an extraneous-conductive-parts (EXCPs), which is not bonded to the grounding system (Fig. 7.2).

In this case, like in TT systems, both the resistance to ground of the EXCP and R_{N} limit the fault current. The overcurrent device might not operate in a timely fashion, as this current may be too low. The risk of dangerous touch potentials, therefore, may arise. A proper main equipotentialization, that is, a sound connection via the protective conductor between EXCPs and the system ground, prevents this hazard and, therefore, is necessary.

7.1.1 Why Earthing the Transformer?

As said, in TN systems the fault-loop does not comprise the actual earth; however, the user must earth the center of its transformer's wye. The purpose of the system ground is to allow the operating voltage-to-earth to remain stable and to limit overvoltages in fault conditions.

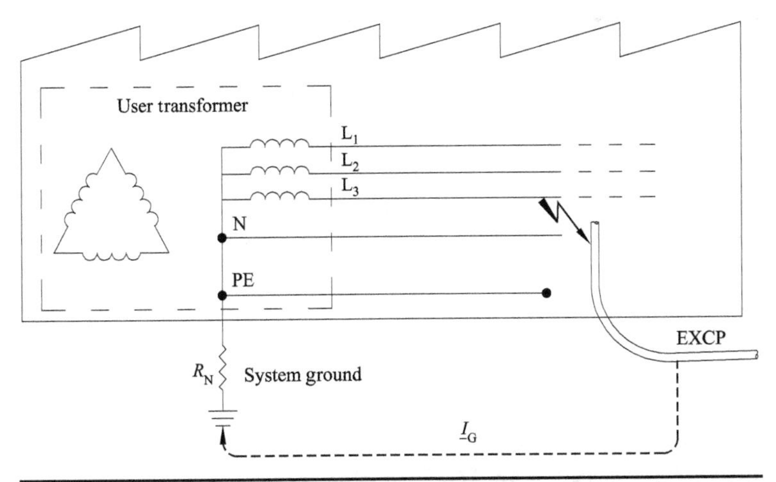

Figure 7.2 Ground fault toward an EXCP not bonded to the grounding system in TN-S.

The earth connection is an important requirement in TN systems, as upon its loss users would be exposed to the risk of electric shock (Fig. 7.3).

At the occurrence of a fault to earth, such as a fault toward a grounded metal structural element, the current cannot reclose to the source if the system is not earthed, therefore, protective devices cannot clear the fault. Persons touching any ECP, even if healthy, may then become the return path to the fault current, with risk of electrocution. In Chap. 9, we will discuss the issues with ungrounded systems and the related safety requirements.

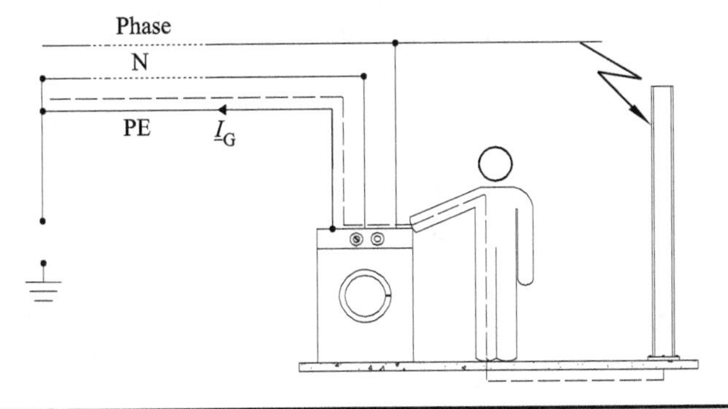

Figure 7.3 Risk of electric shock in the absence of the earth connection in TN system.

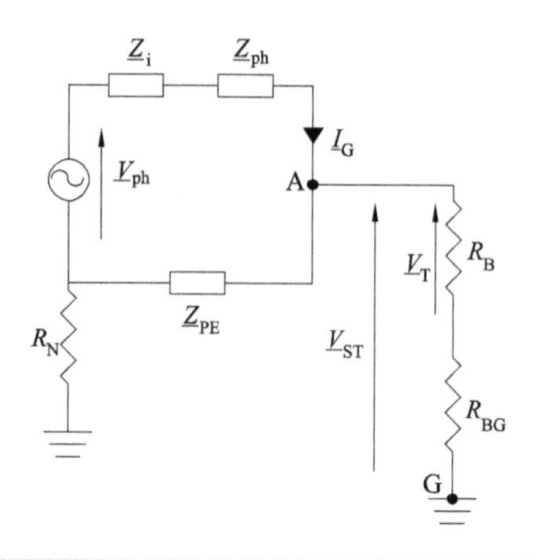

FIGURE 7.4 Fault-loop in TN-S systems.

7.2 Voltage Exposure in TN-S Systems

Let us conservatively assume that persons are exposed to the prospective touch voltage $\underline{V}_{ST}$.[1] The fault-loop, not involving the actual earth, is diagrammatically represented in Fig. 7.4, where the symbols have the same meaning as in Fig. 6.2.

In order to obtain the prospective touch voltage to which persons will be exposed, we apply the Thevenin's theorem between point *A* (i.e., point of contact with the faulted ECP) and *G* (i.e., the ground). We can reasonably neglect the internal impedance $\underline{Z}_i$ of the transformer if the fault occurs at a sufficient distance from it. Modules of $\underline{Z}_{ph}$ and $\underline{Z}_{PE}$, in fact, proportionally increase with the quantity of conductor interposed between the source and the fault, and they both become much larger than $\underline{Z}_i$.

Figure 7.5 shows the Thevenin's equivalent circuit as seen between the two aforementioned points, where

$$\underline{V}_{Th} = \frac{\underline{V}_{ph}\underline{Z}_{PE}}{\underline{Z}_{ph} + \underline{Z}_{PE}} = \frac{\underline{V}_{ph}}{(\underline{Z}_{ph}/\underline{Z}_{PE}) + 1} \cong \underline{V}_{ST} \tag{7.1}$$

$$\underline{Z}_{Th} = \frac{\underline{Z}_{ph}\underline{Z}_{PE}}{\underline{Z}_{ph} + \underline{Z}_{PE}} + R_N \tag{7.2}$$

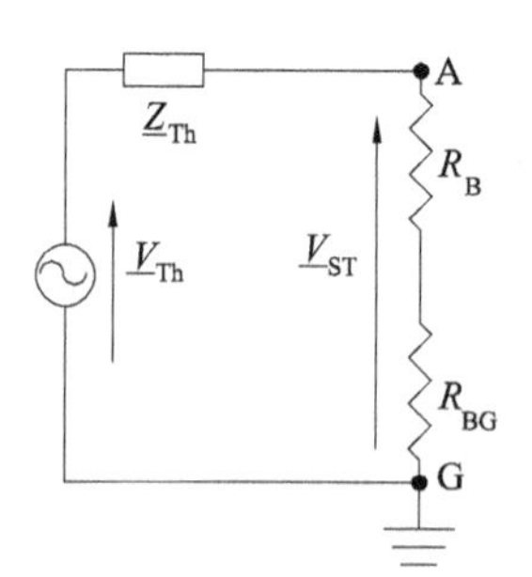

FIGURE 7.5 Thevenin's equivalent circuit between point *A* and point *G*.

It should be noted that $\underline{Z}_{Th}$ (magnitude of fractions of ohms) could be neglected, as it is much smaller than the series ($R_B + R_{BG}$). As a consequence, there is almost no voltage drop on $\underline{Z}_{Th}$ and the fault-loop is virtually an ideal voltage source. The person's total resistance does not significantly change the voltage output $\underline{V}_{Th}$ in case of a contact and, therefore, $\underline{V}_{Th}$ equals $\underline{V}_{ST}$.

As phase and protective conductor belonging to a same circuit usually run together, they are characterized by the same length and type of installation. Therefore, the ratio of their impedances is constant along their entire route up to the fault location. Ergo, Eq. (7.1) shows that the magnitude of $\underline{V}_{ST}$ remains constant regardless of the location of the fault as long as the ratio of Z_{ph} to Z_{PE} is constant.

If the protective conductor has the same cross section as the phase, Eq. (7.1) yields: $V_{ST} = V_{ph}/2$. If the PE has half section of the phase conductor[2] (i.e., $\underline{Z}_{PE} = 2\underline{Z}_{ph}$), we obtain a larger prospective touch voltage: $V_{ST} = 2V_{ph}/3$. Thus, the touch voltage increases where the PE has a lower section than the phase conductor.

7.2.1 Ground Fault in the Vicinity of the User's Transformer

Ground faults might occur within an ECP in the vicinity of the user's transformer, for example, at the main low-voltage panel (Fig. 7.6).

In this case, the internal impedance $\underline{Z}_i$ of the transformer (Fig. 7.3) cannot be neglected, as it may be even larger than the conductors' impedances to the panel because of their short run. By assuming, therefore, that $\underline{Z}_i \gg \underline{Z}_{PE}$ and $\underline{Z}_i \gg \underline{Z}_{ph}$, we obtain

$$\underline{V}_{ST} = \frac{\underline{V}_{ph}\underline{Z}_{PE}}{\underline{Z}_i + \underline{Z}_{PE} + \underline{Z}_{ph}} \cong \frac{\underline{V}_{ph}\underline{Z}_{PE}}{\underline{Z}_i} \cong 0 \qquad (7.3)$$

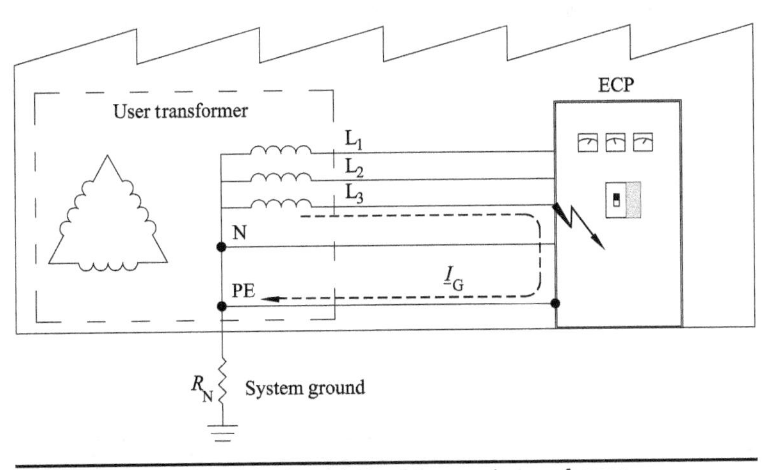

FIGURE 7.6 Ground fault in the vicinity of the user's transformer.

In the above conditions of proximity to the transformer, the indirect contact is hardly ever dangerous.

7.3 Potential Differences Between ECPs, and Between ECPs and EXCPs in TN Systems

Equation (7.1) shows that the prospective touch voltage $\underline{V}_{ST}$ has a constant magnitude regardless of the location of the faulty ECP, if the ratio of $\underline{Z}_{ph}$ to $\underline{Z}_{PE}$ is constant along the circuit. The difference with the TT systems is that both the other healthy ECPs, located upstream of the fault, and the EXCPs acquire for the duration of the fault prospective touch voltages, which decrease moving toward the source. These potentials approach zero as the location of the healthy ECPs approaches the origin of the electrical system (i.e., the transformer). All the ECPs located downstream of the fault, instead, acquire the same potential as the faulty equipment. The presence of nonzero potential differences between bonded metal parts is a salient trait of the TN systems and constitutes one major difference with the TT systems.

For a better comprehension of the above concept, let us examine Figs. 7.7 and 7.8.

Assuming that the protective conductor's cross section varies during its course, as exemplified in Fig. 7.7, the protective conductor, which bonds the faulted ECP 2, is made of three runs of impedances $\underline{Z}_{PE0}$, $\underline{Z}_{PE1}$, and $\underline{Z}_{PE2}$. Both ECP 2 and the healthy ECP 3 will attain the potential $\underline{V}_2$ with respect to ground, as there is no circulation of fault

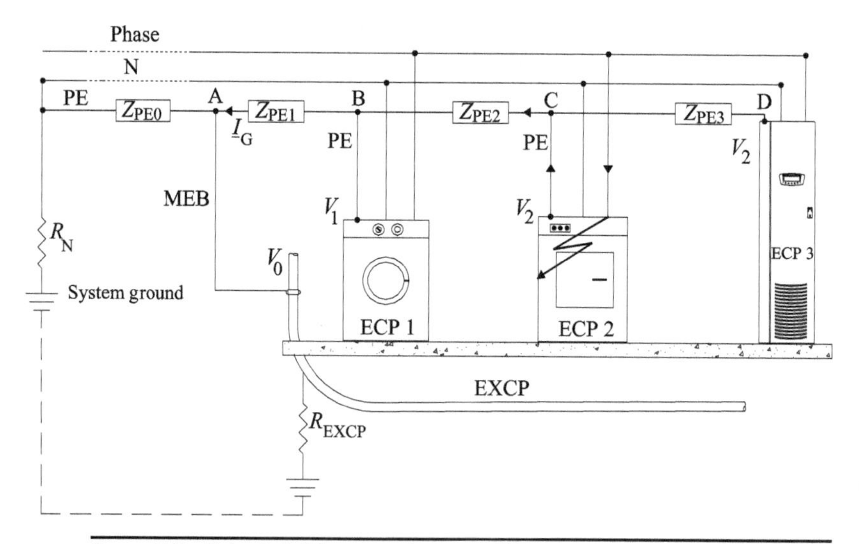

FIGURE 7.7 Potential differences between ECPs and between ECPs and EXCPs in TN systems.

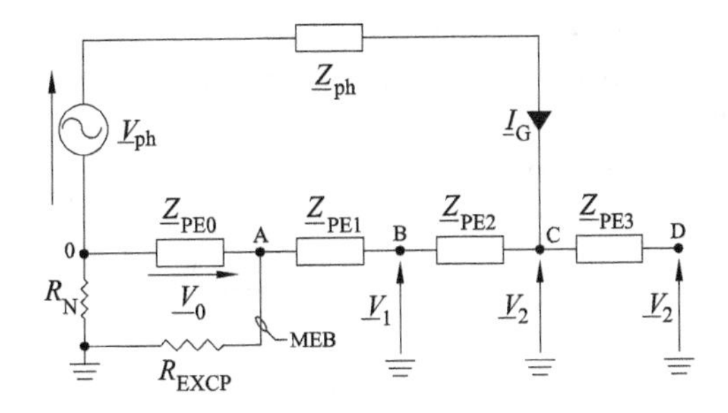

FIGURE 7.8 Equivalent fault circuit and potential differences in TN systems.

current through the protective conductor's impedance $\underline{Z}_{PE3}$. The series of R_N and R_{EXCP}, where R_{EXCP} is the "natural" resistance to earth of the metalwork entering the building, is much larger than $\underline{Z}_{PE0}$. For this reason, the series $R_N + R_{EXCP}$ can be considered as an open circuit in Fig. 7.8.

The profile of the prospective touch voltage from the origin of the electrical system to the faulty equipment is exemplified in Fig. 7.9.

In sum, since $V_0 < V_1 < V_2$, we can infer that potential differences do exist between bonded metal parts during the time the protective device takes to clear the fault; the magnitude of such potential differences varies with the location of the ECP as a function of the distance from the supply.

7.4 Protection Against Indirect Contact in TN-S Systems by Using Overcurrent Devices

In TN systems, overcurrent devices (i.e., circuit breakers or fuses) can be successfully employed for protection against indirect contact, because of the large magnitude of the ground currents circulating in the fault-loop, comparable to that of short-circuit currents. Unlike in TT systems, then, overcurrent devices in TN systems are facilitated in detecting, and thereby interrupting, fault currents.

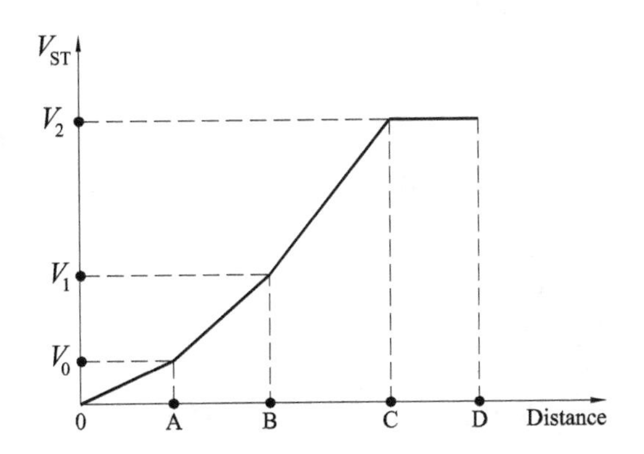

FIGURE 7.9 Prospective touch voltage profile as a function of the distance from the origin of the electrical system.

However, protective devices, RCDs included, must fulfill the following condition applied at the farthest point of the circuit being protected, where the fault current is at its minimum due to the build-up of the wires' impedances:

$$|\underline{I}_G| = \frac{|\underline{V}_{ph}|}{|\underline{Z}_{Loop}|} \geq I_a \tag{7.4}$$

or equivalently, by solving for $|\underline{Z}_{Loop}|$,

$$|\underline{Z}_{Loop}| \leq \frac{|\underline{V}_{ph}|}{I_a} \tag{7.5}$$

where $|\underline{I}_G|$ is the minimum phase-to-protective conductor fault current and $\underline{Z}_{Loop}$ is the series of the impedances of the components that form the fault-loop, specifically the source, the line conductor up to the farthest point of the fault from the source, and the protective conductor up to the farthest point of the fault. $\underline{V}_{ph}$ is the system nominal voltage to ground. I_a is the current causing the automatic operation of the overcurrent protective device within the time specified in Table 7.1.

If RCDs are employed, I_a represents the residual operating current, which provides the disconnection of supply within the time specified in Table 7.1.

For example, an overcurrent device operating at 230 V, with continuous rating $I_n = 16$ A, may trip in 0.4 s in correspondence with $I_a = 4I_n$; therefore, by applying Eq. (7.5), $Z_{Loop} \leq 3.6\ \Omega$. This inequality is fairly easy to fulfill.

As said, Eq. (7.4), applicable to any point of the circuit being protected, should, indeed, be verified for ground faults at its end because that is the point where the smallest amount of fault current is generated. If the overcurrent device can sense the lowest current in the circuit and trip within the safe times, the protection against indirect

Voltage Range (V)	Maximum Disconnection Times t_a (s)
$50 < V_{ph} \leq 120$	0.8
$120 < V_{ph} \leq 230$	0.4
$230 < V_{ph} \leq 400$	0.2
$V_{ph} > 400$	0.1

TABLE 7.1 Maximum Disconnection Times as a Function of the Nominal A.C. Voltage of the TN System

contact upon faults occurring in any other location will be accomplished.

Table 7.1, although generally applicable, has been conceived by assuming standard operating conditions. Such conditions are based on the following considerations:

1. The circulation of fault currents causes large voltage drops.
2. Persons within buildings are not subject to the whole ground potential, thanks to the main equipotential bonding connection that ties together all the EXCPs.
3. The cross-sectional areas of protective and phase conductors are the same (i.e., $\underline{Z}_{PE} = \underline{Z}_{ph}$).

In the aforementioned conditions, IEC standards conventionally estimate a reduction in the postfault driving voltage of 20% of the nominal value of the system voltage.

By applying this reduced voltage in Eq. (7.1) and, for example, assuming $|\underline{V}_{ph}| = 230$ V, we obtain for V_{ST} the value of 92 V. As per the *time–voltage safety curve* (Fig. 5.19), persons can withstand this touch voltage for a maximum time of 0.4 s, which is the value listed in Table 7.1.

Table 7.1 applies to final circuits not exceeding 32 A. In some countries (e.g., China), the maximum disconnecting times stated in Table 7.1 applies to final circuits supplying hand-held or portable equipment. Mobile equipment (e.g., drills, hairdryers, or any piece of electrical equipment that is required to be moved by persons during its use), in fact, are considered to be more dangerous than stationary ones (e.g., light fixtures, air conditioners, etc.). This assumption is justified by the greater mechanical stress mobile equipment normally undergoes, which may expose its live parts. In these countries, a longer disconnection time of 5 s is permitted for stationary loads.

According to IEC, a maximum disconnection time of 5 s is allowed in distribution circuits as long as the fault potentials appearing on them do not affect any final circuit supplied by the corresponding distribution panel (see Sec. 7.6).

For phase conductors with cross-sectional areas exceeding 16 mm^2, IEC standards allow the reduction of the cross section of the PE to half of the phase conductor. The protective conductor's impedance, therefore, double (i.e., $\underline{Z}_{PE} = 2\underline{Z}_{ph}$); hence, V_{ST} reaches the value of 123 V, and the *time–voltage safety curve* is not fulfilled anymore.

In addition, outside of the building, where the benefits of the equipotentialization hardly exist, the diminution in the driving voltage can no longer be applied and the disconnection time of 0.4 s may be excessive.

In the above nonstandard circumstances, to ensure protection against indirect contact, RCDs may be employed together with a fully rated protective conductor.

7.4.1 Calculation of the Approximate Minimum Value of the Phase-to-Protective Conductor Fault Current

As said, ground faults cause short circuits due to the highly conductive nature of the fault-loop in TN systems. The minimum ground-fault current can be approximately calculated at the farthest fault point, under the following assumptions:

1. The supply voltage reduces to 80% of its nominal value due to the voltage drop caused by the ground-fault current. This assumption is deemed acceptable only if the combined impedance of the fault-loop conductors is much greater than the source's impedance.
2. The resistance of conductors included in the fault-loop increases by 50% with respect to their 20°C value because of the heat from the ground-fault current.
3. The reactance of conductors included in the fault-loop can be neglected for cross-sectional areas not exceeding 90 mm^2. Hence, $\underline{Z}_{\text{Loop}} = R_{\text{Loop}}$. For larger sections, a multiplying factor $k < 1$, which takes into account the limiting effect of conductor's reactance to the fault current, can be obtained from Table 7.2.

We define the factor m as the ratio of the phase conductor's cross-sectional area to the protective conductor's cross-sectional area:

$$m = \frac{S_{\text{ph}}}{S_{\text{PE}}} \tag{7.6}$$

k	Cross-Sectional Areas (mm^2)
0.90	120
0.85	150
0.80	185
0.75	240
0.72	300

Table 7.2 Multiplying Factor k, Which Takes into Account the Limiting Effect of Conductor's Reactance of Large Cables to the Fault Current

By neglecting the source impedance, the resistance of the fault-loop is given by the series of the resistances of phase and protective conductors, as follows:

$$R_{\text{Loop}} = R_{\text{ph}} + R_{\text{PE}} = \frac{1.5\rho L}{S_{\text{ph}}} + \frac{1.5\rho L}{S_{\text{PE}}} = 1.5\rho L\left(\frac{S_{\text{PE}} + S_{\text{ph}}}{S_{\text{PE}} S_{\text{ph}}}\right) \tag{7.7}$$

where L and ρ are, respectively, length and resistivity at 20°C of cables at fault point. The multiplier 1.5 accounts for the 50% increase in conductors' resistance.

By combining Eqs. (7.6) and (7.7), we obtain

$$R_{\text{Loop}} = \frac{1.5\rho L\,(1 + m)}{S_{\text{ph}}} \tag{7.8}$$

Thus, by applying the Ohm's law, the minimum ground-fault current I_{G} is

$$I_{\text{G}} = \frac{0.8V_{\text{ph}}}{R_{\text{Loop}}} = \frac{0.8V_{\text{ph}} S_{\text{ph}} k}{1.5\rho L\,(1 + m)} \tag{7.9}$$

7.5 Protection Against Indirect Contact in TN-S System by Using RCDs

When Eq. (7.4) cannot be fulfilled through overcurrent devices (i.e., the loop impedance $\underline{Z}_{\text{Loop}}$ is too high), or the user is not within the equipotential area, RCDs may constitute the only way of protection against indirect contact. However, in some particular circumstances, residual current devices cannot protect persons.

For instance, let us consider a ground fault occurring on the primary side of the user's substation in a TN-S system, where the earthing system is shared by high- and low-voltage ECPs[3] (Fig. 7.10).

If the transformer's enclosure is linked to the same system ground as the low-voltage system, the neutral wire becomes energized at the ground potential $\underline{V}_{\text{G}}$. The protective conductor conveys $\underline{V}_{\text{G}}$ to ECPs, and persons touching them will be exposed to dangerous touch potentials. The RCD, installed on the low-voltage side of the supply system, cannot trip, because it cannot sense the fault current, that does not circulate through it.

In TN-C systems, the RCD cannot work at all, as the ground-fault current is returned to the source by the PEN conductor, which is encircled by the toroid as a neutral wire. This would cause no unbalance in the case of a ground fault and the operation of the RCD is prevented.

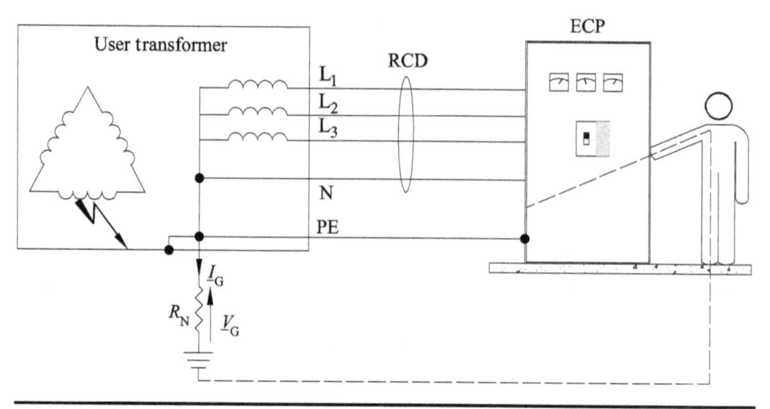

FIGURE 7.10 Indirect contact with ground potential rise in TN-S systems.

7.6 Transferred Potentials Between Distribution and Final Circuits in TN Systems

Within electrical systems, touch potentials may be transferred away from the fault location. If a fault originates on a distribution circuit, it is allowed to persist for 5 s before it is cleared. This potential can also affect other remote metal parts within the system (Fig. 7.11).

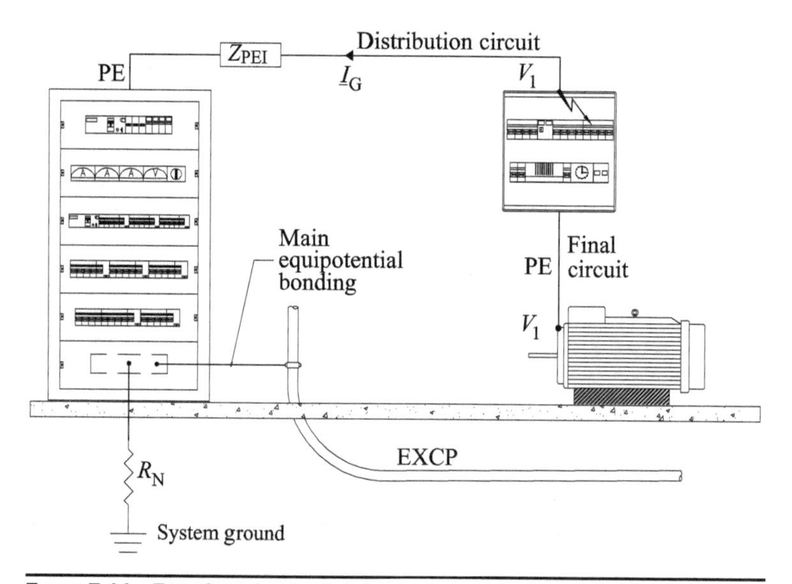

FIGURE 7.11 Transferred potential due to ground faults on distribution circuits in TN systems.

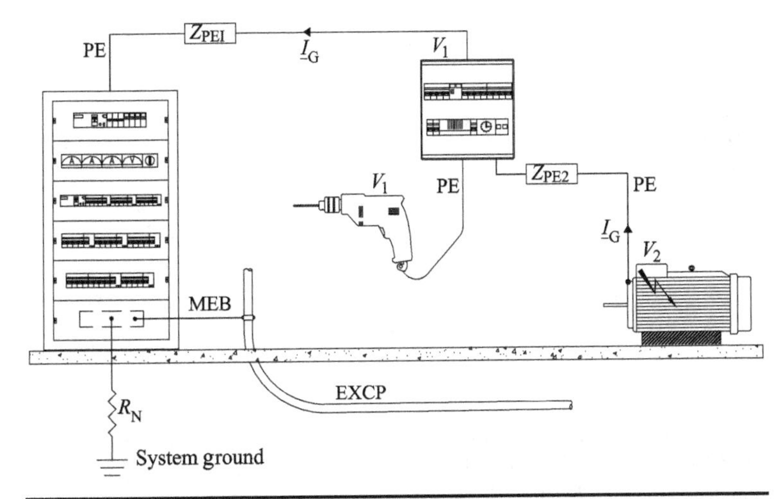

Figure 7.12 Transferred potential due to ground faults on final circuits in TN systems.

As shown in Fig. 7.11, the prospective touch voltage $V_1 = (V_{ph} Z_{PE1})/(Z_{ph} + Z_{PE1})$ is transferred to the healthy stationary equipment (i.e., the electrical motor), whereas the EXCP remains at the earth potential.

Between the motor and the EXCP, therefore, a potential difference V_1 arises for a time not exceeding 5 s, even if the load is healthy. Persons, therefore, are exposed to the risk of electric shock even though not in contact with the faulty distribution circuit. This risk, though, is deemed acceptable[4] by international standards, as the probability of faults in distribution circuits is assumed low.

Should the fault occur on the stationary equipment, exemplified in Fig. 7.12 as a motor, the prospective touch voltage V_1 is also transferred to the healthy mobile equipment (i.e., the drill).

Also in this case, a potential difference V_1 arises between any EXCP within reach and bonded equipment. The probability of a ground fault on stationary loads, though, is considerably higher than on a distribution circuit; therefore, a clearing time of 5 s would pose an unacceptable hazard to persons. For this reason, European standards prescribe this clearing time only if the calculated value V_1 does not exceed 50 V, a value conventionally not dangerous if sustained for no more than 5 s.

7.6.1 Supplementary Equipotential Bonding

If touch voltages cannot be cleared within the safe time, supplementary equipotential bonding connections (SB) between EXCPs and ECPs may be employed. Such bonding connections are locally realized between the enclosures of loads and EXCPs (Fig. 7.13).

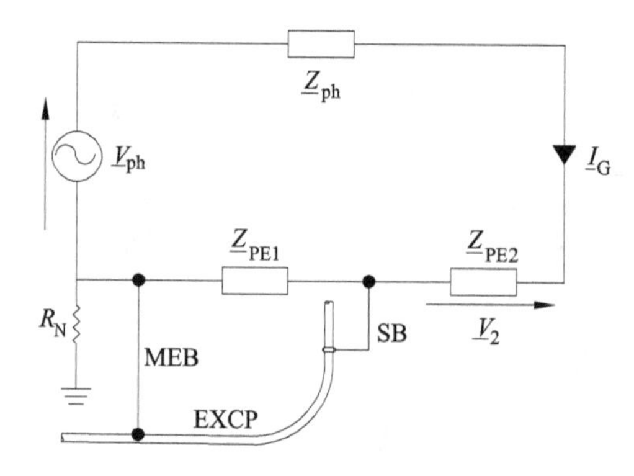

FIGURE 7.13 Supplementary equipotential bonding (SB) in TN systems.

The supplementary bond reduces the potential difference between the EXCP and the faulty ECP to the voltage drop across the protective conductor employed to realize this equipotential connection, which must not exceed 50 V.

In addition, the presence of SB decreases the prospective touch voltage offered by the faulty equipment by "short circuiting" part of the protective conductor's impedance (i.e., $\underline{Z}_{PE1}$).

It is important to note that the equipotentiality in TN systems is even more crucial than in TT systems because the safety of the installation depends on it.

7.7 Local Earth Connection of ECPs in TN Systems

In TN systems, ECPs are not required to be locally earthed, for example, via grounding rods. Nonetheless, the presence of local supplemental grounding electrodes may improve the electrical safety of the installation in areas where the conventional decrease in the driving voltage (see Sec. 7.4) cannot be applied and the permissible disconnection times of Table 7.1 may be excessive. This situation may occur outside of the equipotential area, for example, around equipment installed outside of the building, but supplied by a circuit originating within it. In these zones, the benefits of the main equipotential bonding cannot be enjoyed.

Let us examine Figs. 7.14 and 7.15. The ground fault on the ECP causes circulation of current through the protective conductor of impedance $\underline{Z}_{PE}$ and through the earth, via the series of R_N and R_G. The prospective touch voltage $\underline{V}_{ST}$ is the result of the voltage divider of $\underline{V}_0$ between R_N and R_G:

$$V_{ST} = \frac{V_0 R_G}{R_G + R_N} = \frac{V_0}{1 + (R_N/R_G)} \tag{7.10}$$

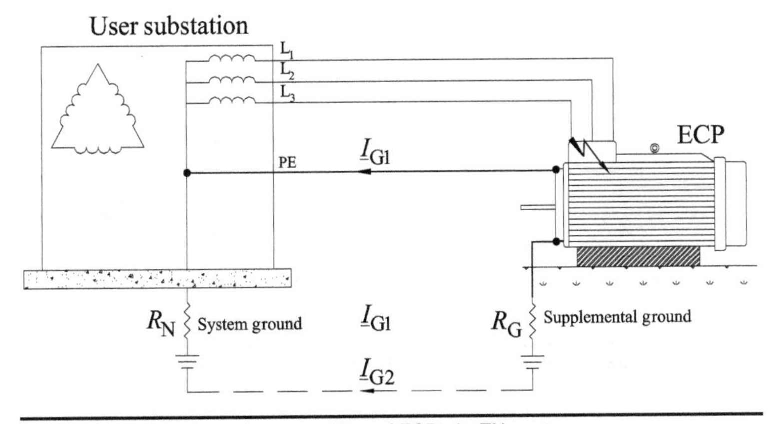

FIGURE 7.14 Local earth connection of ECPs in TN systems.

It appears clear that V_{ST} approaches zero as the ratio of R_N to R_G approaches infinity. This means that R_G should be extremely low with respect to R_N. In reality, the opposite generally happens; therefore, the reduction in the prospective touch voltage due to the supplemental earthing connection is not always substantial.

A better result can be obtained by additionally connecting the ECPs to a local grounding grid, if present throughout the facility (Fig. 7.16).

The main purposes of the grounding grid are to earth the user substation's transformer and provide an equipotential area throughout the zone below which it is embedded. By using this as an additional protective conductor, the prospective touch voltage is reduced in the case of ground faults occurring at low-voltage equipment.

To better clarify this concept, let us examine the equivalent circuit in Fig. 7.17. Persons touching the faulted enclosure are subject only to the voltage drop across the bonding connection, of impedance $\underline{Z}_{PE2}$,

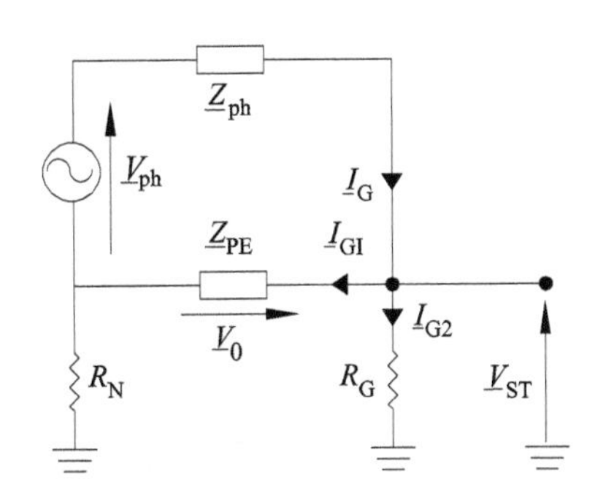

FIGURE 7.15 Equivalent fault circuit with local earth connection of ECPs in TN systems.

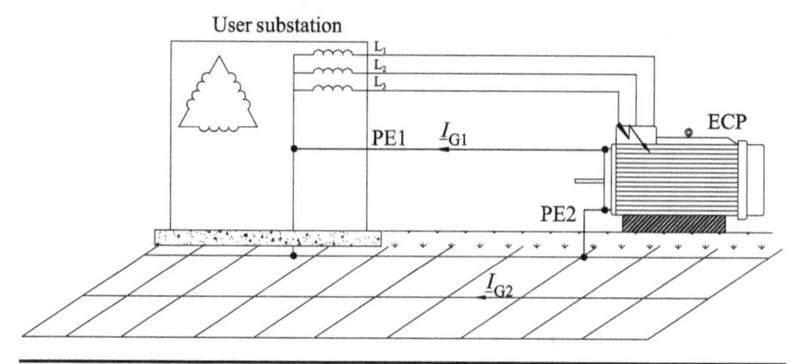

FIGURE 7.16 Use of the grounding grid as protective conductor in TN systems.

between the equipment and the grid, as calculated in Eq. (7.11):

$$\underline{V}_{ST} = \frac{\underline{V}_{PE1}\underline{Z}_{PE2}}{\underline{Z}_{Grid} + \underline{Z}_{PE2}} = \frac{\underline{V}_{PE1}}{(\underline{Z}_{Grid}/\underline{Z}_{PE2}) + 1} = \underline{G}\,\underline{V}_{PE1} \tag{7.11}$$

It is clear that the bonding connection to the grid lessens the prospective touch voltage $\underline{V}_{PE1}$ of a factor $\underline{G}$ depending on the ratio of $\underline{Z}_{Grid}$ to $\underline{Z}_{PE2}$, which is generally <1 in magnitude.

7.8 TN-C Systems and the PEN Conductor

In TN-C systems, and under specified conditions, the functions of neutral wire and protective wire may be combined in a single conductor, referred to as PEN conductor. Of course, the PEN conductor, which besides the neutral current also carries the ground-fault current, must never be switched off in order to preserve the continuity of the fault-loop. Basically, by combining two functions in a single wire, a conductor can be avoided, thereby allowing cost reductions, especially for distribution circuits with large cross-sectional areas. However,

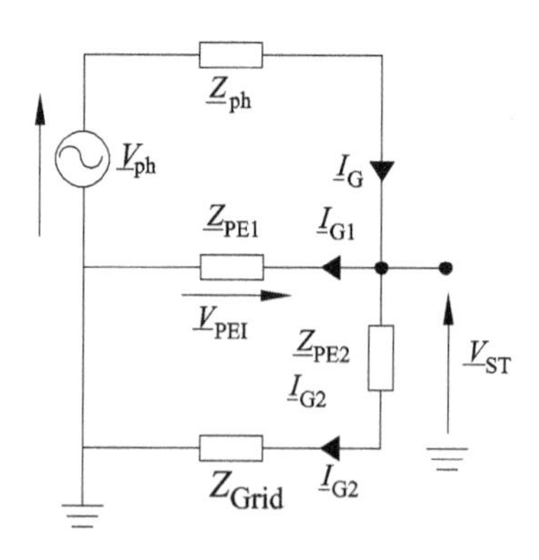

FIGURE 7.17 Equivalent circuit including the bonding connection of ECP to the grounding grid.

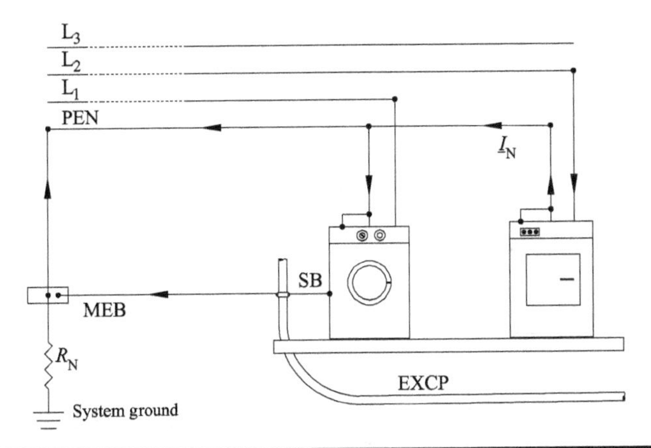

FIGURE 7.18 TN-C system and PEN conductor.

this practice may cause several technical issues. Let us now examine Fig. 7.18.

If the electrical three-phase system is not balanced, the neutral current is nonzero and will not only legitimately circulate through the PEN conductor, but also, inevitably, through the ECPs and the EXCPs that are part of the main and supplementary equipotential bonding system. Neutral currents, which in the ordinary functioning of the installation return to the source via paths not specifically designed to carry them, may be defined as "stray currents." In addition, in ground-fault conditions, both EXCPs and healthy ECPs will carry fault currents.

Nonlinear loads,[5] which draw currents that are no longer sine waves, also cause the existence of nonzero neutral currents; therefore, the neutral current is no longer zero, even if the load is balanced among the phases.

The presence of stray currents on metal parts, by increasing their superficial temperature or even by causing sparks, worsens the fire and explosion hazards. This is crucial in locations where accumulation of dust, both on enclosures and EXCPs, and explosive atmospheres are expected.

Additionally, the neutral currents flowing through ECPs and EXCPs may induce disturbances to sensitive electronic equipment, such as computers, control systems, and similar, preventing their regular operation.

The accidental interruption of the PEN conductor creates a hazardous situation even in the absence of any fault situation (Fig. 7.19).

The enclosures of the single-phase equipment in Fig. 7.19, in fact, become energized, even if healthy, by acquiring the line-to-ground potential V_{ph}. The continuity of the PEN conductor is crucial and, therefore, must be guaranteed.

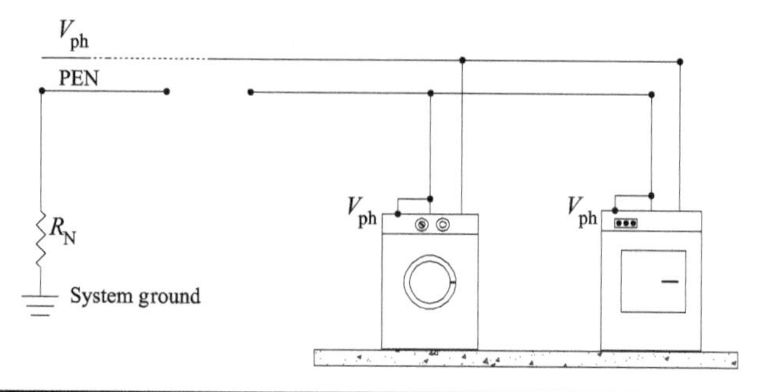

FIGURE 7.19 Interruption of the PEN conductor.

In sum, the TN-C system within a facility must be used only in areas where none of the aforementioned safety issues can occur. A sensible rating of the zones within any facility, with regard to the presence of dust and/or explosive atmospheres, is, therefore, an important prerequisite to evaluate if the PEN conductor may be employed.

7.9 The Neutral Conductor in TN Systems

As already mentioned, in TN-C systems, the neutral conductor must never be switched off. In TT systems, instead, the neutral must be interrupted with the phase wires, as, upon faults caused by the utility, it may become energized; therefore, in TT systems, the neutral wire must be treated as a "live" conductor.

In TN-S systems, where the user owns and maintains the supply source, the risk of energizing the neutral conductor can and must be analyzed, thereby, allowing an "educated" decision about its possible simultaneous interruption with the line wires.

A hazardous situation in the case of "passing-through" neutral is exemplified in Fig. 7.20.

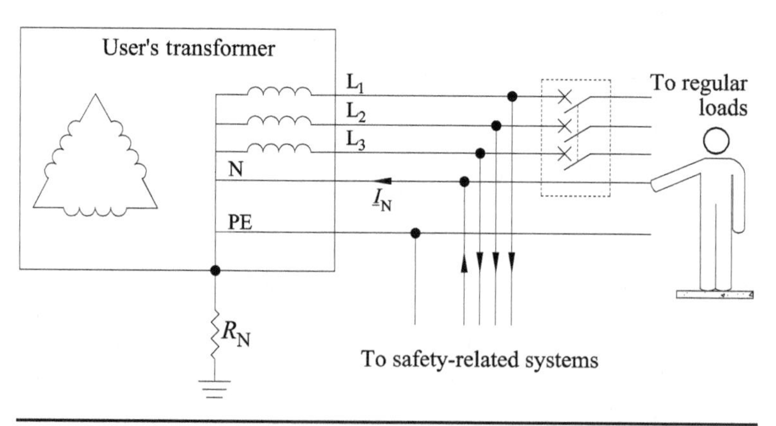

FIGURE 7.20 Hazardous situation in the case of nonswitched neutral.

Safety-related systems (e.g., fire alarm, fire pumps, etc.), that is, systems whose correct operation is necessary for ensuring or maintaining safety, may be supplied ahead of the main protective device. If the corresponding loads are not balanced among the phases, neutral current will flow. The neutral conductor may result, therefore, energized, even if the main breaker has been opened. In an emergency situation, which requires the safety-related systems to be running, persons (e.g., first responders) may be exposed to the risk of electrocution if in contact, directly or indirectly, with the neutral wire, even if the main breaker is in the open position. In this case, the simultaneous switching off of neutral and line conductors is a safety requirement.

7.10 The Touch Voltage in TN Systems

As already anticipated in Chap. 4, in electrical systems exceeding 1 kV, defined as *high-voltage* systems, international standards conventionally define the touch voltage as the potential difference between faulty equipment touched with one hand and the two feet of the person.[6] Ground faults occurring on the primary side of substations cause circulation of currents through the earth and therefore also step potentials. In reality, part of the fault current $\underline{I}_F$ may be prevented from circulating through the ground, with benefits for the safety. Upon permission of the local utility, in fact, the metallic sheaths of their incoming power cables, or the overhead ground wires of their transmission lines, may be used to drain off part of the fault current provided that the sheaths are adequately sized to withstand such current (Fig. 7.21).

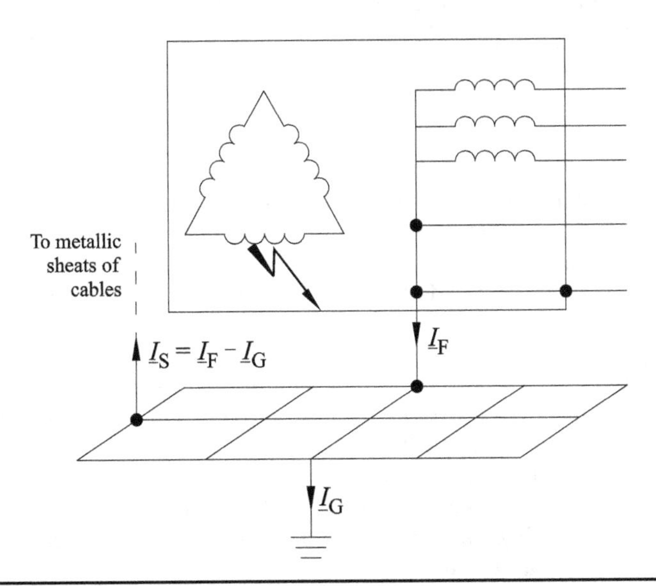

FIGURE 7.21 Earth current due to primary side faults.

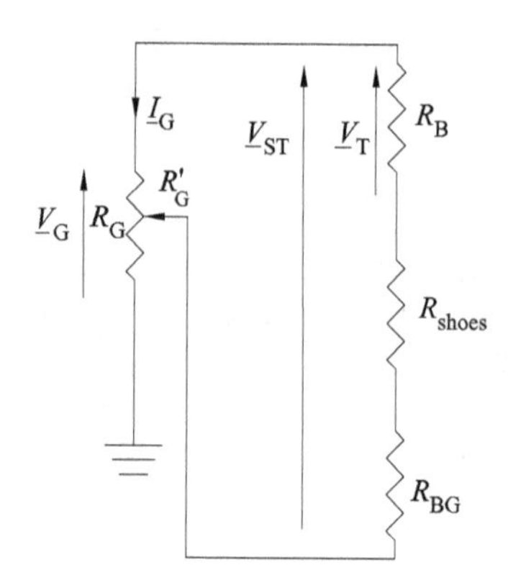

FIGURE 7.22 Touch voltage in TN systems due to high-voltage faults.

As a result, the component $\underline{I}_S = \underline{I}_F - \underline{I}_G$, by not circulating through the grounding system, lowers both touch and step voltages. This circumstance might permit less stringent parameters for the design of the electrode (e.g., the ground grid), by allowing larger earth resistances for it. In the normal practice, though, one can conservatively consider the larger value $\underline{I}_F$, instead of $\underline{I}_G$, as a design parameter for sizing the electrode.

In high-voltage systems (e.g., industrial facilities), and unlike in low-voltage systems, European standards,[7] in establishing the permissible limits for touch and step voltages, assume in series to the body resistance R_B, conventionally 1 kΩ, and the person resistance-to-ground R_{BG} (Fig. 4.15), the resistance of standard footwear of 1 kΩ[8] (Fig. 7.22).

For faults on the secondary side of substations, the following inequality is true:

$$|\underline{V}_T| \leq |\underline{V}_{ST}| \leq |\underline{V}_G| \tag{7.12}$$

where $|\underline{V}_G| = R_G\,|\underline{I}_F|$.

The touch voltage persists in the system during the time the protective device takes to clear the fault. However, primary ground faults may occur outside of the zone of protection of the user's protective device. In this case, the duration of the touch voltage depends on the utility's clearing time, which is out of the user's control and cannot be changed.

Thus, electrical utilities should communicate the fault clearing time as well as the magnitude of the earth current so that strategies to minimize the hazards through the proper design of the ground electrode can be implemented.

Permissible limits of touch voltages V_{TP} compatible with the standard person in Fig. 7.22 have been elaborated in the aforementioned European standard as a function of the fault duration (Table 7.3).

Clearing Time t_f (s)	Permissible Touch Voltage V_{TP} (V)
0.04	800
0.06	758
0.08	700
0.10	660
0.14	600
0.15	577
0.20	500
0.25	444
0.29	400
0.30	398
0.35	335
0.39	300
0.40	289
0.45	248
0.49	220
0.50	213
0.55	185
0.60	166
0.64	150
0.65	144
0.70	135
0.72	125
0.80	120
0.90	110
0.95	108
1.00	107
1.10	100
3.00	85
5.00	82
7.00	81
10.00	80
>10	75 (asymptotic)

Table 7.3 Permissible Touch Voltages as a Function of the Clearing Time

These values, or their interpolation, allow the application of the following basic safety criterion:

$$R_G I_F = |\underline{V}_G| \leq V_{TP} \tag{7.13}$$

where V_{TP} is chosen based on the communicated clearing time and R_G is the measured value of the earth resistance of the electrode.[9]

It is evident that Eq. (7.13), if solved for R_G, provides the criterion to determine the maximum acceptable value of the electrode earth resistance that makes the installation safe.

If Eq. (7.13) is true, the risk of indirect contact is acceptable, as $|\underline{V}_T| \leq |\underline{V}_G|$; if it is false, it might still be possible that $|\underline{V}_T| \leq |\underline{V}_{TP}|$, and it is necessary to perform the actual measurement of the touch voltage, thereby assessing the effectiveness of the grounding system.

The above procedure is shown as the flowchart in Fig. 7.23.

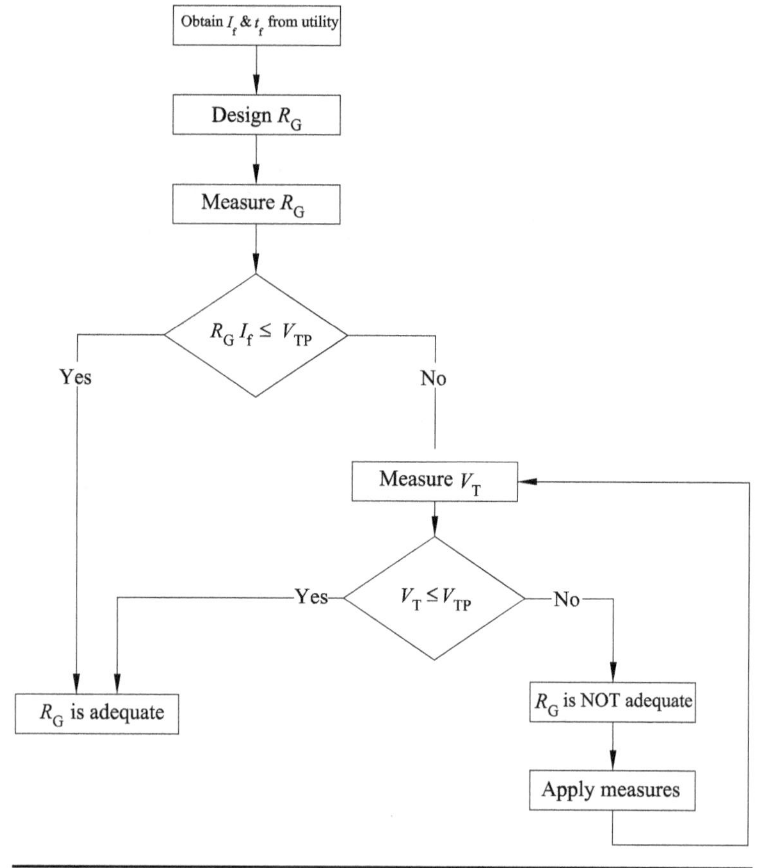

FIGURE 7.23 Procedure to deem adequate the resistance R_G of the grounding electrode.

If the touch voltage exceeds the permissible values given in Table 7.3, corrective mitigation measures must be taken. One might be to spread a layer of asphalt on the soil around the substation so as to increase the person's resistance to earth.

7.11 Step Voltage

As discussed in the previous sections, at the occurrence of high-voltage ground faults, high currents may flow through the actual earth. In addition to touch voltages, this circulation of current exposes persons to step voltages, that is, potential differences between two distinct points of the earth, conventionally taken 1 m apart.

We define the prospective step voltage V_{SS} as the potential difference between two points on the surface of the earth, displaced by the distance of 1 m, when the earth is not being stepped on by the person. The step voltage V_S is defined as the potential difference, which a person may be subject to, between the two feet, conventionally displaced by 1 m. It is always $V_S \leq V_{SS}$.

The difference between V_{SS} and V_S is due to the resistance of the person to ground R_{BG}, which is in series to the person's body resistance R_B, and limits the circulation of current through the person (Fig. 7.24).

In Fig. 7.25, prospective and the actual step voltages are shown.

As examined in Chap. 4, in outdoor locations, R_{BG} can be calculated by considering a person's feet as two round plate electrodes, each with a ground resistance approximately equal to 4ρ, where ρ is the superficial soil resistivity. By assuming the two feet as parallel electrodes, R_{BG} equals 2ρ; therefore, V_S decreases with respect to V_{SS}, when the superficial resistivity of the soil increases.

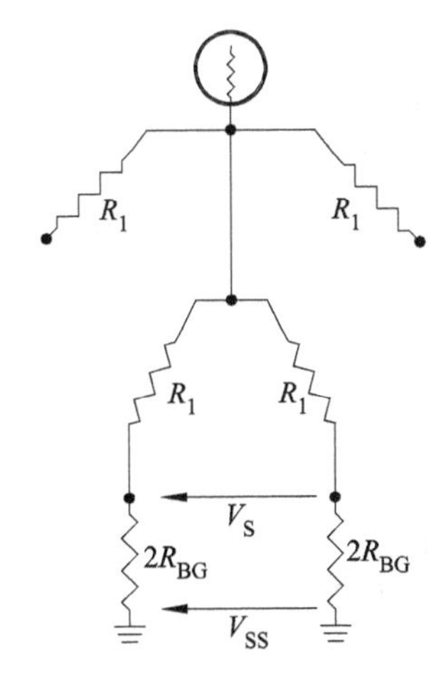

FIGURE 7.24 Diffrence between prospective step voltage V_{SS} and step voltage V_S.

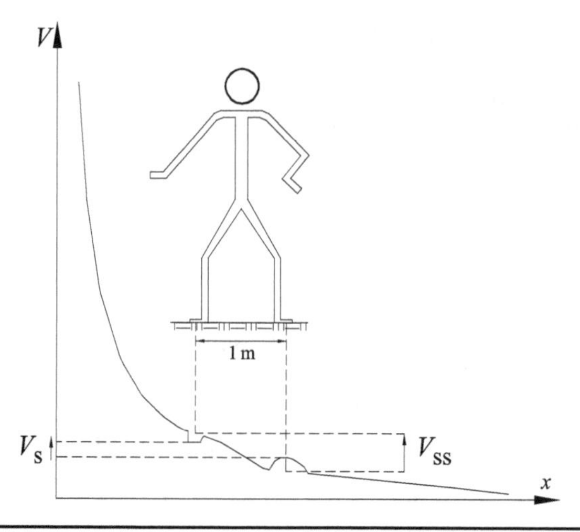

FIGURE 7.25 Prospective step voltage V_{SS} and step voltage V_S.

7.11.1 A Comparison Between the Dangerousness of Touch and Step Voltages

In high and medium voltage, the standard contact is assumed to be applied between a person's hand and both feet. With regard to the touch voltage, therefore, the body resistance is composed of the series of the upper limb's resistance with the parallel of the resistance of two lower limbs (i.e., total resistance equals $1.5R_l$).

With reference to the step voltage, instead, the body resistance is the series of the resistances of the lower limbs (i.e., total resistance equals $2R_l$). As a result, step voltages of the same magnitude as touch voltages are less dangerous, as the larger body resistance limits the magnitude of the current through the person, benefiting the safety.

In addition and as already substantiated in Chap. 5, the probability of ventricular fibrillation depends not only on the current's intensity and its duration but also on the pathway through the person's body.

As per Table 5.1, the heart-current factor F for the left-foot-to-right-foot current path equals 0.04. Therefore, as per Eq. (5.4):

$$I_F^{\text{LF-RF}} = \frac{I_{\text{LH-2F}}}{F} = 25 I_{\text{LH-2F}} \tag{7.14}$$

Thus, a left-foot-to-right-foot current, as caused by step voltages, minimally involves the cardiac region and, therefore, must be 25 times larger than the left-hand-to-feet reference current in order to have the same probability to cause ventricular fibrillation. The left-foot-to-right-foot current path is, therefore, less hazardous and so is the step voltage, if compared to an equal value of touch voltage.

A European standard[10] takes into account the above considerations, obtaining the permissible values of the step voltage by multiplying by the factor 3 the permissible values of the touch voltage, in correspondence with the same clearing time.

FAQs

Q. What is the difference in the status of the neutral wire in TT and TN systems?

A. The neutral wire is generally always energized, in some case even at dangerous potentials. In TT systems, the magnitude of the neutral potential depends on faults caused by the utility, therefore, out of customers' control. For this reason, the TT neutral wire is conventionally treated as a live conductor, and always switched off together with the phase conductors.

In TN-S systems, users own and maintain the supply source; therefore, they can evaluate the risk of the dangerous energization of the neutral wire and decide how to treat it. In TN-C systems, the neutral is also a protective conductor (PEN) and, therefore, must not be switched off.

Q. If the fault-loop in TN systems does not comprise the actual earth, why do we have to ground the source?

A. TN systems require the system ground. Its purpose is to guarantee the stability of the operating voltage-to-earth under regular and fault conditions. In addition, the absence of the earth connection of the source would expose persons to the risk of electric shock, in the conditions depicted in Fig. 7.3.

Endnotes

1. As earlier mentioned, we herein indicate with the underlined symbols the complex numbers as representative of sinusoids. The same symbols, but with no underline, or between bars, signify the magnitude of the complex number. All the network theorems can be applied to sinusoids by means of complex numbers, as it is illustrated in the final appendices of the book.
2. This sizing is permitted by international standards and codes, as further examined.
3. In Chap. 12, the effects of high-voltage faults on low-voltage equipment will be examined.
4. See Chap. 3 for the definition of acceptable risk.
5. Common nonlinear loads may include variable frequency drives and uninterruptible power supplies.

6. In low-voltage systems, the touch voltage is defined as the potential difference across both hands and both feet.
7. Cenelec HD 637 S1–1998-12, *"Power Installations Exceeding 1 kV A.C."*
8. It is assumed that the presence of shoeless workers at industrial job sites is unrealistic.
9. See Chap. 14 for measurement techniques.
10. Cenelec HD 637 S1–1998-12, *"Power Installations Exceeding 1 kV A.C."*

CHAPTER 8

Protective Multiple Earthing (TN-C-S Grounding System)

Marking dynamos for repair $10,000.00,
2 hours labor $10.00
knowing where to mark $9,990.00.

CHARLES P. STEINMETZ (1865–1923)

8.1 Introduction

The *protective multiple earthing* (PME) is a TN-C-S system (Fig. 1.5) employed as the grounding method for low-voltage public supply, and is in use in several countries of the world, for example, Australia, Canada, China, Germany, South Africa, Sweden, Switzerland, U.S.A., U.K., etc.

The utility supply neutral conductor is solidly grounded at the source and at intervals along its distribution. At the dwelling unit's service entrance, the neutral wire (PEN) is connected to the customer's protective conductor (PE) (Fig. 8.1).

In these conditions, the ground-fault currents, arising at the user's installation, will basically return to the supply through the distributor's neutral conductor, which therefore acts also as a protective conductor and thus designated as PEN.

In PME, the main equipotential bonding conductors (MEBs) of a building may be subject to currents, even if the installation is switched off. This is the contribution of other faulty units supplied by the same neutral network (Fig. 8.2).

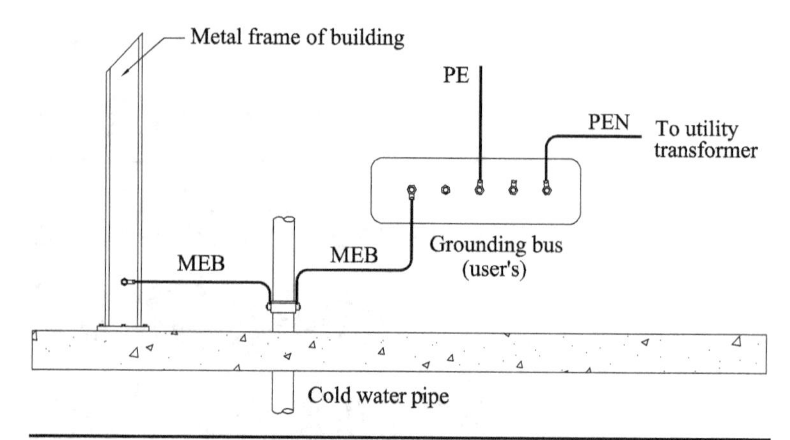

FIGURE 8.1 Connection at the service entrance between utility's PEN and user's PE.

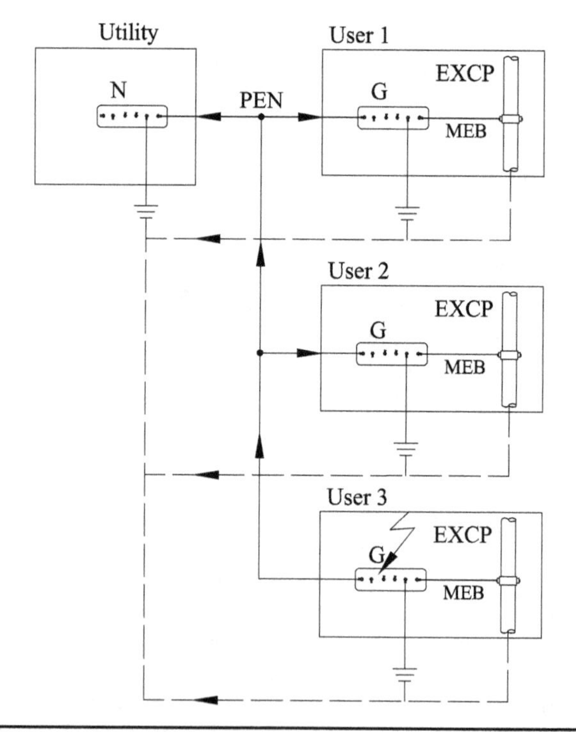

FIGURE 8.2 Faults currents through the MEBs due to the neutral network.

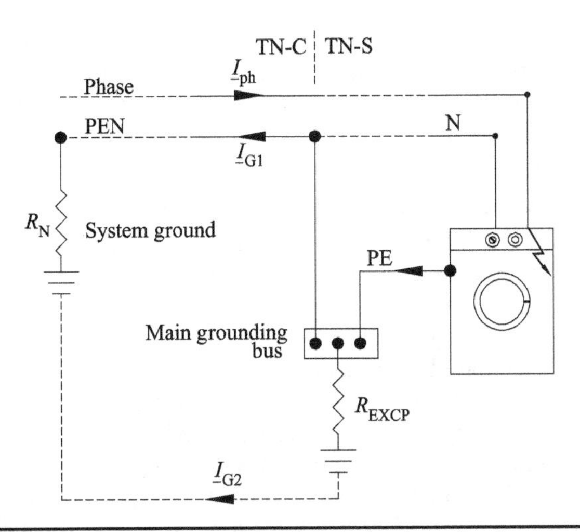

Figure 8.3 Fault-loop in PME systems for single-phase loads.

Unlike in TT systems, the PME customer need not install its own grounding system. In fault conditions, the earth will be partially involved as a return path to the source because of the connection of the EXCPs[1] to the main grounding bus (Fig. 8.3). As already substantiated, this bond is essential to guarantee equipontiality between simultaneously accessible metal parts.

Moreover, in the absence of ground faults, the above earthing arrangement may cause part of the neutral current to return to the source through the earth. In PME systems, therefore, the ground becomes a path to the utility transformer and the circulation of "stray currents" is unavoidable.

In PME systems, ground-fault currents arising at the user's location have large magnitude, as their return path is a conductor, that is, the PE and the PEN to which the earth is in parallel. Therefore, the total impedance of the fault-loop $|\underline{Z}_{\text{Loop}}|$ may be low enough to allow users to use overcurrent devices as protection against indirect contact. A large ground-fault current, in fact, can easily fulfill Eq. (7.4). Hence, the presence of RCDs in PME systems is not a mandatory safety requirement, as is in TT systems.

The combined neutral and protective conductors (PEN) may exist throughout the utility low-voltage distribution system, as this allows some savings in the length of cables to be employed (i.e., four-wire distribution system), whereas, within the user's installation, neutral and protective conductors may be separate (i.e., five-wire distribution system). In some countries (e.g., Norway, U.S.A.), the use of a PEN conductor downstream of the user's main distribution panel is forbidden.

In PME systems, the bonding connection is of utmost importance, at the service entrance of the dwelling unit, between the PEN and the PE. It is, therefore essential to periodically inspect and maintain such connection. Upon loss of this bond, the building would become a TT system, wherein in the absence of RCDs, which are not strictly necessary in PME, users are exposed to electric shock hazards.

8.1.1 Fault-Loop Impedance in PME Systems

As said, low fault-loop impedances guarantee the safe operation of the user's overcurrent protective device with respect to the permissible times of Table 7.1.

In TN systems, $\underline{Z}_{Loop}$ exclusively depends on parameters known to the user, such as the impedances of circuits and transformers. In PME, instead, the fault-loop impedance includes the impedance $\underline{Z}_e$ of the utility low-voltage distribution system, which is usually unknown to the customer. $\underline{Z}_e$, which increases with the distance of the fault's location from the supply source, may also change in time without the user knowing it because of modifications in the utility distribution system. If the total fault-loop impedance $\underline{Z}_{Loop} = \underline{Z}_e + \underline{Z}_{user}$ is excessive, the ground fault current might be so low that Eq. (7.4) cannot be fulfilled, and there would be no effective protection of persons against indirect contact.

The installation of RCDs in dwelling houses, even in PME systems, although redundant in the case of low value of $\underline{Z}_{Loop}$, can, indeed, guarantee safety when $\underline{Z}_{Loop}$ is too high.

8.2 Energization of the PEN Conductor in PME Systems

PME systems imply a considerable responsibility of the local utility, since, together with the electric energy, the distributor provides the users with an earth connection, which must ensure public safety.

In fault conditions, the utility PEN, although multiple grounded, may assume a voltage, with respect to the earth, as is substantiated later on. Such neutral-to-ground voltage can be transferred as a shock potential to the users' ECPs and EXCPs. If utilities cannot "certify" the neutral potential as harmless to persons, a TT system should be employed, instead.

8.2.1 Ground Fault on the Low-Voltage Utility Distribution System

The PEN conductor may become live due to a ground fault occurring along its distribution system, for example, as a result of the fall to earth of overhead cables (Fig. 8.4) or of a contact of the line with an EXCP not connected to a protective conductor.

R_N represents the ground resistance of the utility's earth electrode system: the neutral conductor is earthed not only at intervals along its run (e.g., at the transmission poles) but also at the customers' dwelling

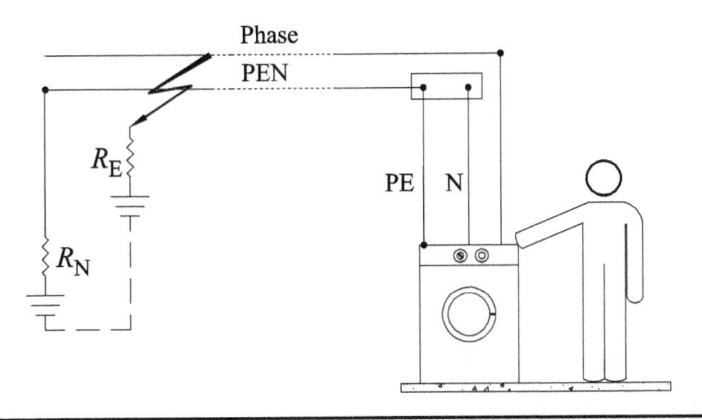

FIGURE 8.4 Ground fault on low-voltage distribution system.

units by means of their ground electrodes (e.g., cold water pipe). R_E is the minimum earth resistance of EXCPs not connected to an equipotential system, through which a fault may occur.

The earth current, by circulating through R_N and R_E energizes the PEN conductor and, therefore, the user's ECPs. The contact resistance with earth also limits this ground current and in some cases can be very high (e.g., line in contact with snow or sand). The distributor's overcurrent devices, therefore, may not be able to clear the fault within the maximum permissible times of Table 7.1, exposing persons to the risk of electric shock.

To identify safe values for the PEN potential and the maximum earth resistance R_N, let us examine Fig. 8.5 and Eq. (8.1):

$$V_N = V_{ph} \times \frac{R_N}{R_E + R_N + Z_i + Z_{ph}} \cong V_{ph} \times \frac{R_N}{R_E + R_N} \leq 50 \qquad (8.1)$$

In Eq. (8.1), we can ignore both the phase conductor impedance $\underline{Z}_{ph}$ and the internal impedance of the source $\underline{Z}_i$ because they are generally negligible with respect to R_N and R_E. As a safety criterion, we can assume as "safe" the PEN conductor if its potential V_N does not exceed the safety limit of 50 V.

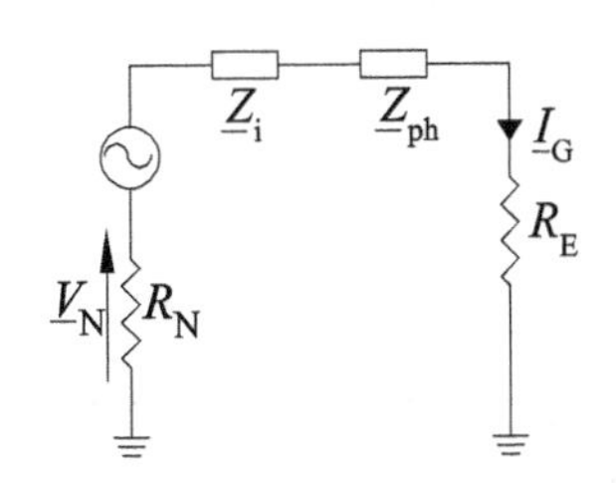

FIGURE 8.5 Fault-loop for a ground fault on the low-voltage PME distribution system.

From Eq. (8.1) we can derive the condition R_N must comply with to keep the PEN potential rise below 50 V:

$$\frac{R_N}{R_E} \leq \frac{50}{V_{ph} - 50} \tag{8.2}$$

In some countries (e.g., Germany), fulfilling Eq. (8.2) is a compulsory requirement for the supply network operators.

8.2.2 Ground Fault on the Medium-Voltage Utility Distribution System

The enclosure of the utility transformer, being an ECP, needs to be earthed and therefore may be connected to the system ground R_N. Low- and medium-voltage systems, then, share the same earth terminal where the PEN conductor originates.

At the occurrence of a ground fault at the transformer primary, the fault current by circulating through the earth and reclosing toward the upstream source of the supply network energizes the system ground. R_N reaches the potential V_N and so does the PEN conductor (Fig. 8.6).

Consequently, persons in contact with low-voltage ECPs, remotely supplied by the "live" PEN, are exposed to the whole earth potential $\underline{V}_N$ during the utility fault-clearing time.

In PME, utilities, in order to ensure that the neutral potential due to primary faults is not dangerous, must accordingly lower the neutral resistance R_N. When it is impossible to decrease R_N, the distributor must alternatively separate the service PEN conductor from the transformer enclosure's earth by creating two distinct grounds (Fig. 8.7).

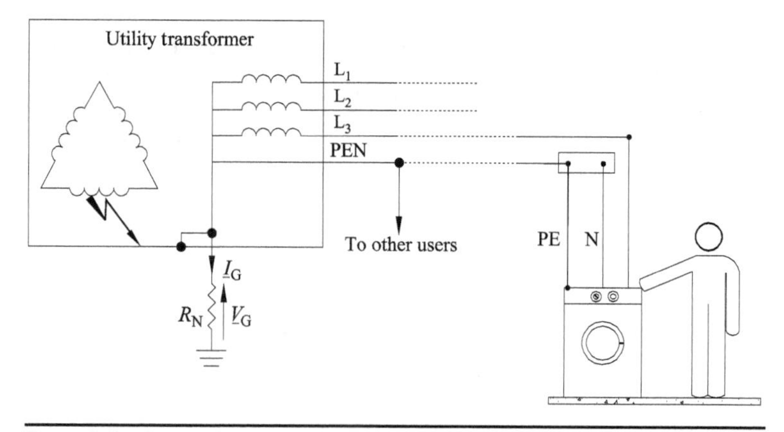

FIGURE 8.6 Ground fault at the transformer primary.

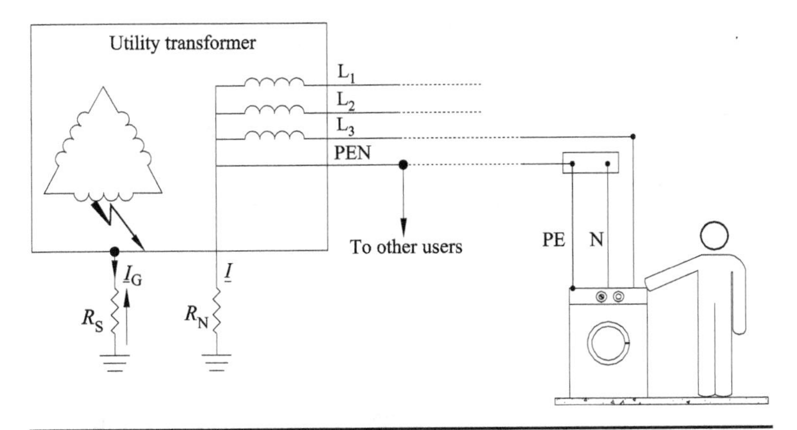

FIGURE 8.7 Distinct grounds at the utility transformer.

This strategy, although effective for the customers, is not free of risk for the utility's operators engaged in maintenance activity within the substation. The PEN conductor, in fact, becomes an EXCP, liable to introduce a "zero" potential or a fault potential into the premises. With this arrangement, workers must take precautions and treat the PEN as a "live" conductor.

8.2.3 Faults Phase-to-PEN in Low-Voltage PME Networks

Another cause of energization of the PEN conductor may be the accidental contact with the phase conductor in low-voltage distribution networks (Fig. 8.8).

The resulting short circuit causes a circulation of current back to the source through the PEN conductor. We assume to neglect the fault current derived by the EXCPs at the user's location, connected for

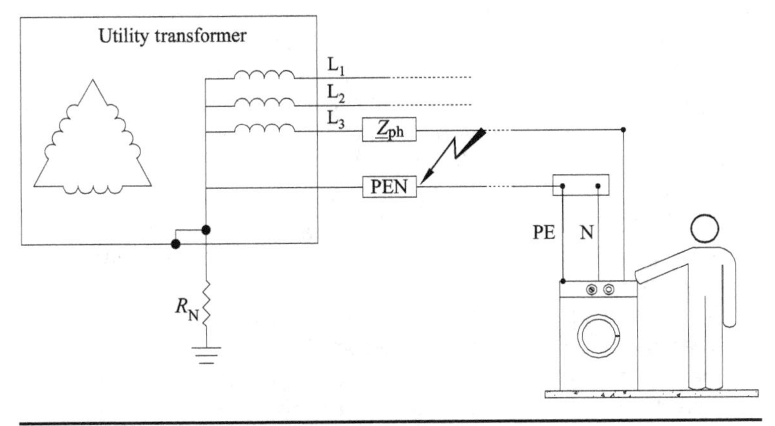

FIGURE 8.8 Short circuit phase-to-PEN in PME.

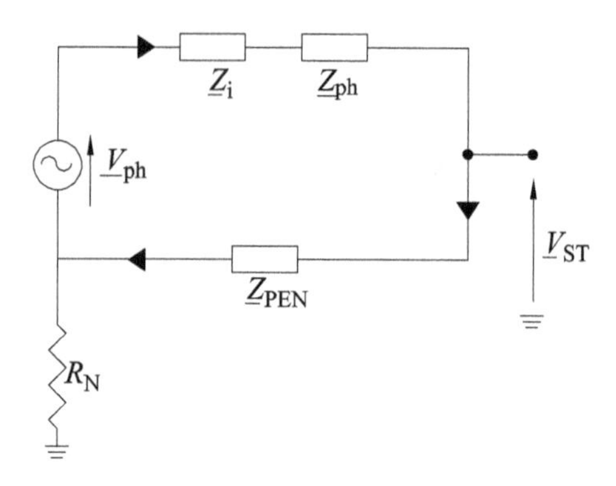

FIGURE 8.9 Equivalent fault-loop for a short circuit phase-to-PEN in PME.

equipotential reasons to the PEN. Such current, in fact, is greatly limited by the EXCPs' resistance-to-ground, which is much larger than the impedance of the PEN conductor. The equivalent fault-loop circuit is shown in Fig. 8.9.

By assuming both the internal resistance of the source to be negligible and the cross-sectional area of the PEN as half of the phase conductor (common situation), the user's ECPs will reach the following prospective touch voltage:

$$\underline{V}_{ST} = \underline{V}_{ph} \times \frac{\underline{Z}_{PEN}}{\underline{Z}_{PEN} + \underline{Z}_{ph}} = \underline{V}_{ph} \times \frac{2\underline{Z}_{ph}}{2\underline{Z}_{ph} + \underline{Z}_{ph}} = \frac{2}{3} \times \underline{V}_{ph} \qquad (8.3)$$

Users are exposed to this voltage for the time the distributor's protective device takes to clear the fault.

8.3 Interruption of the PEN Conductor in PME

As already substantiated in Sec. 7.8 for TN systems, the accidental interruption of the PEN conductor causes all the ECPs supplied downstream of the interruption to be energized at the line-to-line potential, even if healthy. PME have a much larger geographical extension than TN systems and therefore the risk of the interruption of the PEN conductor and of the energization of the ECPs of more than one customer is higher. Hence, the installation of the PEN conductor should be in such a way as to minimize the probability of its break.

The loss of the PEN conductor also triggers overvoltages. Let us examine Fig. 8.10, where two users are supplied by two different phases and the same PEN.

The absence of the PEN as a return path causes a voltage divider between the two users' single-phase loads, which are now supplied by the line-to-line voltage. This may cause the supply to each load to exceed the nominal value, with great risk of overheating of the equipment and therefore of initiating fire.

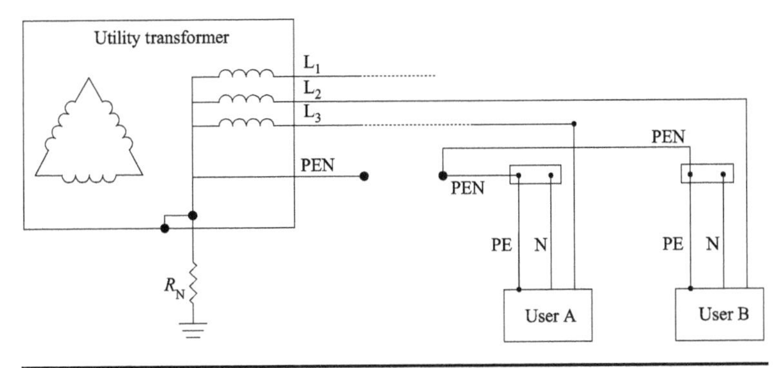

Figure 8.10 Interruption of the PEN conductor in PME.

In order to limit the risk of accidental loss of the PEN, the distributor must develop a redundancy by establishing a network of conductors instead of using a single conductor. In addition, the mechanical strength of the PEN must be assured by using cables of cross-sectional area of at least 10 mm^2 (copper) or 16 mm^2 (aluminum).

In some countries (e.g., the Netherlands), the utility cables are equipped with insulated metal sheaths. The sheaths and the PEN are linked together at the utility and at the customer's service entrance. With this arrangement, even if the PEN is lost, the sheaths of the intact cables will act as a return path to the source, safeguarding the person's safety.

The installation of an intentional ground electrode at the user's service entrance, connected to the main ground bus, as mandated in some countries (e.g., U.S.A.), would limit the prospective touch voltage to the potential difference across the user's electrode in case of a broken PEN (Fig. 8.11). However, the system would become now TT without necessarily having the associated safety requirements (i.e., residual current devices), and the users would still be at risk of electrocution.

8.4 Stray Currents

As already anticipated, in PME systems, unavoidable stray currents continuously circulate through the actual earth. As they depend on the supplied loads, they are likely to escalate over time. Stray currents may produce interferences among electrical systems by transferring potential rises to healthy systems,[2] thereby exposing persons to touch voltages; these may also trigger corrosion phenomena, involving underground metalwork, if they have a d.c. component.

In the presence of ascertained high stray currents, a solution might be to isolate the user by installing a transformer in the dwelling unit.

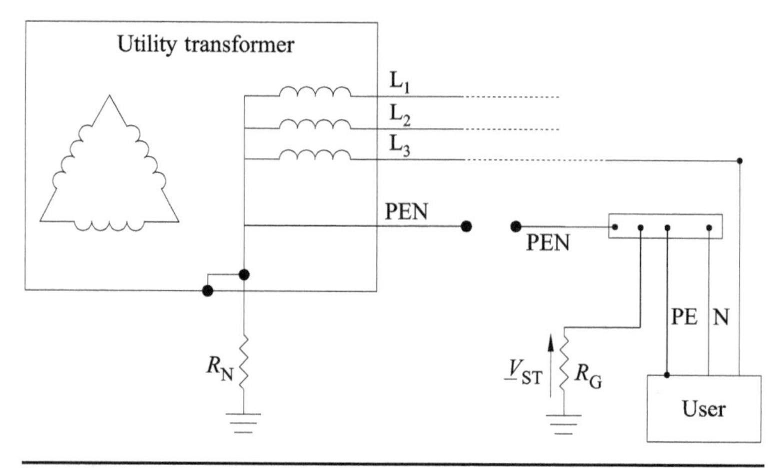

FIGURE 8.11 Interruption of the PEN conductor in PME with intentional user's ground.

Such a transformer, with turn-ratio of 1:1, must be grounded at the secondary winding, preferably at the midpoint (Fig. 8.12).

This transformer breaks the electrical system into independent "islands"[3] and confines both return and fault current in a metallic return path (i.e., the PE) by creating a TN-S system within the PME.

8.5 Stray Voltages

The term stray voltage indicates permanent neutral-to-ground potentials and not temporary potentials due to fault conditions at the customer's location. The neutral conductor, although earthed at the substation, may, in fact, be energized above ground, where it enters the user's premises.

The reason being that supply cables employed to power up the customers, composed of phase and PEN conductors, have finite impedance, which may cause voltage drops along their runs (Fig. 8.13).

The voltage drop along the PEN increases with its length and the contributions from the customers; therefore, $\underline{V}_{NG}$ at User B is greater

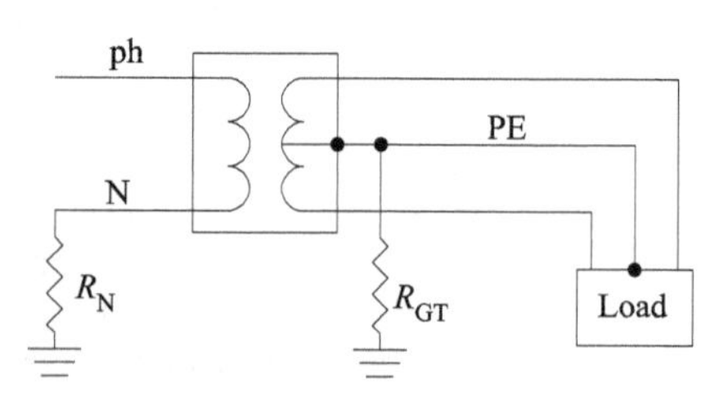

FIGURE 8.12 Transformer inside of the dwelling unit.

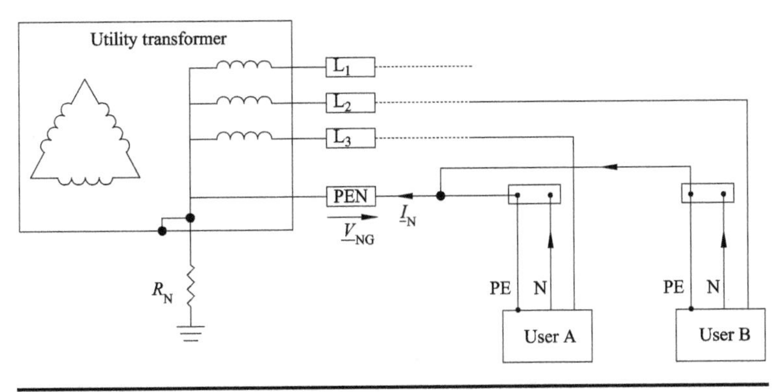

FIGURE 8.13 Neutral-to-ground potentials.

than at User A. The neutral-to-ground voltage, which may reach several volts, also exists between the neutral and the protective conductors. In the case of a defect in the insulation between these two conductors, a low-resistance object might bridge the gap between them, with the possible result of setting on fire any flammable material eventually present.

Endnotes

1. As a rule of thumb, the percentage of ground-fault current flowing through the earth may be estimated as large as 20% of the return current.

2. M. Mitolo, M. Tartaglia, G. Gruosso, and A. Canova, "*Evaluation of Voltage Exposures Due to AC/DC Stray Currents*," Proceedings of the IEEE-IAS Industry Application Society 42nd Annual Meeting, New Orleans, LA, September 2007.
3. M. Mitolo, G. Parise, "*TN—Island Grounding System*," Proceedings of the IEEE-IAS Industrial & Commercial Power Systems Technical Conference, Dearborn, MI, May 2006.

CHAPTER 9

IT Grounding System

The lecturer should give the audience full reason to believe that all his powers have been exerted for their pleasure and instruction.

MICHAEL FARADAY (1791–1867)

9.1 Introduction

In IT (Isolation Terre) systems, the power source, for example, transformers, is not solidly connected to earth, and therefore, is defined as ungrounded. Enclosures of ECPs, though, must be grounded, individually, in groups or collectively (see Chap. 1).

The insulation of secondary sides of supply transformers from the earth may be obtained through a high-resistance grounding resistor, typically in high-/medium-voltage systems. It is not advisable, although not forbidden by technical standards, to distribute the neutral conductor in order to facilitate its insulation from ground.

As already anticipated in Chap. 2, electrical systems cannot be completely isolated from ground, even in the absence of any intentional connection to the earth. In IT systems, ground-fault currents can circulate through the distributed system capacitance to ground C_0 (Fig. 9.1).

At the system frequency, in fact, cables and earth can be seen as armatures of a capacitor, whose dielectric is the surrounding air. In addition, the resistance offered by the cable insulation to ground, which has the magnitude of a few megaohms per kilometer, constitutes another leaking path in parallel to the system net capacitance. In addition, connected equipment and devices (e.g., electrical motors and surge arresters), although to a lesser amount, increases both the system capacitance and the leakage resistance to ground.

The leakage resistance is, indeed, very high and, therefore, may be considered as an open circuit in parallel to the system capacitance.

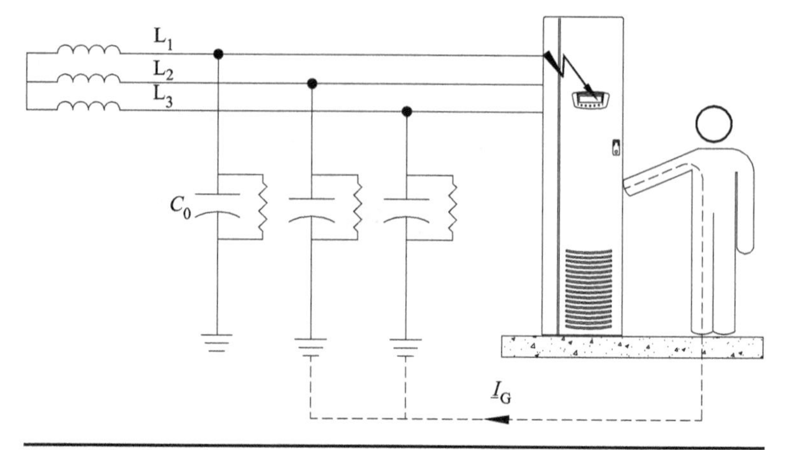

Figure 9.1 Ground-fault currents circulating through the system capacitance to ground.

The current circulating after the first fault, thus, is mainly capacitive. In the case of indirect contact with a faulty ECP, therefore, persons are exposed to the risk of electrocution (Fig. 9.1). C_0 increases with the cables' extension and so does the ground-fault current flowing through the person. Such current, relatively small in absolute value (i.e., tens of amperes), are large enough to be lethal to persons. To avoid this hazard, the ECPs must be earthed, as previously anticipated (Fig. 9.2).

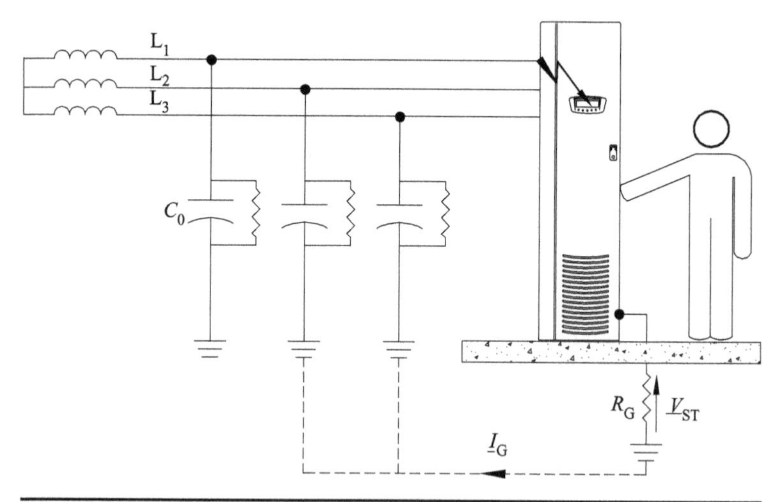

Figure 9.2 Earth current flowing through the ECPs ground and the capacitive system impedance to ground.

The presence of the intentional ground lowers the prospective touch voltage to the potential drop across R_G, with evident benefit for the safety against indirect contact. If $V_{ST} = R_G I_G \leq 50$ V, where I_G is the first-fault current to ground, the automatic disconnection of supply is not necessary, as the ground fault does not cause any hazard to persons and may persist within the ECP. However, for the reasons explained later on, it is recommended to clear this fault as soon as practically possible.

The most valuable aspect of the IT systems is the possibility to maintain the supply to a circuit even in the presence of a first fault caused by a live part in contact with enclosures or earth. This feature is of paramount importance when the loss of the electrical service can compromise the safety of persons or disrupt a costly industrial process.

During an unresolved first fault to ground, a second fault involving a different phase might take place. In this case, the phase-to-phase voltage drives the fault current, and an actual short circuit occurs (Fig. 9.3).

In this situation, at least one of the protective devices safeguarding the circuits will trip and disconnect the supply. Even the two faulty circuits might be simultaneously tripped off and the safety of the installation, where the continuity of supply is essential, may be further compromised.

On the other hand, the second fault exposes persons to risk of electrocution in the time frame the overcurrent devices take to trip. With reference to Fig. 9.3, let us calculate the prospective touch voltage

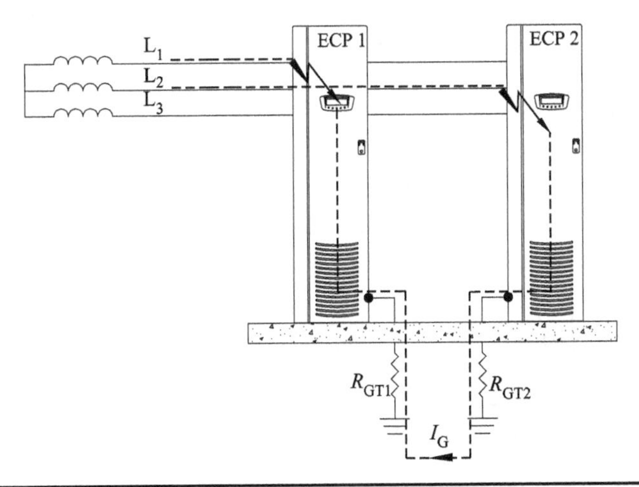

FIGURE 9.3 Second fault to ground driven by the phase-to-phase voltage (ECPs are shown independently grounded).

V_{ST1} on ECP 1. We obtain

$$I_G = \frac{\sqrt{3}V_{ph}}{R_{G1} + R_{G2}} \tag{9.1}$$

$$V_{ST1} = \frac{\sqrt{3}V_{ph}}{R_{G1} + R_{G2}} \times R_{G1} \tag{9.2}$$

The fault-loop is the same as in the TT systems, but the driving potential is not the voltage between the line and the neutral (e.g., 230 V), but the voltage between the phase conductors (e.g., 400 V). It is clear from Eq. (9.2) that if R_{G2} were low when compared to R_{G1}, persons in contact with ECP 1 would be exposed to nearly the whole line-to-line voltage.

To prevent these hazards, and as already anticipated, the first fault should be resolved in the shortest possible time by the maintenance team. To this purpose, in IT systems an insulation monitoring device (IMD) must be employed to detect the presence of the first fault to ground.

9.1.1 Insulation Monitoring Device

The IMD supervises the insulation reactance and/or resistance between the power lines and the earth (Fig. 9.4).

The IMD continuously monitors the impedance to ground (i.e., resistance and capacitive reactance) by injecting both a d.c. and an a.c. current through the neutral point of the system.[1] If such impedance decreases below a predetermined value, due to a first fault to ground, an audible/visual alarm will be initiated. Such alarm will alert the maintenance crew and will stay on for the entire duration of the fault. Once the faulty circuit has been located and fixed, operators will manually switch it off.

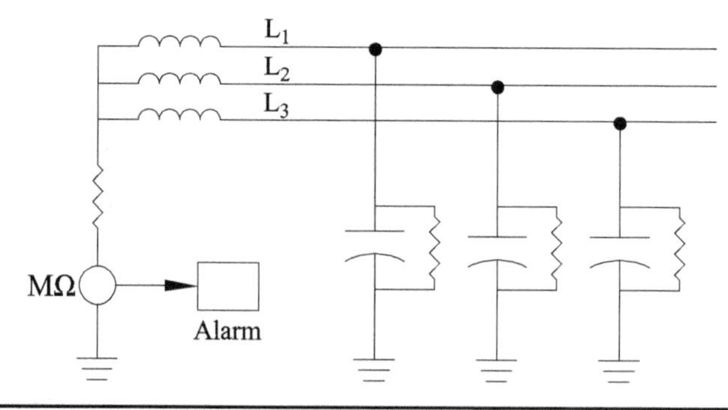

FIGURE 9.4 IMD monitors the insulation reactance/resistance to ground.

A simpler method of ground-fault detection consists of three lamps connected between each line and the earth. At the occurrence of a ground fault, there will be a reduction in the voltage across the lamp linked to the faulted phase, which therefore will dim, while the others will become brighter.

9.1.2 Equipotential Bonding

In IT systems, the equipotential bonding, consisting of connections between enclosures of fixed equipment and EXCPs, when simultaneously accessible, add safety to the installation. The equipotential bonding should also connect, if practicable, the metal frame of the building or the reinforced bars embedded in the structure's concrete.

The resulting equipotential system converging to the main grounding bus should "generate" from there all the protective conductors, including, of course, those to receptacles.

The criterion to assess the safety of the bonding connection between ECPs and EXCPs is based on the resistance of the connection itself, as follows:

$$R_B I_a \leq 50 \text{ V} \tag{9.3}$$

where R_B is the resistance of the bonding connection and I_a is the operating current of the protective device in correspondence with the maximum disconnection time as per Table 6.1 or Table 7.1, according to the way the ECPs are earthed, singularly or collectively, as further explained in Sec. 9.5.

9.2 Overvoltages Due to Faults in IT Systems

The main issue in IT systems is the possibility of overvoltages induced by ground faults. With reference to Fig. 9.2, where the system leakage resistance is neglected (i.e., it is considered as an open circuit), the equivalent three-phase fault circuit is shown in Fig. 9.5.

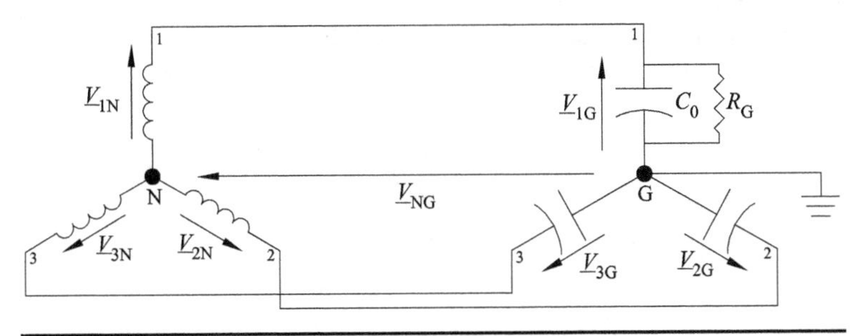

FIGURE 9.5 Equivalent circuit in the case of ground fault.

The occurrence of a ground fault causes the system capacitance to become an unbalanced load (i.e., the impedances of the three branches 1, 2, and 3 are no longer equal). The earth electrode resistance R_G, in fact, is in parallel to the capacitance of the faulty phase (e.g., line 1 in Fig. 9.5). Upon the first fault the system evolves from a balanced three-phase capacitive load, with no neutral wire, to an unbalanced capacitive-resistive load.

Because of this unbalance, a potential difference $\underline{V}_{NG} = \underline{V}_N - \underline{V}_G$, also referred to as *neutral potential rise*, appears between the point of neutral N at the source and the ground G at the faulty ECP. The presence of $\underline{V}_{NG}$ changes the voltage between the line conductors and the ground (i.e., $\underline{V}_{1G}$, $\underline{V}_{2G}$, and $\underline{V}_{3G}$) with respect to the systems voltages (i.e., $\underline{V}_{1N}$, $\underline{V}_{2N}$, and $\underline{V}_{3N}$). The two sets of vector quantities identical in normal conditions will now differ.

By applying Kirchhoff's voltage law to each one of the closed loop in Fig. 9.5, we obtain

$$\underline{V}_{1G} = \underline{V}_{1N} + \underline{V}_{NG} \tag{9.4}$$

$$\underline{V}_{2G} = \underline{V}_{2N} + \underline{V}_{NG} \tag{9.5}$$

$$\underline{V}_{3G} = \underline{V}_{3N} + \underline{V}_{NG} \tag{9.6}$$

Let us calculate $\underline{V}_{NG}$ by applying Millman's theorem to the three-phase circuit in Fig. 9.5 (see App. B for more details):

$$\begin{aligned}\underline{V}_{NG} &= \frac{-\underline{V}_{1N}[(1/R_G) + j\omega C_0] - \underline{V}_{2N}\, j\omega C_0 - \underline{V}_{3N}\, j\omega C_0}{(1/R_G) + j3\omega C_0} \\ &= \frac{-\underline{V}_{1N}(1/R_G) - j\omega C_0\left(\underline{V}_{1N} + \underline{V}_{2N_0} + \underline{V}_{3N}\right)}{(1/R_G) + j3\omega C_0} \\ &= -\frac{\underline{V}_{1N}}{1 + j3\omega C_0 R_G}\end{aligned} \tag{9.7}$$

The above result has been obtained by knowing that the vectorial summation ($\underline{V}_{1N} + \underline{V}_{2N} + \underline{V}_{3N}$) is zero, since these vectors are equal in magnitude and equally displaced by 120°.

We can now rewrite Eqs. (9.4) through (9.6), as follows:

$$\underline{V}_{1G} = \underline{V}_{1N} - \frac{\underline{V}_{1N}}{1 + j3\omega C_0 R_G} = \frac{j3\omega C_0 R_G \underline{V}_{1N}}{1 + j3\omega C_0 R_G} \tag{9.8}$$

$$\underline{V}_{2G} = \underline{V}_{2N} - \frac{\underline{V}_{1N}}{1 + j3\omega C_0 R_G} \tag{9.9}$$

$$\underline{V}_{3G} = \underline{V}_{3N} - \frac{\underline{V}_{1N}}{1 + j3\omega C_0 R_G} \tag{9.10}$$

Equations (9.8) through (9.10) show that in the absence of ground faults (i.e., $R_G = \infty$), there is an identity between the system voltages at the supply and at the load.

If R_G equals zero (i.e., bolted fault to ground), we obtain

$$\underline{V}_{1G} = 0 \tag{9.11}$$

$$\underline{V}_{2G} = \underline{V}_{2N} - \underline{V}_{1N} \tag{9.12}$$

$$\underline{V}_{3G} = \underline{V}_{3N} - \underline{V}_{1N} \tag{9.13}$$

In the above case, the voltage between each healthy phase conductor and the ground is the result of a vector difference, which assumes a magnitude as large as the line-to-line potential (e.g., 400 V vs. 230 V). The basic insulation of single-phase loads, eventually present when the neutral conductor is distributed, may be overstressed and punctured, if loads are not rated to withstand this overvoltage.

With reference to Fig. 9.2, where we neglect the system leakage resistance, we can now calculate the representative phasor $\underline{I}_G$ of the ground current, together with its magnitude $|\underline{I}_G|$, by using Eq. (9.8):

$$\underline{I}_G = \frac{\underline{V}_{1G}}{R_G} = \frac{j3\omega C_0 \underline{V}_{1N}}{1 + j3\omega C_0 R_G} \tag{9.14}$$

$$\begin{aligned} |\underline{I}_G| &= \frac{|\underline{V}_{1G}|}{R_G} = \frac{3\omega C_0 \, |\underline{V}_{1N}|}{\sqrt{1 + (3\omega C_0 R_G)^2}} \\ &= \frac{|\underline{V}_{1N}|}{(1/3\omega C_0)\sqrt{1 + (3\omega C_0 R_G)^2}} = \frac{|\underline{V}_{1N}|}{\sqrt{[1/(3\omega C_0)^2] + R_G^2}} \end{aligned} \tag{9.15}$$

where $|\underline{V}_{1G}|$ is the magnitude of the phase-to-ground voltage of the system as per Eq. (9.8) in correspondence with a generic value of R_G.

9.3 Resonant Faults in IT Systems

A further technical drawback of IT systems, besides overvoltages, is the possible occurrence of resonant faults-to-ground. Such faults can cause very high earth potentials, thereby destroying the insulation of equipment, as well as originating fires.

Resonant ground faults may be set off by contacts of the line conductor with ground by means of an inductance (e.g., the winding of a transformer). In this case, an inductive reactance $X_L = \omega L$, where L is the self-inductance, is in parallel to the line capacitance of the faulty phase (Fig. 9.6).

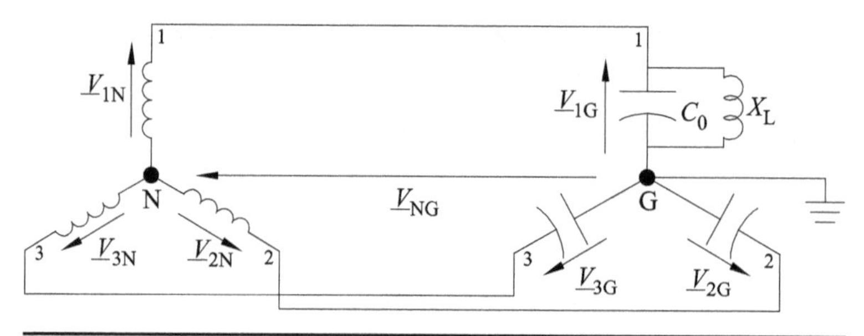

FIGURE 9.6 Resonant ground fault.

Also in this case, we will calculate the neutral potential rise $\underline{V}_{NG}$ by applying Millman's theorem to the system in Fig. 9.6.

$$\begin{aligned}\underline{V}_{NG} &= \frac{-\underline{V}_{1N}[(1/j\omega L) + j\omega C_0] - \underline{V}_{2N}\, j\omega C_0 - \underline{V}_{3N}\, j\omega C_0}{(1/j\omega L) + j3\omega C_0} \\ &= \frac{-\underline{V}_{1N}(1/j\omega L) - j\omega C_0\left(\underline{V}_{1N} + \underline{V}_{2N_0} + \underline{V}_{3N}\right)}{(1/j\omega L) + j3\omega C_0} \\ &= \frac{-\underline{V}_{1N}}{1 - 3\omega^2 LC_0} \end{aligned} \tag{9.16}$$

The magnitude $\underline{V}_{NG}$ approaches infinity (i.e., resonant condition), with disruptive consequences, when the denominator of Eq. (9.16) approaches zero:

$$1 - 3\omega^2 LC_0 = 0 \tag{9.17}$$

As the system frequency f is fixed (e.g., 50/60 Hz), the system will resonate when the product of the distributed capacitance and the fault inductance satisfies Eq. (9.18):

$$\frac{1}{3\omega^2} = LC_0 \tag{9.18}$$

To prevent the accidental fulfillment of Eq. (9.18), the point of neutral of the source may be earthed via a grounding resistor R_s of appropriate high value (Fig. 9.7).

Let us calculate the neutral potential rise $\underline{V}_{NG}$ by applying Millman's theorem to the system in Fig. 9.7.

$$\underline{V}_{NG} = \frac{-\underline{V}_{1N}(1/j\omega L)}{(1/j\omega L) + j3\omega C_0 + (1/R_s)} = \frac{-\underline{V}_{1N}}{1 - 3\omega^2 LC_0 + j(\omega L/R_s)} \tag{9.19}$$

In resonant conditions, by substituting Eq. (9.18) in Eq. (9.19), we will obtain

$$\underline{V}_{NG} = \frac{-\underline{V}_{1N}}{j(\omega L/R_s)} = \frac{-\underline{V}_{1N} R_s}{j(1/3\omega C_0)} = j\underline{V}_{1N} R_s \times 3\omega C_0 \tag{9.20}$$

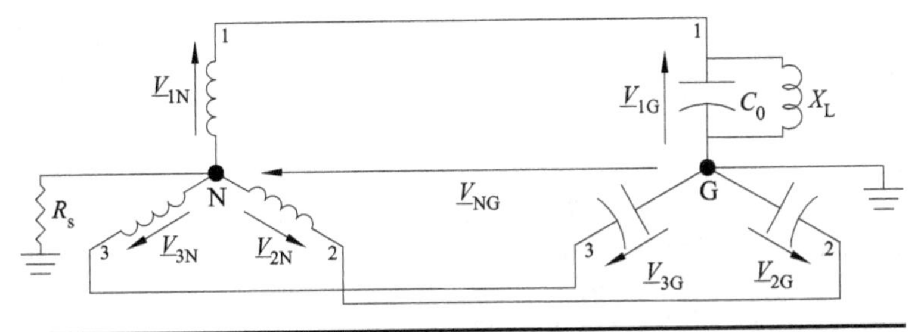

Figure 9.7 IT system earthed via a neutral grounding resistor R_S.

In order for $|\underline{V}_{NG}|$ not to exceed $|\underline{V}_{1N}|$, R_s must fulfill the following condition:

$$R_s \leq \frac{1}{3\omega C_0} \tag{9.21}$$

Equation (9.21) constitutes the criterion to size the grounding resistor in IT systems, once C_0 is estimated. However, it is important to note that R_s must be sufficiently high to prevent first fault currents from tripping protective devices. As long as the system is high-resistance grounded, the benefits of the continuity of the service can be enjoyed without the risk of overvoltages.

9.4 Protection Against Direct and Indirect Contact by Using RCDs in IT Systems

The direct contact can cause the circulation of dangerous current through the human body even in IT systems. The additional protection, normally provided by RCDs in TT and TN, is not effective in IT systems. The fault current, in fact, cannot activate the RCD, because it flows entirely back through its toroid via the system distributed impedance. The RCD does not sense any unbalance and, therefore, cannot intervene (Fig. 9.8).

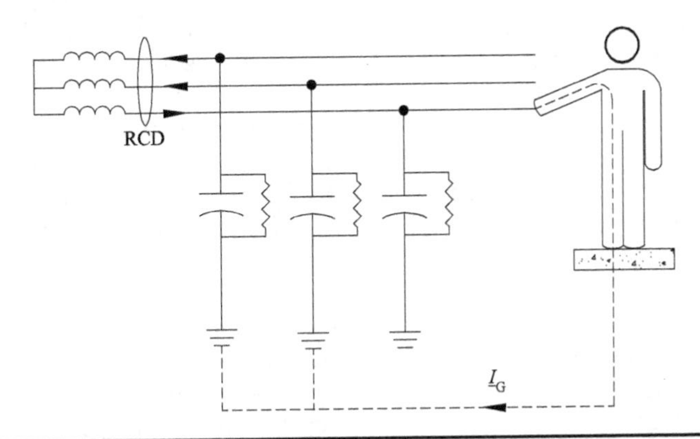

Figure 9.8 Direct contact in IT systems and RCDs.

A similar situation also occurs in the case of first fault to ground caused by the failure of the basic insulation of an ECP. The RCD cannot protect against indirect contact, as the circulation of the fault current is, once again, entirely allowed through the toroid's windings by the system distributed impedance.

In sum, in IT systems RCDs neither can function nor can be blamed for not intervening as a protection for both direct and indirect contacts. The nature of the fault-loop, in fact, prevents their proper operation and renders their installation ineffective.

9.5 Protection Against Indirect Contact in the Event of a Second Fault to Ground

After the occurrence of the first fault to earth, the IT system is no longer ungrounded, because of the accidental connection of the faulty phase to earth. In the event of a second fault involving a different phase, the IT system "evolves" into TT or TN according to the earthing arrangement of ECPs (i.e., individually or collectively).

9.5.1 ECPs Earthed Individually or in Groups

If ECPs are earthed individually, or in groups, in the event of a second fault, the system becomes TT and we are in the case in Fig. 9.3. Protection against indirect contact is achieved if the following condition, applied to the generic ith ECP, is fulfilled:

$$R_{Gi} I_a \leq 50 \text{ V} \tag{9.22}$$

where I_a is the current causing the automatic operation of the disconnection device within the maximum permissible time specified in Table 6.1 for TT systems for final circuits, or in a time not exceeding 1 s for distribution circuits. As we have already substantiated in Chap. 6, the optimum protection against indirect contact in TT systems is carried out by RCDs. The fault current circulating through the earth, due to the second fault may in fact be too low to operate promptly the overcurrent devices. RCDs can clear the fault within the safe time required by Table 6.1, generally in correspondence of a ground current of at least five times their residual operating currents.

The additional costs due to the necessity of RCDs, effective under second fault conditions in the previous arrangement, and due to the installation of individual ground electrodes, usually induce designers to collectively earth the ECPs to a single ground electrode.

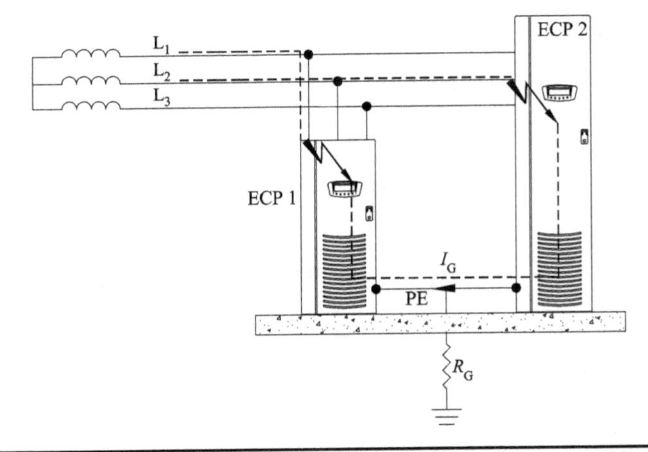

FIGURE 9.9 ECPs earthed collectively to a single grounding system (neutral not distributed).

9.5.2 ECPs Earthed Collectively to a Single Grounding System

The ECPs may be earthed collectively, that is, connected to one single grounding system. If the neutral is not distributed, at the occurrence of a second fault, involving a different live conductor, the system becomes TN (Fig. 9.9).

In this case, the voltage between the line conductors, equal to 1.732 times the voltage between line conductor and neutral, drives the fault current. The second fault may randomly occur in a different circuit, which, for example, supplies an ECP remotely located with respect to the location of the first fault. Thus, the fault impedance may be due to the contributions of line conductors and protective conductors of different cross-sectional areas. This possibility renders extremely challenging the prediction of the total fault-loop impedance in IT systems "evolved" into a TN. As seen in Chap. 7, in TN systems created as such in industrial facilities, the fault-loop exclusively comprises one circuit at the time.

To take into account the second fault in a different circuit, the protection against indirect contact of persons touching one faulty enclosure is effective if the following condition is fulfilled:

$$\frac{\sqrt{3}V_{\text{ph}}}{2Z_{\text{S}}} \geq I_{\text{a}} \tag{9.23}$$

or equivalently, by solving for Z_S,

$$Z_{\text{S}} \leq \frac{\sqrt{3}V_{\text{ph}}}{2I_{\text{a}}} \tag{9.24}$$

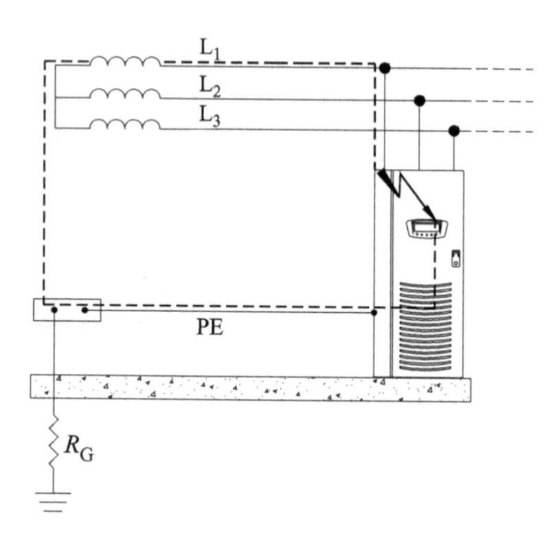

FIGURE 9.10 Fault-loop impedance Z_S.

where Z_S is defined as the impedance of the fault-loop composed of the line conductor and the protective conductor of only one of the faulty circuits (Fig. 9.10). I_a is the current causing the automatic operation of the overcurrent protective device within the time specified in Table 7.1 for TN systems. If RCDs are employed, I_a represents the residual operating current, which provides the disconnection of supply within the times of Table 7.1. A disconnection time of 5 s is allowed in distribution circuits. The factor 2 at the denominator in Eq. (9.23), by doubling the above-defined impedance of the fault-loop, takes into account the limiting effect of the additional impedance of the second circuit involved in the fault.

Although defined as a "loop" in international standards, in reality the fault-loop to be considered in Eqs. (9.23) and (9.24) is "open," as it misses the side that connects the source to the protective conductor. However, one can imagine a conductor with zero impedance as the missing "side" of the loop, as indicated by the dotted line in Fig. 9.10. Both faulty circuits at ECP1 and ECP2, instead, form the true fault-loop (indicated by the dotted line in Fig. 9.9).

If the neutral conductor is distributed to loads, a first or second fault may involve this conductor (Fig. 9.11).

In this case, the voltage V_{ph} between the faulty line and neutral conductor will drive the fault current. However, this is the most conservative case that the designer should take into consideration. In correspondence with a lower current, in fact, the clearing time of protective overcurrent devices increases and might dangerously exceed the maximum permissible disconnection time.

Condition for safe automatic disconnection of supply is

$$\frac{V_{ph}}{2Z'_S} \geq I_a \tag{9.25}$$

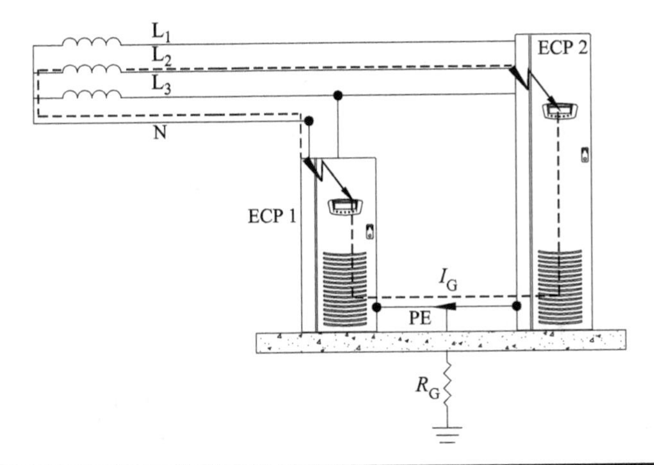

FIGURE 9.11 First or second fault involving the neutral conductor in IT systems.

or equivalently, by solving for Z'_S,

$$Z'_S \leq \frac{V_{ph}}{2I_a} \tag{9.26}$$

where Z'_S is the impedance of the fault-loop composed of the neutral conductor and the protective conductor (Fig. 9.12). All the other terms of the previous equations have the analogous meaning as those in Eqs. (9.23) and (9.24).

9.6 Role of the Fault Resistance in TT and IT Systems

TT and IT systems have in common the inclusion of the actual earth in the fault-loop. In the presence of the same value of R_G, the voltage exposure in TT systems depends on the resistance R_N of the utility neutral [Eq. (6.1)], whereas in IT systems it depends on the distributed capacitance to ground C_0 [Eq. (9.8)]. The prospective touch voltage, therefore, greatly differs between the two systems.

V_{ST} as per Eqs. (6.1) and (9.8) and as a function of R_G is shown in Fig. 9.13 for both grounding systems. It has been assumed that $R_N = 1\ \Omega$ and $C_0 = 2$ nF.

The chart clearly shows that in IT systems even in the presence of relatively high value of the ground resistance, the prospective touch voltage is still below the threshold of danger. In addition, the rate of change of V_{ST} with R_G is much lower in IT systems.

Another important difference between TT and IT systems is the influence of the eventual resistance of the fault R_F, which is in series to

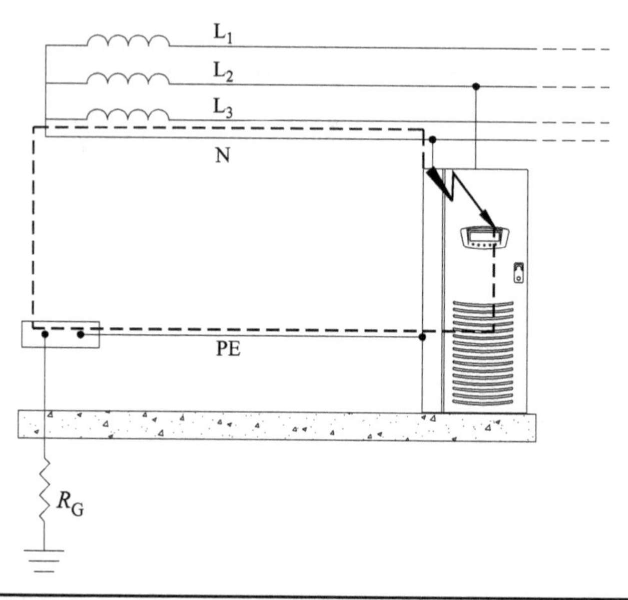

FIGURE 9.12 Fault-loop impedance Z'_S including the neutral conductor.

the ground resistance, in the determination of the prospective touch voltage V_{ST}.

With reference to Fig. 6.2, in TT systems the prospective touch voltage, taking into account R_F, becomes

$$V_{ST}^{TT} = V_{ph} \times \frac{1}{1 + [R_N/(R_G + R_F)]} \tag{9.27}$$

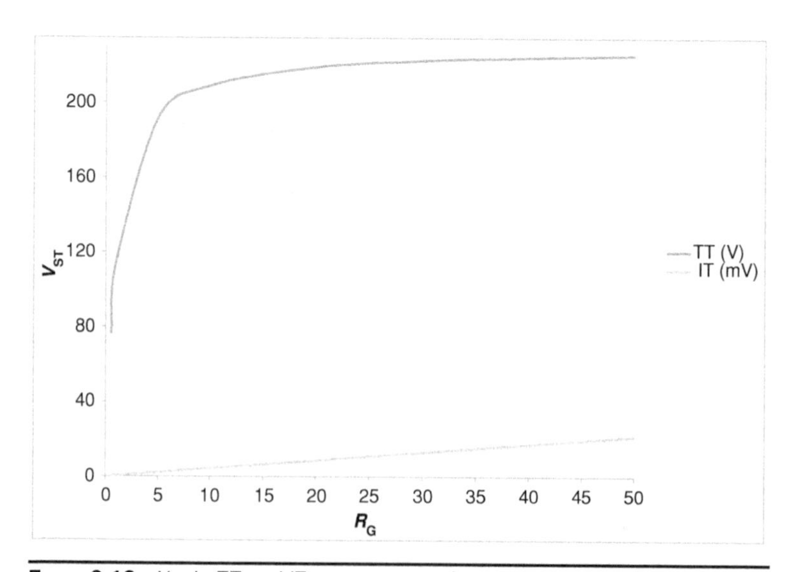

FIGURE 9.13 V_{ST} in TT and IT systems as a function of R_G.

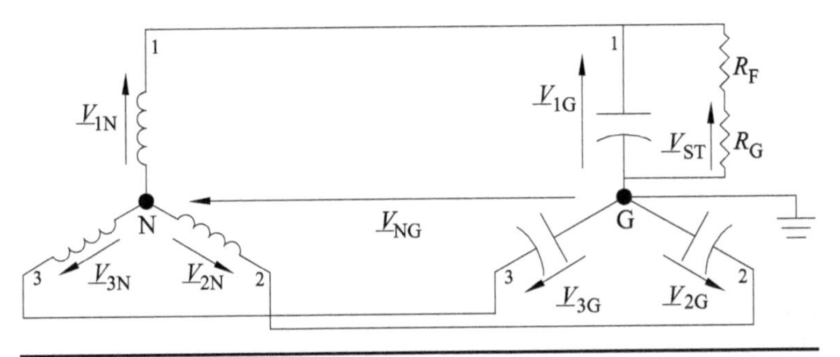

Figure 9.14 Fault resistance in IT systems.

With reference to Fig. 9.14, in IT systems the prospective touch voltage becomes

$$V_{ST}^{IT} = V_{1G}\frac{R_G}{R_G + R_F} = \frac{V_{1N}(R_G + R_F)}{\sqrt{[1/(3\omega C_0)^2] + (R_G + R_F)^2}} \times \frac{R_G}{R_G + R_F}$$
$$= \frac{V_{1N} R_G}{\sqrt{[1/(3\omega C_0)^2] + (R_G + R_F)^2}} \tag{9.28}$$

In order to graph Eqs. (9.27) and (9.28) as a function of R_F, let us assume $R_N = 1\ \Omega$, $C_0 = 2$ nF, and $R_G = 10\ \Omega$. The result, for a nominal voltage line-to-ground equal to 230 V, is shown in Fig. 9.15.

In IT systems, the presence of a fault resistance R_F plays in favor of safety. The prospective touch voltage, in fact, decreases with R_F, unlike in TT systems.

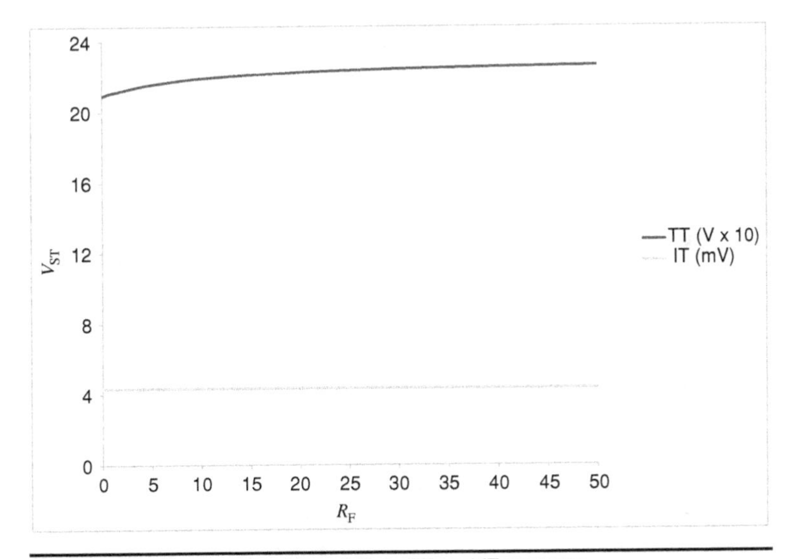

Figure 9.15 V_{ST} as a function of R_F in TT and IT systems.

FAQ

Q. What is the difference between IT systems and electrically separated systems?

A. In IT systems, the power source is not solidly connected to earth, but enclosures of ECPs must be grounded. In electrically separated systems, the power source is still not connected to earth, but ECPs must not be grounded.

As seen, in IT systems, the fault currents through the human body might reach lethal magnitudes of tens of amperes in the absence of the grounding of the ECPs.

In electrically separated systems, there is no need for earthing ECPs, as the fault currents to ground are limited to harmless values by definition: the product of the nominal voltage of the separated circuit (in volts) and its length (in meters) must not exceed 10^5 V · m, and the length of the wiring system must not exceed 500 m. These two conditions actually define electrically separated systems and, thereby, dictate the number of transformers necessary to fulfill them.

Endnote

1. In the absence of a neutral point (e.g., delta connection of the system), an artificial neutral can be created.

CHAPTER 10
Extra-Low-Voltage Systems

Despite our best efforts to achieve complete electrical safety, against stupidity, ignorance, and negligence, there is no defense.

M. MITOLO

10.1 Introduction

Protection by extra-low voltage, realized by supplying electrical systems with nondangerous voltages, is a measure against direct and indirect contact, suitable in all situations, but especially indicated in wet locations or in restrictive conductive locations.[1] The extra-low voltage must not exceed 50 V a.c. or 120 V ripple-free d.c. between conductors or between any conductor and the earth.[2] The aforementioned values are deemed not hazardous to persons in standard conditions; therefore, persons cannot undergo electric shocks even if in contact with live parts. Typical applications of extra-low voltages may be lighting systems in particular locations and electrical equipment of machines.

Extra-low-voltage systems are grouped in three different categories: *separated extra-low voltage* (SELV), *protective extra-low voltage* (PELV), and *functional extra-low voltage* (FELV).

A constant voltage to which may be superimposed a sinusoidal ripple, whose r.m.s. value does not exceed 10% of the d.c. voltage itself is conventionally defined as ripple-free (Fig. 10.1).

If the ripple is not a sine wave, the maximum peak of the total voltage must be less than 140 and 70 V, respectively, for a nominal 120 and 60 V ripple-free d.c. system (Fig. 10.2).

To prevent neighboring electrical systems at higher voltages from accidentally coming in contact with extra-low-voltage circuits, the

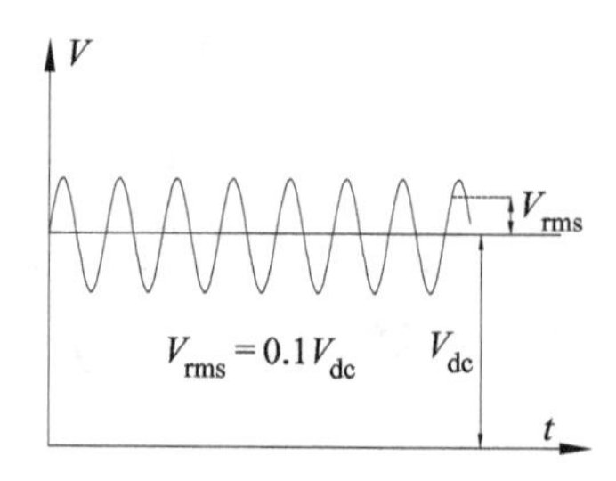

Figure 10.1 Ripple-free voltage.

interposition of double, or reinforced, insulation[3] is required between them. To facilitate this separation, extra-low-voltage circuits often have dedicated conduits and junction boxes. Extra-low-voltage systems may be insulated from each other via the basic insulation. For SELV systems, basic insulation between live parts and the earth is also required.

10.2 Separated Extra-Low-Voltage (SELV) Systems

10.2.1 Protection Against Indirect Contact

Protection against indirect contact in SELV systems is fulfilled by a harmless extra-low-voltage supply. This solution is acceptable only if an effective electrical separation from higher voltages during both normal operations and fault conditions is in place. Acceptable sources to supply SELV systems, which guarantee sufficient electrical insulation from other non-SELV circuits, are listed as follows:

- A safety isolating transformer with no intentional connection to earth and secondary voltage not exceeding 50 V.
- A source providing the same degree of protection as the safety isolating transformer (e.g., electric motor-driven generator).
- An electrochemical source (e.g., a battery) or another independent source (e.g., an engine-driven generator).

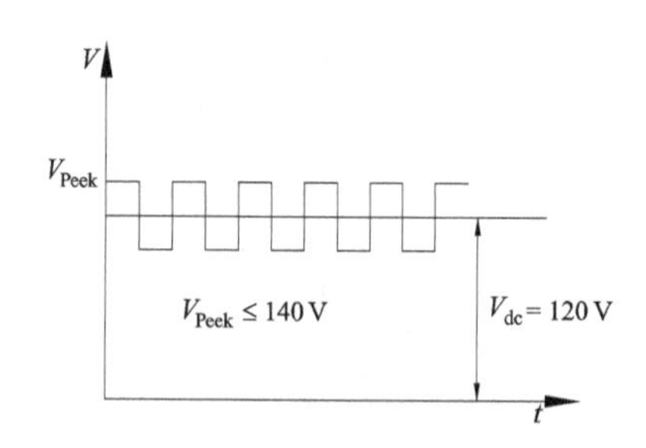

Figure 10.2 Nonsinusoidal ripple-free voltage.

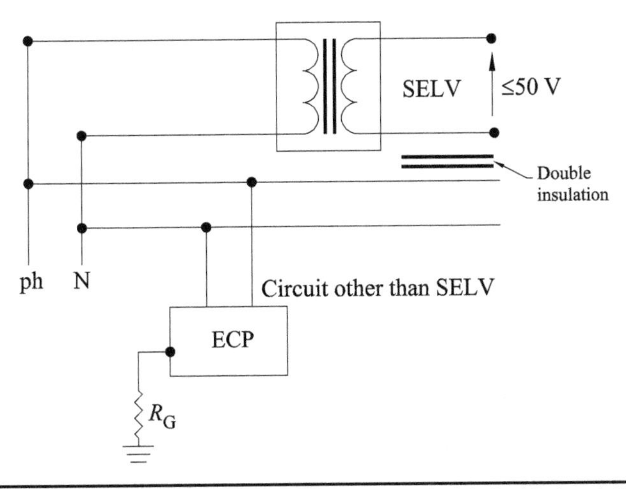

FIGURE 10.3 Safety isolating transformer equipped with double insulation between primary and secondary windings.

The separation between primary and secondary windings is achieved through double, or reinforced, insulation (Fig. 10.3) or through basic insulation and the interposition of a grounded metallic screen, or sheath, between the windings.

As explained in Chap. 2, the electric separation might as well be carried out by ordinary transformers, characterized by the same primary and secondary voltages and by basic insulation between input and output. However, SELV systems do require safety isolating transformers with the double protective insulation. If, in fact, the safety transformer only had the basic insulation between the primary (e.g., at 230 V) and secondary windings, its failure would connect the input voltage to loads insulated only for extremely low voltages, and cause their immediate failure. In this case, a single fault occurring at the transformer might expose persons in contact with metal enclosures of SELV equipment to the risk of electric shock. The presence of the double insulation (or of the grounded metallic screen) between the windings fulfills the general rule of having at least two layers of protection safeguarding persons against indirect contact.

The grounded metallic screen, although functional as a protective separation, is less reliable than the double insulation. The failure of the insulation between the secondary winding and the metal screen, in fact, would earth the system (Fig. 10.4).

In this situation, a ground fault occurring on any circuit supplied at low voltage would pose a threat to the SELV system (Fig. 10.4). The earth potential V_G, in the worst-case scenario, is additive to the safety transformer output voltage, defying the purpose of the SELV system. This overvoltage would persist for the time the protective device of

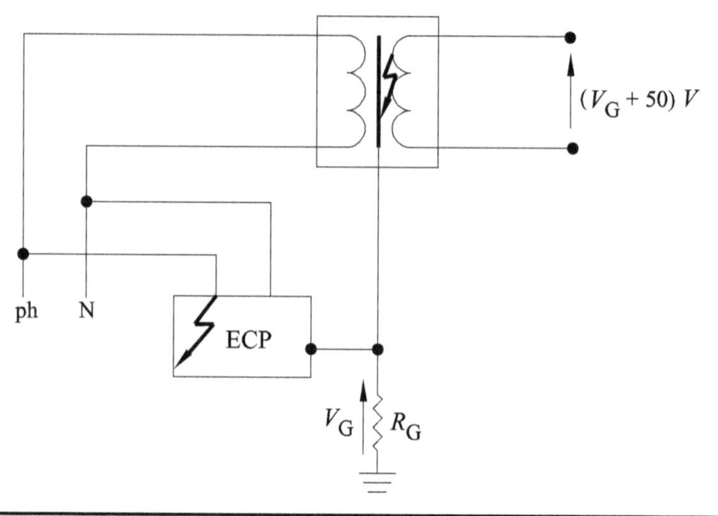

FIGURE 10.4 Failure of the insulation between the secondary winding and the grounded metal screen.

the higher voltage faulty circuit takes to clear the fault. Thus, in this arrangement the protection against electric shock does not solely depend on the SELV system, but also on the protective provisions of the low-voltage system.

In general, ECPs in SELV systems must not be connected to earth, to ECPs of other earthed systems, or to EXCPs.[4] The absence of the earthing connection will prevent the transfer of dangerous potentials originating in other locations to the SELV equipment. In addition, plugs and receptacles must have no protective conductor contacts, as well as should not be able to enter/admit any receptacles/plugs assigned to other non-SELV electrical systems.

10.2.2 Protection Against Direct Contact

The direct contact with one live terminal in SELV systems is harmless, because the supply, having no reference to earth due to the electrical separation, causes ground currents of very small magnitude to circulate. Even the simultaneous direct contact with both source's terminals is not dangerous, as the resulting touch voltage is considered harmless (i.e., $V_T \leq 50$ V a.c.). In addition, in dry conditions if the nominal voltage of the SELV does not exceed 25 V a.c., or 60 V d.c., the basic insulation of live parts is not deemed necessary for protection against direct contact. However, in wet conditions the extra-low voltage may be dangerous in the case of fault to ground of one of the output terminals (Fig. 10.5). In this case, the direct contact may cause electrocution.

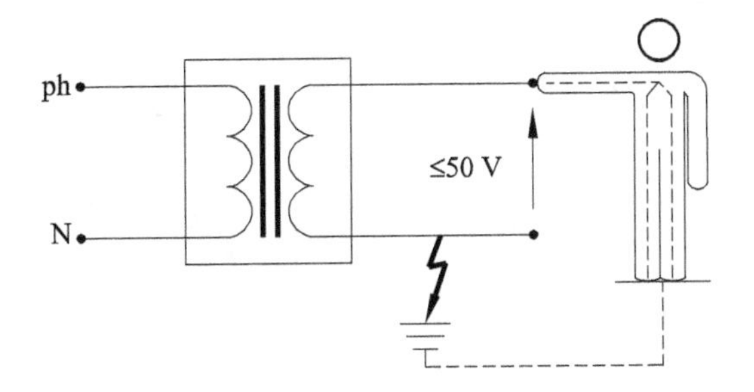

FIGURE 10.5 Fault to ground of an output terminal and consequent touch voltage.

10.3 Protective Extra-Low-Voltage (PELV) Systems

Like SELV systems, PELV circuits must be supplied by safety isolating transformers whose secondary output does not exceed 50 V. The distinct difference with SELV consists of the permitted connection to earth of a point of the system (Fig. 10.6).

The connection to earth may be necessary due to safety reasons, as explained in Sec. 10.3.1, and be achieved through a link to the main earth terminal, as well as to a grounded ECP. PELV systems offer a greater risk of electrocution than SELV. In fact, during ground faults occurring in the higher voltage system, the output voltage of the safety transformer may exceed the nominal voltage of the PELV by the ground potential rise V_G. In the worst-case scenario, persons standing in a zero potential area would be exposed to the touch voltage $V_G + 50$. For this reason, to lower the risk of electrocution it is important to

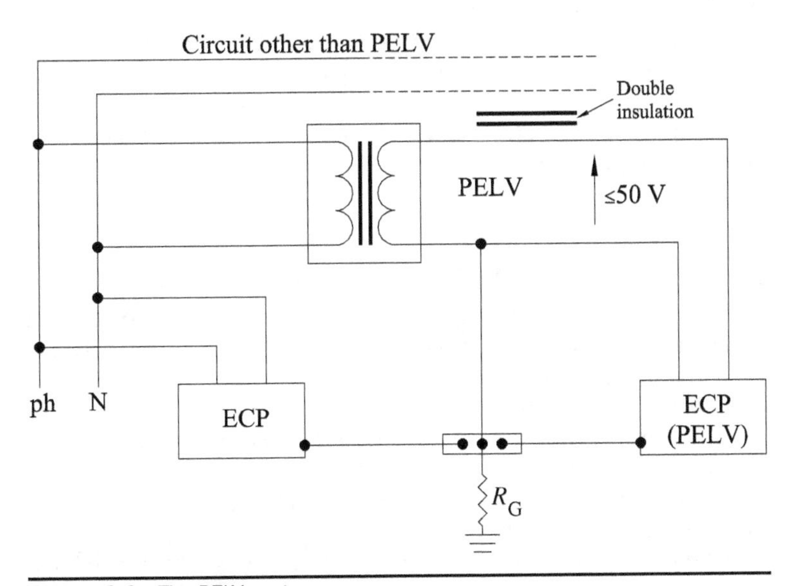

FIGURE 10.6 The PELV system.

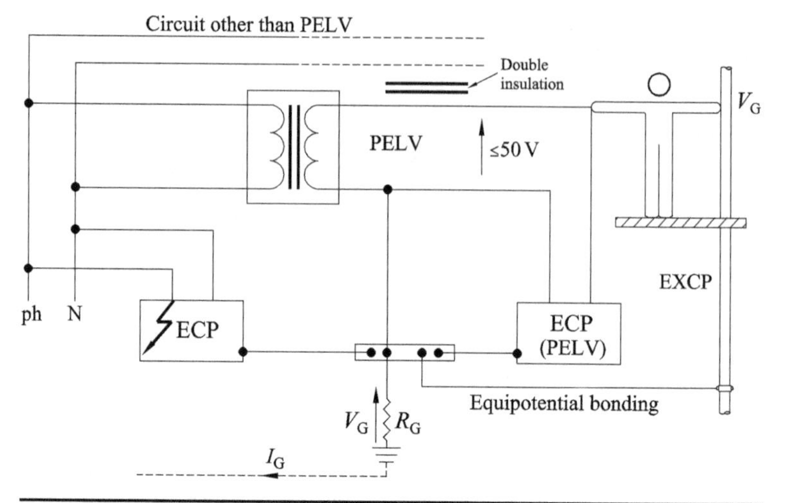

FIGURE 10.7 Equipotential bonding in PELV systems.

create an equipotential area by connecting all the ECPs and EXCPs of the PELV systems to the main earth terminal.

In the case of a ground fault on the higher voltage system, a person in simultaneous contact with a phase conductor and an EXCP would be subject only to the extra-low voltage (i.e., $\leq$50 V), as in the equipotential area the EXCPs "elevate" their potential to V_G (Fig. 10.7).

Thus, the potential difference between the phase conductor and the EXCP will not exceed 50 V.

In PELV circuits, in dry conditions and in the equipotential area, the basic insulation of live parts is a safety requirement only if the nominal voltage exceeds 25 V a.c., or 60 V d.c. Plugs and receptacles, which must have a protective conductor terminal, must not be able to enter/admit any receptacles/plugs assigned to non-PELV electrical systems.

10.3.1 Application of PELV Systems to Control Circuits

The PELV system may be required in circuits that need an earth point for safety reasons, for example, control circuits of machines. Faults in control circuits should cause prompt disconnection of supply, as it cannot be tolerated that, by not clearing them, functionality of controls is lost or unexpected start-ups or turn-offs of machines occur.

To better understand this safety issue, let us examine Fig. 10.8, where the case of multiple faults in a machine's control system supplied by SELV is exemplified.

The two consequent faults toward the metal control box bypass contact 2, and because of the electrical separation are not cleared.

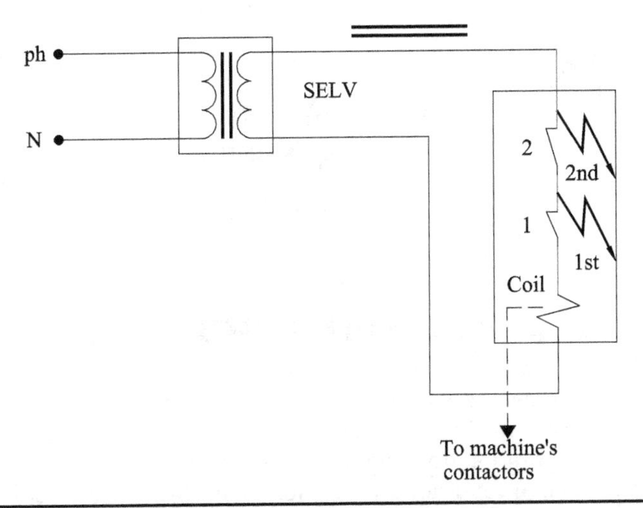

Figure 10.8 Ground faults in SELV control circuit.

Contact 2 loses its control capability and even if it is opened the coil stays energized and so might the machine.

Disconnection of supply can be achieved by turning the first fault into a short circuit by using PELV systems (Fig. 10.9).

The first fault toward the enclosure determines a short circuit caused by the earthing connection of the safety insulating transformer's pole. This provokes the tripping of the protective device, typically a fuse, which clears the fault and puts the machine in safety.

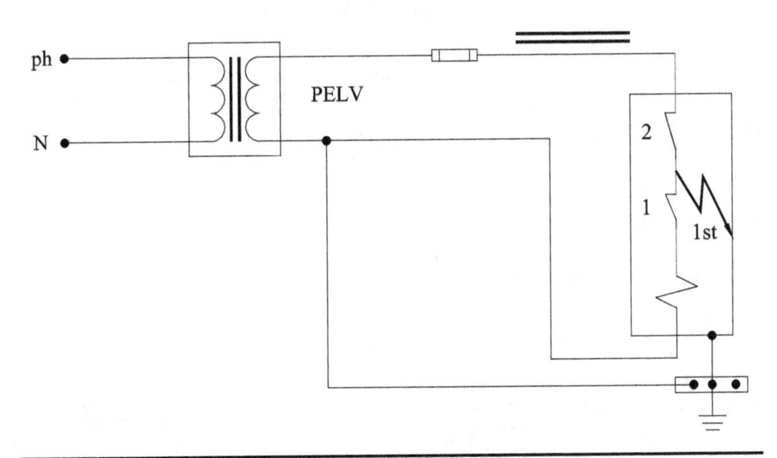

Figure 10.9 Ground faults in control circuits supplied by PELV systems.

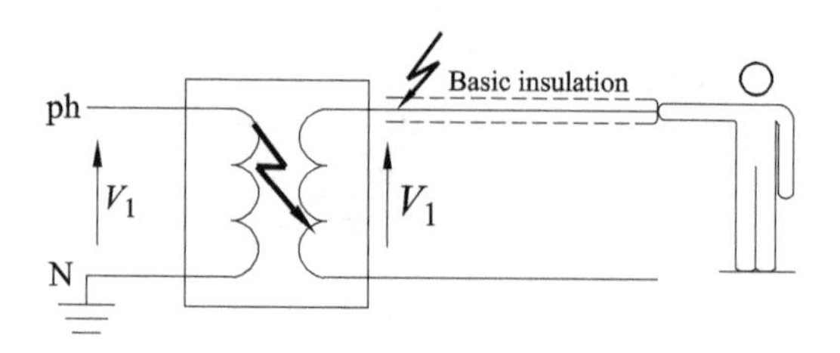

Figure 10.10 Fault between windings and direct contact.

10.4 Functional Extra-Low-Voltage (FELV) Systems

FELV systems work at extra-low voltages, when these are required for their functioning. FELV do not fulfill the fundamental safety requirements of SELV or PELV, such as safety insulating transformers and protective separation from higher voltage systems.

In the above conditions, protection against direct contact is necessary because the rupture of the basic insulation between the windings, or of neighboring circuits at higher voltages, would cause the input voltage to transfer to the secondary side of the transformer, with the result to also puncture the insulation of the FELV circuits (Fig. 10.10).

The secondary winding, in fact, is insulated to operate at extra-low voltage and cannot withstand the primary voltage-to-ground. This event exposes persons to the touch voltage $V_1 > V_{FELV}$. For this reason, protection against direct contact is fulfilled if the basic insulation of the FELV system is adequate to withstand the nominal voltage of the primary circuit.

As to indirect contact, failure of the basic insulation between input and output windings will also cause the failure of the FELV equipment, whose insulation is "punctured" by the primary voltage V_1. Thus, persons in contact with the failing FELV ECP would be exposed to the prospective touch voltage V_1.

Protection against indirect contact caused by faults in both the low- and extra-low-voltage systems is achieved if the FELV ECPs are connected to the same grounding system as the low-voltage ECPs (Fig. 10.11).

If the primary circuit is properly protected against indirect contact by automatic disconnection of supply, according to its earthing system (e.g., TT or TN), so will be the FELV circuit. In the aforementioned conditions, in fact, the first and subsequent second fault will be cleared by the primary protective device, and persons will be exposed to the ground potential $V_G \leq V_1$ during the time it takes to intervene.

FELV plugs and receptacles must have a protective conductor terminal, but should not be able to enter/admit any receptacles/plugs assigned to other non-FELV electrical systems.

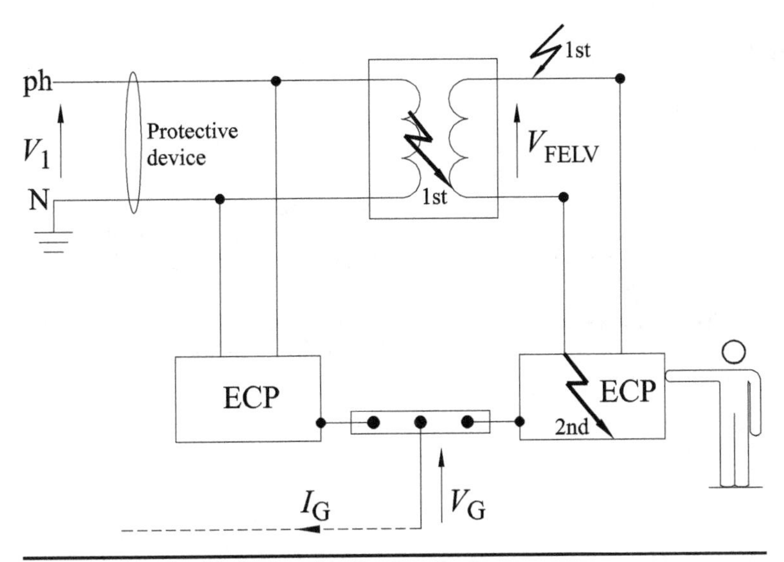

FIGURE 10.11 Double-fault and indirect contact.

In FELV systems, while it is important to ground the ECPs, it is not advisable to earth the secondary winding of the source, because the accidental loss of protective conductors may cause hazardous situations under double-fault conditions (Fig. 10.12).

In the worst-case scenario, persons are exposed to a prospective touch voltage equal to $V_G + V_{FELV}$.

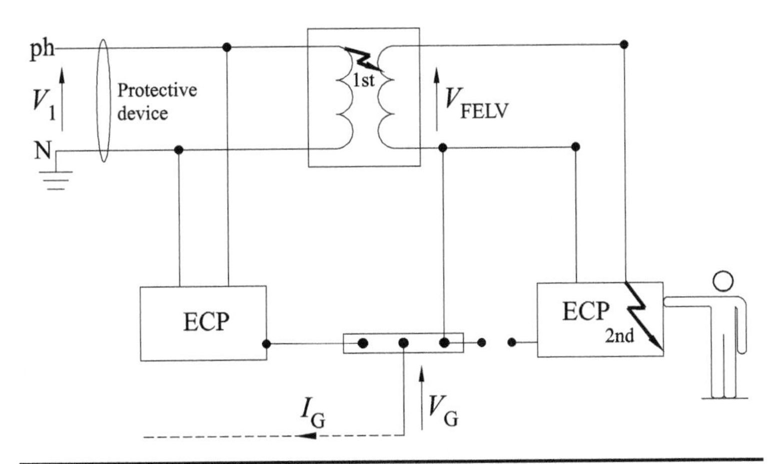

FIGURE 10.12 Earth connection of the supply.

Endnotes

1. See Chap. 15 for more details.
2. Voltages exceeding the above limits, up to 1 kV a.c. and 1.5 kV d.c., are considered low voltages.
3. As already defined in Chap. 2, the double insulation comprises both basic and supplementary insulations. The reinforced insulation is a single insulation system, which provides the same degree of protection against electric shock as the double insulation.
4. The protective separation between the windings of the safety transformer renders the SELV system equivalent to Class II equipment for which bonding is not permitted.

CHAPTER 11

Earth Electrodes, Protective Conductors, and Equipotential Bonding Conductors

The heating of a conductor depends upon its resistance and the square of the current passing through it. JAMES P. JOULE

11.1 Introduction

Earth electrodes, protective conductors (PE), and equipotential bonding conductors, both main (MEB) and supplementary (SB), are the fundamental components of the earthing arrangements. A failure in any of these elements can compromise the electrical safety of the installation as well as its functionality. Earthing arrangements, in fact, may be used for both functional reasons (e.g., system grounding) and safety purposes (e.g., grounding of ECPs in TT systems, bonding of ECPs in TN systems).

Figure 11.1 summarizes the roles of the aforementioned elements in the earthing arrangements.

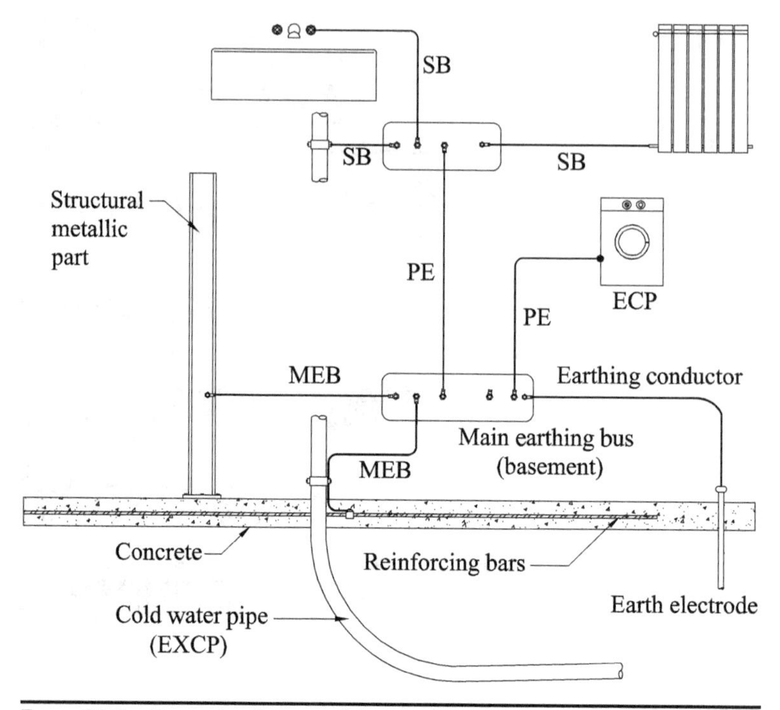

Figure 11.1 Elements of the earthing arrangements.

11.2 Earth Electrodes

Earth electrodes must provide a reliable link to ground, primarily for safety purposes, but also for the proper functioning of equipment. As already substantiated in Chap. 4, electrodes must be able to carry ground-fault currents and dissipate them to ground, without causing hazards caused by thermal effects and/or electric shock. The effectiveness of the earthing system depends upon its ground resistance R_G, which varies with the resistivity of the local soil. Once the characteristic of the soil and the minimum acceptable value for safety of R_G are known, one or more ground electrodes, even of different nature, must be employed.

Approximate values of the earth resistance at 50/60 Hz of typical-made electrodes may be calculated by using the formulas[1] reported in Table 11.1.

L is the length of the electrode in contact with the soil, ρ is the soil average resistivity, and r is the radius of the circle that circumscribes the grid (Fig. 11.2).

In several countries, cold water pipes are not permitted as earth electrodes under any circumstances (e.g., Austria, Belgium, Finland,

Type of Electrode	R_G
Rod	ρ/L
Buried horizontal wire	$2\rho/L$
Grid	$\rho/4r$

TABLE 11.1 Approximate Formulas of the Earth Resistance of Typical-Made Electrodes

France, Germany, Switzerland, and the U.K.). Users, in fact, having no control over the cold water system, cannot rely on its electric continuity to ground. However, only with the explicit consent of the water utility, which guarantees the aforementioned continuity, water pipes may be relied on as earth electrode (e.g., in Italy).

The same prohibition applies to metal pipes for flammable liquids or gases. However, the above rule pertaining to earth electrodes does not preclude the protective bonding of any metalwork entering the building to the earthing system, for example, by connecting pipes downstream of their water or gas meters. As we know, this connection is indispensable in ensuring a safe equipotential area into the user's premises and cannot be omitted.

11.2.1 Corrosion Phenomena

Earth electrodes must have a minimum size in order to have adequate mechanical strength and withstand corrosion.

Corrosion is an electrochemical process that involves two dissimilar metals electrically connected when embedded in electrolytes, such as earth, concrete, seawater, etc. The two metals, respectively, assume the role of cathode and anode of a galvanic cell. When the current leaves the anode, to reclose to the cathode through the electrolyte, corrosion at the expense of the anode occurs. The phenomenon is more pronounced when the ratio of the cathode's surface to the anode's surface is large. The rule of thumb is to expect appreciable corrosion only when the cathode's surface is 100 times bigger than the anode's.

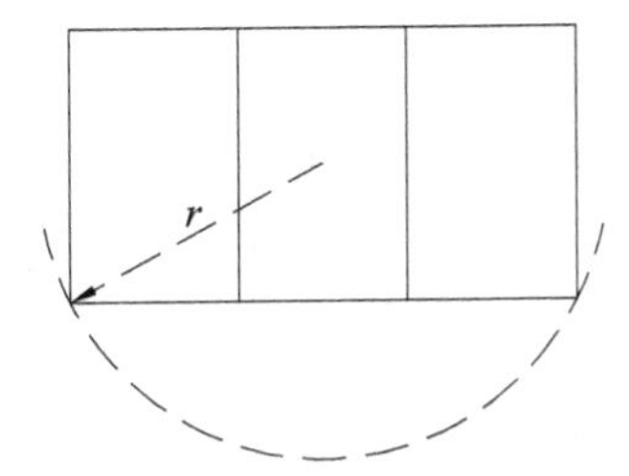

FIGURE 11.2 Radius of the circle circumscribing the grid.

Steel				
Surface	Electrode's Type	Diameter (mm)	Cross-Sectional Area (mm^2)	Thickness (mm)
Hot-dip galvanized or stainless	Strip	—	90	3
	Round rod for deep earth electrodes	16	—	—
	Round wire for surface earth electrodes	10	—	—
	Pipe	25	—	2

TABLE 11.2 Minimum Sizes of Common Electrodes Made of Steel

In this respect, common minimum sizes for earth electrodes, as per IEC 60364–5-54,[2] are reported in Tables 11.2 and 11.3. The two tables allow a comparison between the two commonly used materials, steel and copper, in terms of equivalent withstand capability to corrosion and mechanical strength as a function of their type and dimensions.

We have already mentioned the possibility of using the concrete-encased reinforcing steel bars of a building's footings as an electrode.

Copper			
Electrode's Type	Diameter (mm)	Cross-Sectional Area (mm^2)	Thickness (mm)
Bare strip	—	50	2
Bare rope	1.8 for individual strands	25	—
Bare round wire for surface earth electrodes	—	25	—
Bare pipe	20	—	2
Tin-coated rope	1.8 for individual strands	25	—
Zinc-coated strip	—	50	2

TABLE 11.3 Minimum Sizes of Common Electrodes Made of Copper

The concrete, in fact, by absorbing and retaining moisture provides around the conductive re-bars an even lower resistivity than that of the local soil. As a consequence, materials suitable to be embedded in concrete (e.g., hot-dip galvanized/stainless steel) should not be coated with insulating materials, if they are to be used as electrodes. The concrete-encased electrode makes a very effective earth electrode and at no extra cost for the user. In the presence of more than one concrete-encased electrode in a structure, it is sufficient to bond only one to the main earthing bus, as the entire foundation network is interconnected due to metal re-bars.

However, some may have concerns about the connection of the steel foundation re-bars to other made-electrodes with higher electrochemical potential (e.g., copper rods) eventually employed in the earthing system. This bond, which creates a single electrode system necessary to have an equipotential area, is feared to generate corrosion of the re-bars. Steel re-bars in concrete, in fact, may result anodic to copper rods, and, therefore, undergo corrosion.

In reality, the electrochemical potential of steel, when embedded in concrete, increases and reaches a value close to that of copper. In addition, the surface of earthing rods (i.e., the cathode) is much smaller than the equivalent surface of the network of re-bars embedded in foundations (i.e., the anode). Therefore, only negligible corrosion will occur, especially in residential and commercial power systems, whose earthing electrodes are usually limited in number. However, to completely eliminate the risk of corrosion of elements of foundations, it would be best to use tin-coated copper rods in lieu of bare copper ones, or employ hot-dip galvanized steel rods.

11.3 Protective Conductors

Protective conductors (PEs) provide safety against indirect contact by linking ECPs to the main earthing terminal, thereby creating a clear path for the fault currents. Cross-sectional areas of protective conductors must be adequately large, so that fault currents can promptly activate the protective device and automatically disconnect the supply.

Additionally, protective conductors must be able to withstand the flow of the ground-fault current without reaching dangerous temperatures to the surrounding environment or shorten the life of, or damage, their insulation.

Minimum standard cross-sectional areas deemed adequate for PEs are shown in Table 11.4, when the protective conductor is of the same material as the line conductor.

If a protective conductor is common to more than one circuit, it must be selected in correspondence with the phase conductor's largest

Mimimum Cross-Sectional Area of the Line Conductor, S	Cross-Sectional Area of the Corresponding PE
$S \leq 16$	S
$16 < S \leq 35$	16
$S > 35$	S/2

TABLE 11.4 Minimum Standard Cross-Sectional Areas of PEs

cross-sectional area. In addition, if the PE is not in the same enclosure as the line conductor, its cross-sectional area must not be less than 2.5 mm^2 (copper)/16 mm^2 (aluminum), when it is protected against mechanical damage, or 4 mm^2 (copper)/16 mm^2 (aluminum), when protection against mechanical damage is not being provided.

Protective conductors may not necessarily consist of actual conductors, but metallic layers of cables (e.g., metallic sheaths, armors, concentric conductors, etc.) and metallic conduits can serve the same protective purpose as long as their equivalent cross-sectional area complies with Table 11.2 or 11.3. In some countries (e.g., China, Italy, the U.K., and the U.S.A.), cable trays can also be used as a protective conductor as long as manufacturers guarantee their electric continuity by construction.

Protective conductors may even be bare. This conductor, in fact, is generally at zero potential and, therefore, does not require any dielectric insulation. The PE, in fact, is normally not "hot," but becomes temporarily energized upon faults. There are advantages of having bare protective conductors in the same conduit or tray with insulated conductors. In the case of faults between energized conductors (e.g., short circuits), it is useful that the PE is also involved, so that residual current devices, eventually present, are activated. This would constitute a significant redundancy for safety, as two devices, overcurrent and RCD, initiate the clearing procedure of the short circuit. However, insulation of PE may be required if during its pulling through conduits, damages to other insulated conductors are likely, or feared, to occur.

11.3.1 Analytical Calculation of the Minimum Cross-Sectional Area of PEs

The cross-sectional area of PEs can be analytically calculated, besides being selected from Table 11.3. The advantage of the analytical calculation is that it may lead to less-expensive installations, as it can yield smaller sizes for the protective conductors, and yet are perfectly adequate to assure safety.

At the occurrence of phase-to-ground faults, heat will be developed in protective conductors by the Joule effect. We can assume an adiabatic process, that is, neglect the thermal exchange by convection or radiation between the PE and the surrounding environment. In this case, all the heat developed by the fault accumulates in the protective conductor, as well as in all the components present in the fault loop, with the result to raise their temperatures. This assumption is amply justified since the fault is generally cleared within tens of milliseconds, while the heat transfer requires more time to take place.

In analogy with the procedure described in Chap. 5, the adiabatic process for a conductor of length l, cross-sectional area S (mm^2), resistivity ρ ($\Omega \cdot mm$), and volumetric specific heat capacity c [$J/(°C \cdot mm^3)$] can be described by the thermal balance of Eq. (11.1):

$$\rho \frac{l}{S} i^2 \, dt = Slc \, d\theta \tag{11.1}$$

If a fault current $i(t)$ flows through the protective conductor for the infinitesimal time dt, the conductor undergoes a temperature rise $d\theta$. $d\theta$ represents the difference between the initial temperature θ_0 of the conductor, at the inception of the fault, and its final temperature θ_f, after the fault is cleared. Thus, the left-hand side of Eq. (11.1) represents the heat developed by the fault current during dt, whereas the right-hand side is the heat accumulated in the conductor during the same time.

We may also reasonably assume that the protective conductor's cross-sectional area S does not vary significantly during the temperature variation caused by the fault current. The resistivity of the conductor, instead, cannot be considered constant with the temperature θ. We consider ρ linearly variable with θ according to Eq. (11.2):

$$\rho = \rho_0(1 + \alpha\theta) \tag{11.2}$$

where ρ_0 is the resistivity of the PE at 0°C and α is the temperature coefficient of resistivity. Table 11.5 shows values for α, ρ_0, and c for different conductive materials.

Material	Temperature Coefficient of Resistivity, α ($°C^{-1}$)	Resistivity, ρ_0 ($\Omega \cdot mm$)	Volumetric Specific Heat Capacity, c [$J/(°C \cdot mm^3)$]
Copper	4.26×10^{-3}	15.89×10^{-6}	3.45×10^{-3}
Aluminum	4.38×10^{-3}	25.98×10^{-6}	2.5×10^{-3}
Lead	4.34×10^{-3}	196.88×10^{-6}	1.45×10^{-3}
Steel	4.95×10^{-3}	125.56×10^{-6}	3.8×10^{-3}

TABLE 11.5 Values for α, ρ_0, and c for Different Conductive Materials

If the protective device takes the time t_f to interrupt the fault, in order to calculate the total heat accumulation in the conductor during this time, we integrate Eq. (11.1) and obtain Eq. (11.3):

$$\int_0^{t_f} i^2\,\mathrm{d}t = cS^2 \int_{\vartheta_0}^{\vartheta_f} \frac{\mathrm{d}\theta}{\rho} = \frac{cS^2}{\rho_0} \int_{\vartheta_0}^{\vartheta_f} \frac{\mathrm{d}\theta}{(1+\alpha\theta)} \tag{11.3}$$

where θ_0 is the temperature of the protective conductor at the inception of the fault and θ_f is its final temperature when the fault is cleared.

The left-hand side of Eq. (11.3) is known as *Joule integral* and is measured in A^2s. It is also referred to as "*I square t*, (I^2t)," "*let-through energy*," or "*specific energy*."

Substituting y for $(1+\alpha\theta)$ and differentiating, we obtain

$$\mathrm{d}\theta = \frac{1}{\alpha}\,\mathrm{d}y \tag{11.4}$$

With this substitution, Eq. (11.3) yields

$$\int_0^{t_f} i^2\,\mathrm{d}t = \frac{cS^2}{\alpha\rho_0} \int_{1+\alpha\vartheta_0}^{1+\alpha\vartheta_f} \frac{\mathrm{d}y}{y} = \frac{cS^2}{\alpha\rho_0} \ln \frac{1+\alpha\vartheta_f}{1+\alpha\vartheta_0} \tag{11.5}$$

The adverse effects of fault-to-ground currents to the protective conductor are prevented if the final temperature θ_f reached by the PE does not exceed the maximum values θ_M its insulation can withstand (e.g., the maximum temperature for conductor insulation PVC is 160°C). Therefore, Eq. (11.5) can be rewritten as an inequality. If we define

$$k^2 = \frac{c}{\alpha\rho_0} \ln \frac{1+\alpha\vartheta_M}{1+\alpha\vartheta_0} \tag{11.6}$$

where k depends on the material of the protective conductor, its type of insulation, and the initial and final temperatures that are reached. Through the PE no current normally circulates, therefore, its initial temperature θ_0 corresponds to the standard ambient temperature (i.e., 30°C).[3]

By substituting Eq. (11.6) into Eq. (11.5) and integrating, we obtain

$$\int_0^{t_f} i^2\,\mathrm{d}t \le k^2S^2 \tag{11.7}$$

By solving Eq. (11.7) for S, we obtain cross-section values that guarantee the protection of the PE against damages caused by thermal effects.

In addition, to assure protection against indirect contact, the protective device must open the circuit within a safe time. This is accomplished if the minimum r.m.s. value I_{min} of the prospective fault current is greater than, or equal to, the instantaneous (or magnetic) trip setting I_i of the protective device. The fault current must be in the instantaneous region of the time–current characteristic of the overcurrent device (see chart in Fig. 6.6). In formulas:

$$I_{min} \geq I_i \tag{11.8}$$

The left-hand side of Eq. (11.7) is not of immediate calculation, therefore IEC standards, by conventionally assuming that the fault is cleared within 5 s, allow the following simplification:

$$\int_0^{t_f} i^2 \, dt \cong I^2 t \leq k^2 S^2 \tag{11.9}$$

where I is the r.m.s. value of the prospective fault current circulating through the PE for a fault of negligible impedance, and t is the operating time of the protective device in correspondence with that current. In reality, during the fault the current is not a constant value as it varies with time, but the error caused by this simplification is generally acceptable. It is important to note that at the inception of the fault, the current is asymmetrical due to the development of a transient d.c. component. Equation (11.9), therefore, cannot be applied when the fault duration is extremely short (e.g., when current limiter protective devices are employed), as the d.c. transient might not be over and the current wave might not be yet symmetrical, thereby, accentuating the difference between the left- and right-hand sides of Eq. (11.7).[4]

If a protective conductor is common to more than one circuit, it must be sized for the largest Joule integral.

Calculated values of k from Eq. (11.6) for insulated protective conductors not incorporated or bunched with other cables are shown in Table 11.6.

Conductor Insulation	θ_0 (°C)	θ_M (°C)	*k*		
			Copper	**Aluminum**	**Steel**
PVC	30	160	143	95	52
Rubber	30	200	159	105	58
Thermosetting	30	250	176	116	64
Silicone rubber	30	350	201	133	73

Table 11.6 Values of k for Insulated Protective Conductors Not Incorporated in Cables or Bunched with Other Cables

Conditions of Bare PE	Copper Max Temperature (°C)/*k* Value	Aluminum Max Temperature (°C)/*k* Value	Steel Max Temperature (°C)/*k* Value
Visible and in restricted areas	500/228	300/125	500/82
In normal conditions areas	200/159	200/105	200/58
In fire risk areas	150/138	150/91	150/50

TABLE 11.7 Values of Maximum Temperatures of Bare Protective Conductors

In the case of a bare PE not bunched with other cables, the maximum temperature the conductor can achieve is not dictated by insulation's thermal capabilities, but by the surrounding environment. The bare PE in fault conditions, in fact, may become a temperature hot object and trigger fires or explosive atmospheres. For safety reasons, then, IEC 60364–5-54 lists the maximum temperatures of bare PEs exposed to touch, as a function of the surrounding conditions, assuming these temperatures are not a risk of damage to any neighboring material. Table 11.7 shows the maximum temperatures for different materials and conditions.

11.3.1.1 Iterative Method of Calculation

The following Eq. (11.10), derived by Eq. (11.9), clearly shows that the cross-sectional area S of protective conductors depends on the prospective phase-to-ground current I flowing through it:

$$S \geq \frac{\sqrt{I^2 t}}{k} \tag{11.10}$$

However, the value of I can be calculated only by knowing the impedance of the PE, which varies according to its cross-sectional area S. Thus, I and S in Eq. (11.10) are not independent quantities from each other, especially in TN systems, where ground-fault currents will return to the source through protective conductors.

To solve this quandary, an iterative method to size S, as depicted in Fig. 11.3, should be employed.

One can start by selecting the minimum standard cross-sectional area (i.e., 1.5 mm^2) and calculate maximum and minimum phase-to-ground fault currents accordingly. The maximum ground-fault current originates for a fault at the beginning of the circuit (e.g., contact between phase and protective conductors), as there is virtually no

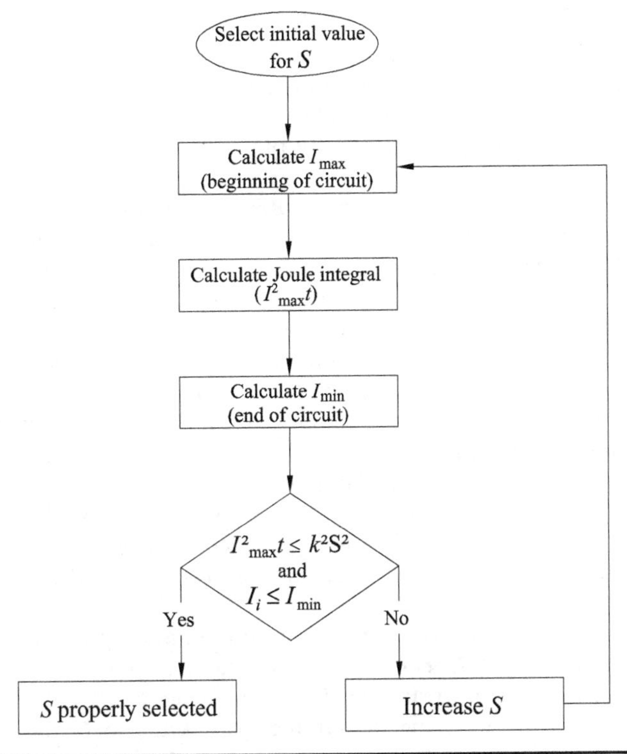

FIGURE 11.3 Iterative method of calculation to size the cross-sectional area S of protective conductors.

limitation due to the impedance of the wires themselves. The minimum fault current, instead, occurs at the end of the circuit (e.g., contact between phase wire and enclosure).

By knowing the aforementioned current values, one must verify the simultaneous fulfillment of Eqs. (11.8) and (11.9). If the equations are not satisfied, the selected cross section must be increased to the consecutive larger standard value, and a successive iteration is required. The iterative process will stop upon fulfillment of both equations, which provides the safe value for S.

As studied in Chap. 6, in TT systems, the ground-fault loop comprises the earth, and the fault current is limited by both the user ground resistance R_G and the system ground resistance of the distributor R_N, whose value may be unknown. The impedance of protective conductors is much less than R_G or R_N, thus affects little the value of I. However, the calculation of the earth current flowing through the PE under fault conditions may be carried out only by postulating the value of R_N, and thus only approximate solutions can be obtained.

11.3.2 Metallic Layers of Cables as Protective Conductor

As already anticipated, metallic sheaths, or armors, of cables may be used to carry the ground-fault current, or a portion of it, when both their ends, supply side and load side, are linked to the earthing system. In this configuration, a multiple return path may be available for fault currents: sheaths/armors and protective conductors are in parallel. The majority of the ground current will circulate through the protective conductor, as they offer a much lesser impedance than metallic sheaths/armors. It is assumed as a general rule of thumb that no more than 15% of the fault current will circulate through metallic sheaths/armors. However, the current-carrying capability of sheaths and armors within the operational time frame of the protective device must be assessed. There may be cases when metal layers of cables cannot sustain any part of the fault current for the duration of the fault, and therefore, should not be included in the fault-loop. This can be achieved by lifting one end of metallic layers of cables.

However, by not bonding both ends of a cable's metallic layers, safety issues may arise. To this regard, let us consider Fig. 11.4 in which a cable is shown whose metallic sheath is bonded only at one end.

A medium-voltage ground fault simultaneously energizes the ground-grid and the metallic sheath of a cable linked to it. The earth potential is, therefore, transferred from the bonded end of the sheath to the floating one. Along the ground grid, whose behavior can be approximated by the combination of buried spherical electrodes, the superficial earth potential is not constant, but decreases and assumes its lowest value almost in correspondence with its edge (Fig. 11.5).

If the sheath is accessible (e.g., at its termination at the equipment box) between the person's hand (at fault potential V_T) and feet (at grid potential), a dangerous potential difference V_{ST} may exist.

Bonding connections between both ends of suitable metallic sheath (i.e., capable to carry the fault current) of medium-voltage

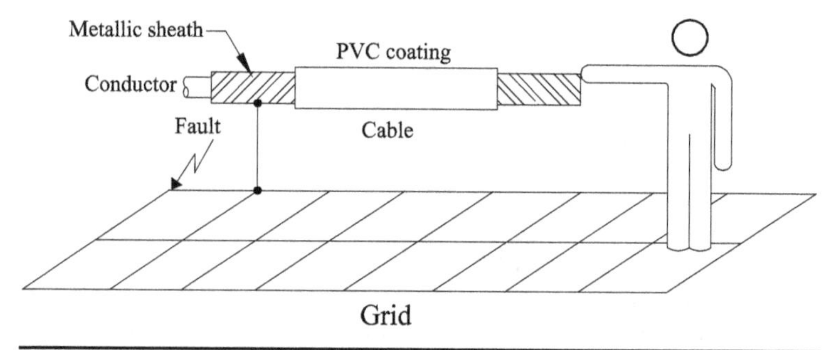

FIGURE 11.4 Metallic sheath bonded only at one end.

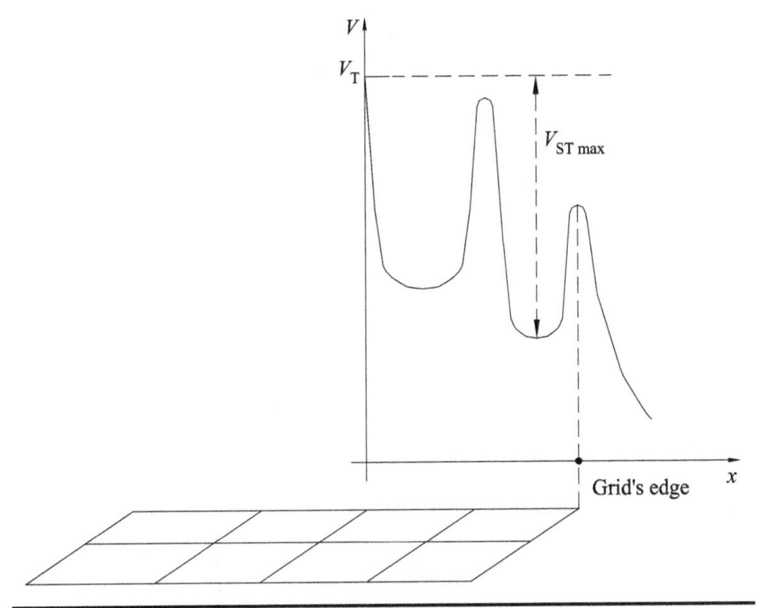

Figure 11.5 Superficial earth potential along the grid.

cables and the grid cancel the above-examined prospective touch voltage, thereby, increasing the electrical safety of the installation.

11.4 Equipotential Bonding Conductors

Equipotential bonding conductors cancel dangerous potential differences between metal parts, for example, ECPs and EXCPs, when ground faults occur. They must be reliable, have negligible resistance and good mechanical strength. As already discussed, main equipotential bonding conductors are employed for connections of EXCPs to the main earthing bus (MEB in Fig. 11.1) and for supplementary bonding (SB in Fig. 11.1) in special locations (e.g., locations containing a bath or shower).

To assist in the decision-making process pertaining to which metal part (MP) needs main bonding, flowchart shown in Fig. 11.6 can be used.

The cross-sectional area of MEBs can be calculated with the same methodology as protective conductors [i.e., Eq. (11.7)], or selected from Table 11.8, as per IEC 60364-5-54, which indicates their minimum standard sizes.

Supplementary bonding conductors must comply with the same requirements as the protective conductors as to their minimum size: their cross-sectional area must not be less than 2.5 mm^2 (copper)/

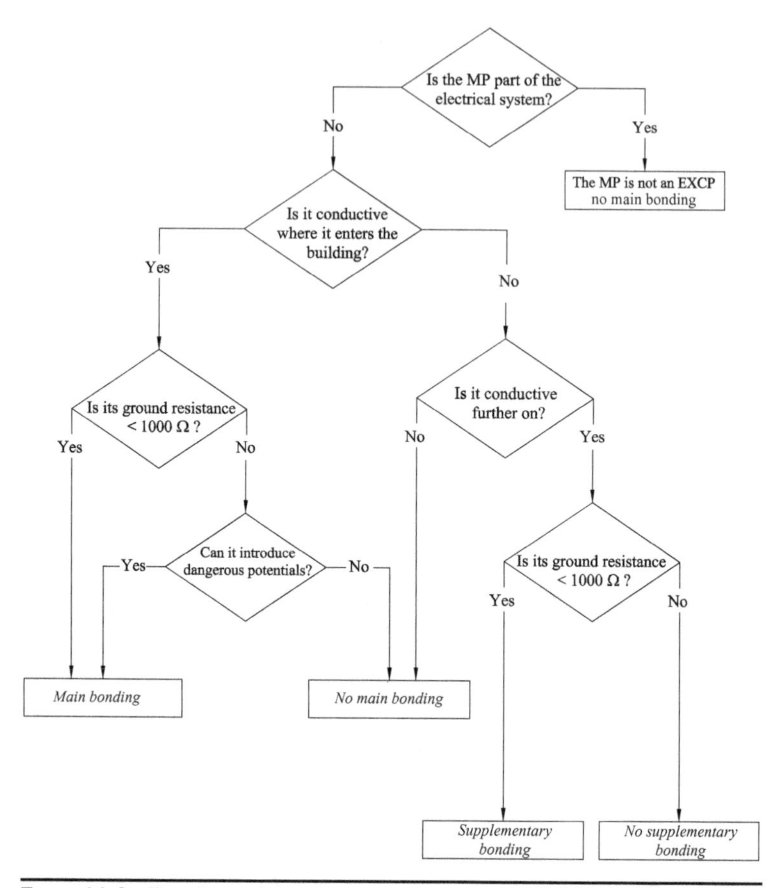

FIGURE 11.6 Flowchart of the decision-making process for main bonding.

16 mm^2 (aluminum), when they are protected against mechanical damage, or 4 mm^2 (copper)/16 mm^2 (aluminum), when protection against mechanical damage is not provided. In addition, a supplementary bonding conductor, which connects two ECPs (Fig. 11.7), must have a cross-sectional area not less than that of the smaller PE serving the ECPs. In the case of ground faults, in fact, this SB will carry part of the fault current due to the current divider taking place between the two PEs.

Copper (mm^2)	Aluminum (mm^2)	Steel (mm^2)
6	16	50

TABLE 11.8 Minimum Standard Cross-Sectional Area for Main Equipotential Bonding Conductors

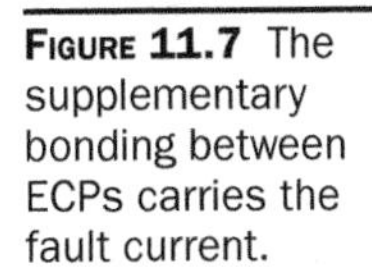

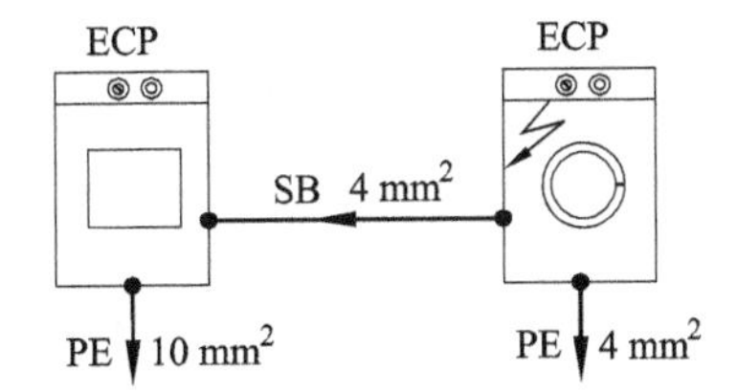

FIGURE 11.7 The supplementary bonding between ECPs carries the fault current.

If the supplementary bonding conductor connects an ECP to an EXCP, as per IEC 60364–5-54, its conductance must be at least half that of the PE serving the ECP. Also, in this case, a current divider takes place and the EXCP carries part of the fault current.

11.4.1 Where Should We Use Equipotential Bonding Conductors?

As we have already substantiated, the only reason to employ an equipotential bonding conductor is to eliminate or reduce potential differences in the case of faults. However, an erroneous application of this concept may induce designers to bond each and every metal part in the vicinity of equipment. An undue link between ECPs and metal objects, which are not required to be bonded, causes the safety to decrease, because the fault potential arising within the ECP may be transferred to that metal part.

As an example of a "legitimate" bonding conductor, let us consider a panelboard, whose door contains live conductors (e.g., door with controls or instrumentation) (Fig. 11.8).

This piece of equipment is composed of a "core" (i.e., the panel itself) and a "satellite" (i.e., the "active" door). Both the core and satellite are ECPs since, by definition, they are normally "dead," but likely to become live upon failure of their basic insulations. Both panelboard and door must be bonded to the grounding system to allow a prompt disconnection of supply upon ground faults.

While the enclosure of the "core" is bonded to the grounding bus within the panel, which is linked to the earthing system, the door might simply employ an equipotential bonding conductor between itself and the "core," as shown in Fig. 11.8. The equipotential bonding conductor, by bypassing the resistance of the hinges,[5] allows a clear path to the ground current and a prompt intervention of the protective device, should the door fail.

On the other hand, a "plain" door (i.e., with no live parts on it) is neither an ECP nor an EXCP and therefore there is no need for bonding jumpers. The presence of a bonding conductor between the door and the frame decreases the safety, as a fault in the panelboard also energizes the door. As a consequence, the probability for persons

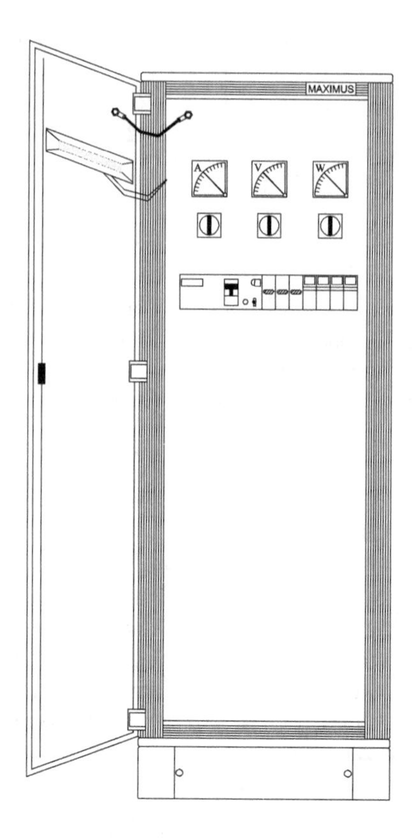

FIGURE 11.8 Panelboard with an "active" door.

to touch a live part of equipment increases, and so does the risk of indirect contact.

In some cases, though, the panel door may directly open to exposed live parts (i.e., bus bars), that is, the door is part of the enclosure and prevents direct contact. In this case, the door, even if it lacks live conductors, must be bonded.[6]

11.5 Earthing Conductors and Main Earthing Terminal

Earthing conductors link the earth electrode(s) to the main earthing bus. The cross-sectional area of these conductors can be selected or calculated as seen for the protective conductors in Sec. 11.3.

If the earthing conductors are embedded in the soil, they must endure mechanical stress and corrosion without compromising their electrical continuity. To this end, their minimum standard cross-sectional areas can be selected from Table 11.9 as per IEC 60364–5-54.

The main earthing terminal links together protective conductors, equipotential bonding conductors, and earthing conductors.

	Minimum Cross-Sectional Areas for Earthing Conductors (mm²)			
	Mechanically Protected		Mechanically Unprotected	
	Cu	Fe	Cu	Fe
Protected against corrosion	2.5	10	16	16
Not protected against corrosion	25	25	25	25

Table 11.9 Minimum Standard Cross-Sectional Area for Earthing Conductors Embedded in the Soil

All the aforementioned conductors must be able to be disconnected individually, but only by means of a tool. This allows both earth resistance tests of the earthing electrode, by isolating the influence of the EXCPs, and the insulation and continuity tests of the PEs.

11.6 The PEN Conductor

As already substantiated in Chap. 7 for TN-C systems, PEN conductors provide both functions of neutral and protective conductor. The accidental interruption of the PEN, when the line conductors are live, energizes the equipment, even if healthy; therefore, its electrical and mechanical continuity must be guaranteed. To this purpose, the PEN should only be used in a fixed installation and have a cross-sectional area not less than 10 mm^2 in copper or 16 mm^2 in aluminum. These conditions limit the probability of its accidental interruption.

In some point of the installation, the PEN conductor may split and originate distinct wires as neutral and PE. The system becomes TN-S and separate bus bars for the neutral wire and the PE may be employed (Fig. 11.9).

In the presence of two bus bars, it is safer to connect the PEN to the PE's bus bar and not to the neutral one, even though there is no electrical difference between the two. This arrangement elevates the

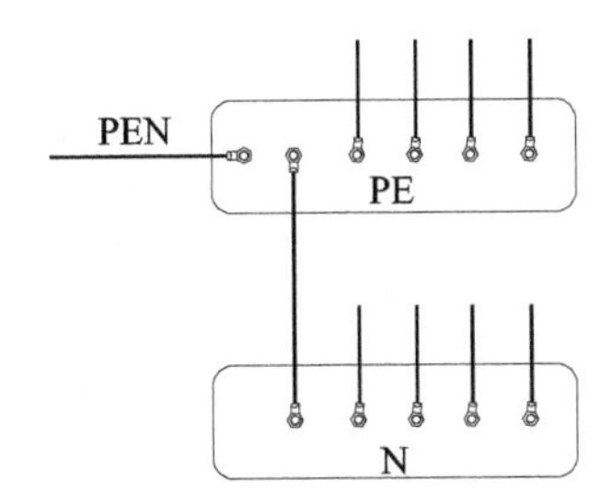

Figure 11.9 Separate bus bars for neutral and PE.

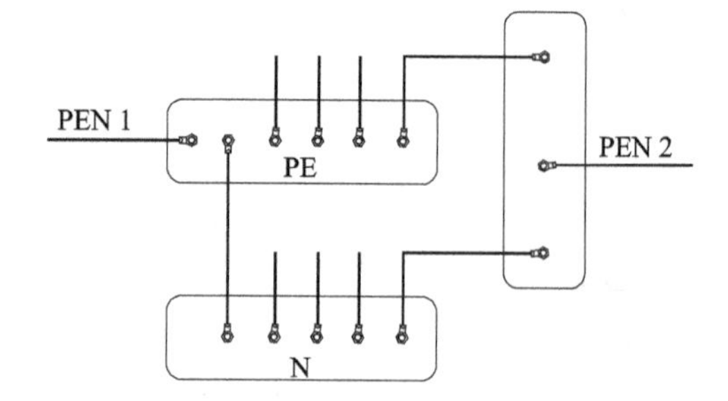

FIGURE 11.10 Neutral and PE must not be combined together to form a second PEN.

reliability of the connection of the PE to the PEN conductor, which is considered crucial for the safety against indirect contact.

Once the system has become TN-S, neutral and protective conductors must not be connected together to form a second PEN (Fig. 11.10).

The reason behind this prohibition is the fact that PE and neutral wire might not reach the high reliability required for a PEN conductor. Among other things, both should be labeled as a PEN conductor to prevent their accidental disconnection, for example, for maintenance purposes.

FAQs

Q. Why in Fig. 11.1 is the cold water pipe indicated as an EXCP and not as an earth electrode?

A. A cold water pipe cannot be relied on as an earth electrode, because the user has no control over its electric continuity to ground. A water utility, in fact, may interpose an insulating insert across the pipe to protect it against corrosion.

However, the cold water pipe may be an EXCP, even in the presence of the insulating insert, as the utility might eliminate it without warning the customer. For these reasons, we must bond the water pipe for equipotential reasons at the customer's side of the meter, but not rely on it as an earth electrode.

Q. Why is the assumption of adiabatic process during ground faults justified?

A. At the occurrence of a fault-to-ground, the heat developed by the Joule effect is proportional to the square of the fault current. The heat released by convention and/or radiation is proportional to the temperature of the conductive material, which increases much more slowly than the fault current (e.g., the temperature can increase four times when the current increases 50 times). This proportion shows the large disparity between the two energies and, therefore, the possibility to neglect that released by convention/radiation during the clearing time.

Endnotes

1. As per *"Residential and Similar Premises—Installation Criteria of Earthing System,"* Norma Italiana CEI 64–12, 1998–02.
2. IEC 60364–5-54, *"Electrical Installations of Buildings—Part 5–54: Selection and Erection of Electrical Equipment—Earthing Arrangements, Protective Conductors and Protective Bonding Conductors."* Second Edition 2002–06.
3. When faults involve line conductors (i.e., phase-to-phase short circuits), the initial temperature of the conductors, at the inception of the fault, is not the ambient one, but the actual temperature is in correspondence of the prefault current. In this case, k can be conservatively calculated by considering, as the initial temperature, the temperature in correspondence of the current-carrying capability of the conductor (e.g., 70°C for PVC insulation).
4. For more details, see: M. Tartaglia, M. Mitolo, *"Evaluation of the Prospective Joule Integral to Assess the limit Short Circuit Capability of Cables and Busways,"* Proceedings of the IEEE Industry Application Society (IAS) 43rd Annual Meeting, Edmonton, October 5–9, 2008.
5. The bonding conductor might be avoided if the hinges are both low-resistance type and protected against corrosion.
6. For further details see M. Mitolo, *"Protective Bonding Conductors: An IEC Point of View,"* IEEE Transactions on Industry Applications, Vol. 44, No. 5, September/October 2008.

CHAPTER 12

Safety Against Overvoltages

Quod Lex non dicit, non vult.
What the Law says not, wants not.
LATIN PROVERB

12.1 Introduction

Overvoltages are defined as the unwanted potentials occurring in electrical systems between one-phase conductor and the earth (referred to as *common-mode* voltage), or between phase conductors (referred to as *differential-mode*, or *transverse*, voltage), having a peak value greater than the peak of the largest nominal voltage of the system itself.

Overvoltages can be triggered by atmospheric lightning discharges, in which case they are defined as *external*, or by a rapid change of system conditions (e.g., ground faults, switching operations, large equipment being turned off, etc.), in which case they are defined as *internal*.

Internal overvoltages caused by switching events are *transient* phenomena of duration of a few microseconds or less, with oscillations usually highly damped; their frequency is in the order of 100 kHz. Ground faults occurring in the primary, or secondary, side of substations may cause internal *temporary* overvoltages at power frequency of relatively long duration in the order of seconds; such temporary overvoltages are usually undamped or weakly damped.

Should the overvoltages exceed the dielectric capability of the insulation across which they are applied, its premature failure may occur and, consequently, current would circulate between live parts, or live parts and earth. The production of heat may cause the escalation of the damage and the extensive destruction of the insulation, unless the supply of electric current is promptly interrupted.

Excessive voltage stress, therefore, can create hazards for persons by triggering fires and explosions and/or compromising the continuity of service of critical apparatus. In addition, overvoltages are a cause of damage to electronic equipment, which usually have lower dielectric strength.

12.2 Temporary Overvoltages and Safety

Ground faults that occur in the primary side of substations, supplied through distribution systems operating at medium/high voltages, may cause circulation of high currents (i.e., tens of kiloampere) in their earthing systems. The intensities of the ground-fault currents depend on how the neutral of the primary side distribution system is "operated." Such neutral, defined as the common point of a polyphase supply system, may be solidly grounded, isolated from ground or grounded through impedances or resistances.

High and low sides of substations, though, may share the same earthing system and, therefore, in primary fault conditions, the low-voltage ECPs connected to it become energized. The duration of the resulting prospective touch voltage in the low-voltage system, in the order of thousand of volts, although temporary, may be excessive and, therefore, unsafe for the low-voltage system. In fact, medium/high voltage protective relays in charge of the protection of the primary side of substations may be time-delayed to facilitate their coordination with other devices.

The high-voltage fault may overstress the low-voltage equipment by imposing an excessive voltage across its insulation and earth. The stress voltages may breakdown the insulation and cause faults as well as start fires.

In the following sections, we will examine the above issues as occurring in typical earthing arrangements.

12.2.1 High-Voltage Ground Faults in TN Systems

In TN systems, the substation's grounding system will connect together the transformer tank, the ECPs of high-voltage and low-voltage equipment, and all the EXCPs eventually present. In Fig. 12.1, the low side and high side of the transformer share the same earthing system.

In the case of earth fault in the high-voltage system, all the aforementioned metal parts become energized due to their connection to the grounding system. This may expose persons to dangerous touch voltages if in contact with low-voltage equipment.

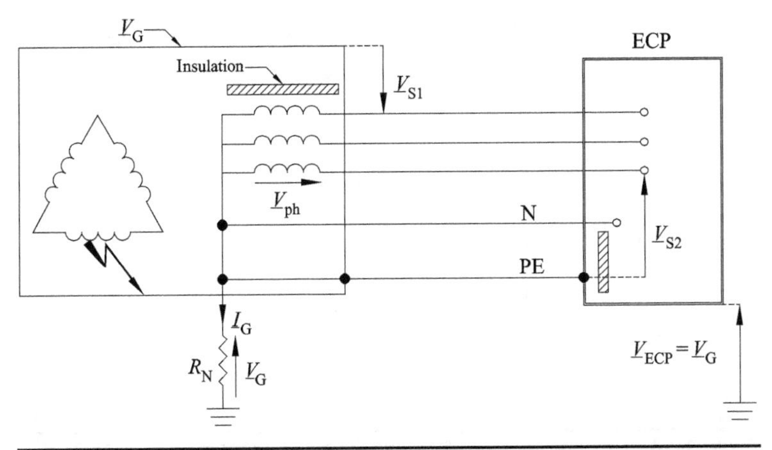

FIGURE 12.1 Low and high sides of substation sharing the same earthing electrode in TN systems.

To ensure safety, magnitude and duration of the touch voltage V_T must fulfill the following equation:

$$V_G = R_N I_G \leq V_T \tag{12.1}$$

where I_G is the earth current caused by a high-voltage fault and V_T is the maximum permissible value for the touch voltage as a function of high-voltage fault duration as shown in the curve in Fig. 12.2, as per IEC 60364–4-44.[1]

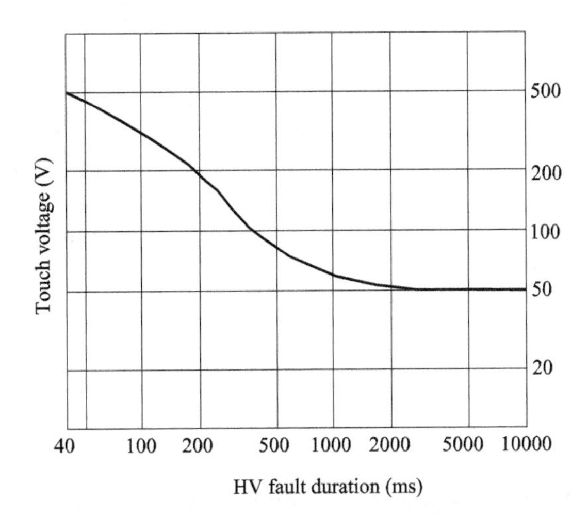

FIGURE 12.2 Permissible touch voltage as a function of the duration of high-voltage faults.

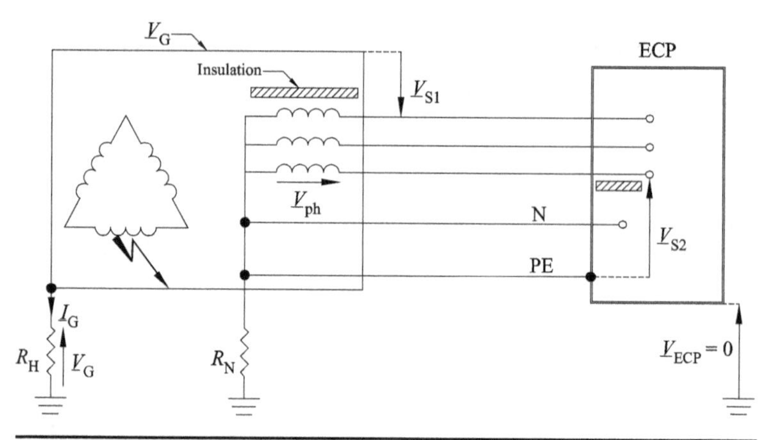

FIGURE 12.3 Low and high sides of substation independently earthed in TN systems.

The curve in Fig. 12.2 relates the maximum duration of faults allowable by the high-voltage protective devices to the permissible touch voltage on ECPs supplied by the secondary side of the transformer.

When Eq. (12.1) is fulfilled, both the neutral and protective conductors of the low-voltage system may be connected to the same earth electrode of the substation.

In TN systems sharing the same grounding electrode, the protective conductor (PE) equalizes the potential between the electrode itself and the low-voltage ECPs. Therefore, even in the presence of a fault potential $\underline{V}_G$ across R_N, the potential difference across the basic insulation of low-voltage equipment (i.e., $\underline{V}_{S1}$ and $\underline{V}_{S2}$) will be equal to $\underline{V}_{ph}$ and no overvoltage will be caused.

If Eq. (12.1) is not fulfilled, the neutral and the protective conductors of the low-voltage system must be earthed independently of the substation's grounding system (Fig. 12.3).

In high-voltage fault conditions, the low-voltage ECPs remain at zero potential (i.e., $\underline{V}_{ECP} = 0$ and $\underline{V}_{S2} = \underline{V}_{ph}$), while the transformer's enclosure in the substation reaches the fault potential $\underline{V}_G = R_H \underline{I}_G$. A potential difference $\underline{V}_{S1}$, whose magnitude may be as large as $V_{ph} + V_G$, appears across the insulation separating the transformer's enclosure and the secondary windings, or any low-voltage system bonded to R_N. This temporary overvoltage stresses the low-voltage insulation and might compromise its integrity, if its dielectric strength is exceeded.

12.2.2 High-Voltage Ground Faults in TT Systems

In TT systems, the neutral conductor serving the low-voltage customer's installation may be connected to same electrode that earths the high-voltage ECPs of the substation (Fig. 12.4).

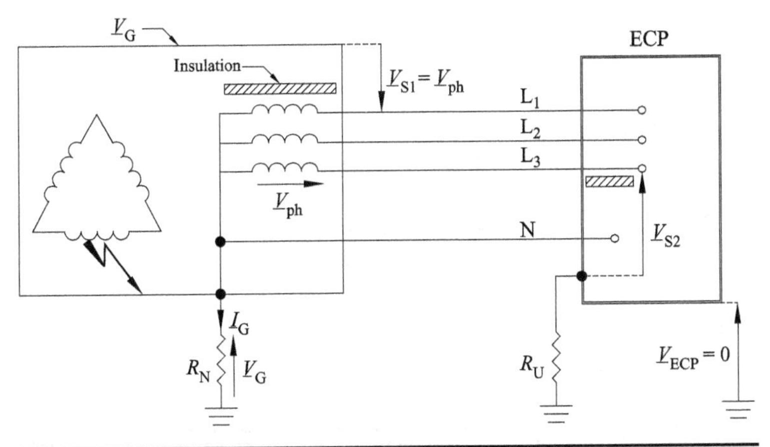

FIGURE 12.4 Neutral conductor connected to the substation earthing system in TT systems.

Upon high-side faults, the neutral will reach, with respect to ground, the earth potential V_G, while the low-voltage ECPs will remain at zero potential. Consequently, a temporary potential difference $\underline{V}_{S2}$, of maximum value $V_{ph} + V_G$, will appear across the insulation of low-voltage equipment.

To limit possible damages to the insulation, permissible values for the stress voltage, as a function of its duration, have been elaborated by IEC 60364-4-44, as shown in Table 12.1.

The disconnecting time is the time the high-voltage protective device takes to clear the fault. A clearing time exceeding 5 s may be typical of inductively grounded HV systems. Table 12.1 provides the criterion to properly rate the insulation-to-ground of low-voltage systems against temporary overvoltages in TT systems.

If the inequalities of Table 12.1 cannot be fulfilled, the neutral conductor of the low-voltage system must be grounded independently of the substation's grounding system (Fig. 12.5).

Also in this case, the maximum stress voltage $V_{S1} = V_{ph} + V_G$ must be interrupted in a time compatible with the minimum insulation rating of the low-voltage equipment present in the substation.

Permissible LV Stress Voltage (V)	Disconnecting Time (s)
$V_{ph} + 250$	>5
$V_{ph} + 1200$	≤5

TABLE 12.1 Permissible Values for the LV Stress Voltage in TT Systems

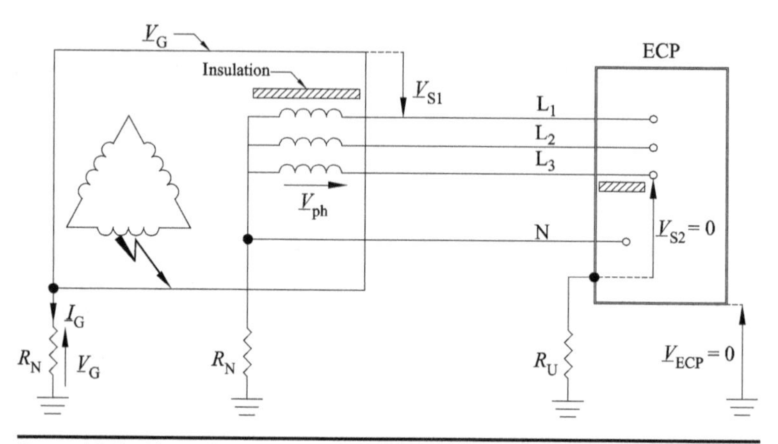

FIGURE 12.5 Neutral conductor grounded independently of substation's earth in TT systems.

12.2.3 High-Voltage Ground Faults in IT Systems

As explained in Chap. 9, in IT systems the power source, that is, the secondary side of the supply transformer, is isolated from the earth, while the ECPs must be independently grounded from the source—individually, in groups, or collectively. The transformer, though, may be subject to primary voltage faults, and therefore its enclosure must be earthed to guarantee a clear return path to the fault currents. The ECPs of the low-voltage system are permitted to be linked to this grounding system (Fig. 12.6), providing that the resulting touch voltage V_T is cleared in a time compatible with the chart in Fig. 12.2.

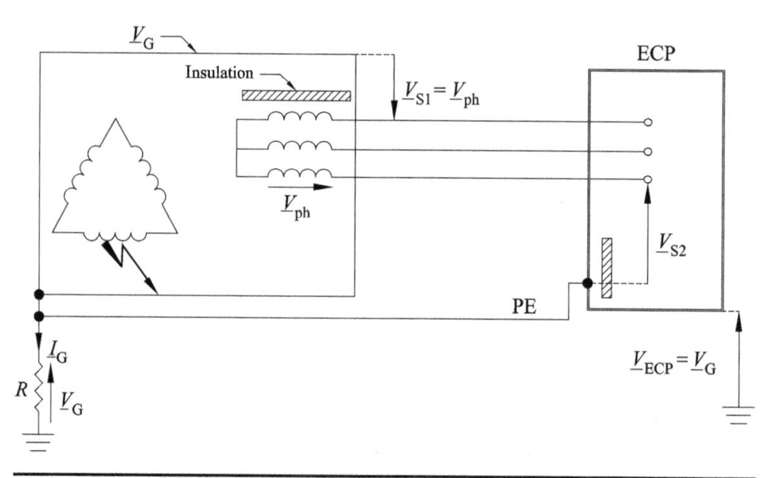

FIGURE 12.6 High-voltage fault in IT system where high- and low-voltage equipment share the same earthing system.

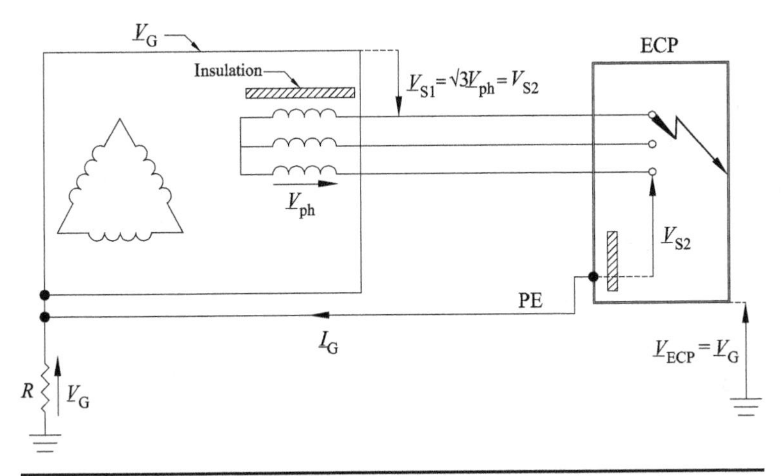

FIGURE 12.7 Low-voltage ground fault in IT system where high- and low-voltage equipment share the same earthing system.

In the above system, the protective conductor PE equalizes the potential between the electrode itself and the low-voltage ECPs. This allows the potential differences $\underline{V}_{S1}$ and $\underline{V}_{S2}$ across the low-voltage basic insulation not to exceed $\underline{V}_{ph}$, and no additional stress voltage is imposed.

As to ground faults of negligible resistance in the low-voltage system (Fig. 12.7), as already explained in Chap. 9, they cause the voltage between each healthy phase and the earth to increase up to the line-to-line potential (e.g., 400 V vs. 230 V). This condition may impose an overstress to the basic insulation of low-voltage equipment, especially to single-phase loads, which may not be rated to withstand the phase-to-phase voltage.

As to high-voltage ground faults, if the touch voltage V_T is not cleared in a time compatible with the chart in Fig. 12.2, the ECPs of low-voltage equipment must be earthed via an independent electrode (Fig. 12.8).

In this arrangement, the stress voltage $\underline{V}_{S1}$ can reach the maximum value of $V_G + V_{ph}$.

Two consecutive ground faults, one in the high-voltage system and the other in the low-voltage system, can occur (Fig. 12.9).

The low-voltage ECPs will be energized at the perspective touch potential $V_{ECP} = R_U I_d$, and the stress voltage $\underline{V}_{S2}$ will equal $\sqrt{3}V_{ph}$. We already know from Sec. 9.1 that if $V_{ECP} \leq 50$ V, there is no need for automatic disconnection of supply, because this voltage does not cause any harm. The high-voltage fault causes the stress voltage $\underline{V}_{S1}$ to be as high as $\sqrt{3}V_{ph} + V_G$.

In both the situations in Figs. 12.8 and 12.9, the stress voltages $\underline{V}_{S1}$ and $\underline{V}_{S2}$ must be interrupted in a time compatible with the low-voltage

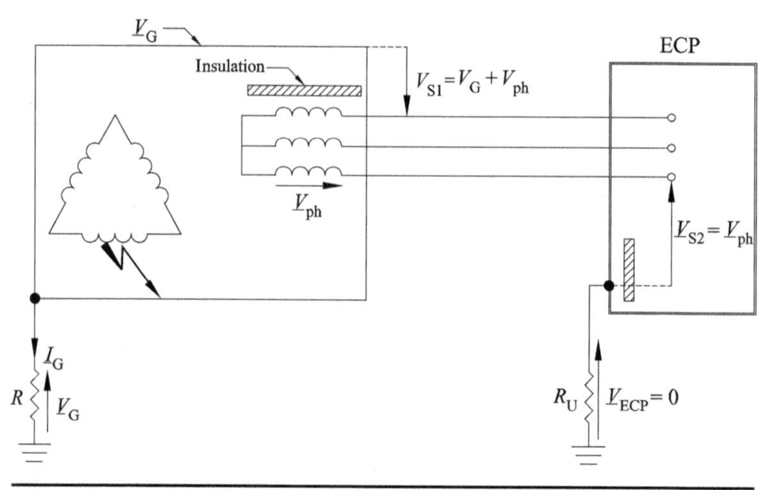

FIGURE 12.8 High-voltage ground faults in IT systems with independent grounds.

insulation rating of the equipment within the substation, which is the most stressed by the ground faults.

12.3 External Overvoltages

External overvoltages are caused by lightning. Lightning is defined by IEEE Standard 100[2] as an electrical discharge that occurs in the atmosphere between clouds or between clouds and ground. In the

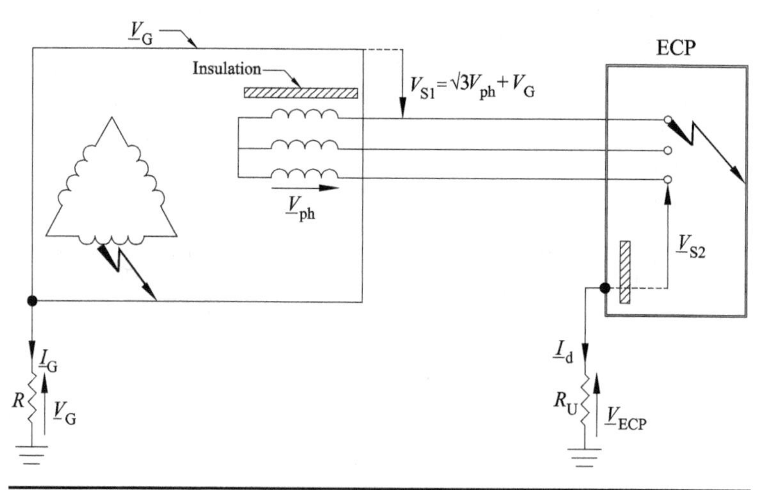

FIGURE 12.9 High- and low-voltage ground faults in IT systems with independent grounds.

latter case, the lightning impulsive current is drained to earth, directly or via the structures that are struck. The consequent release of a large amount of energy (i.e., many hundreds of megajoules) is harmful to persons as well as destructive for equipment.

The cloud-to-earth lighting is initiated by the presence of charges, usually negative, in the lower part of the cloud. The buildup of charges may cause the resulting electric field to exceed the dielectric strength of the air.[3] The breakdown of the air and an initial discharge then occur. The discharge creates a highly conductive channel the charges can use to descend toward ground as if it were a conductor. The channel stops at the point where the dielectric strength of the air equals the electric field caused by the charges. Further charges, though, traveling from the cloud reinforce the field, perturbing the equilibrium and allowing new discharges. Thus, new conductive channels occur in a "zig-zag" fashion toward the earth. Such "stepped" discharge is due to the nonuniformity of the air, caused by the punctual variation of its parameters, such as density, humidity, etc.

The charge from the cloud progressing toward the earth induces an equal amount of charge, but of opposite sign. The electric field increases and an upward-directed discharge from the soil, or a structure, takes place causing an "attachment" between the upward- and downward-directed channels. This "return stroke" causes the circulation of the high-intensity (i.e., hundreds of thousands of amperes) impulsive lightning current to ground. Such a current is characterized by a rapid rise to the peak (i.e., within a few microseconds), a relatively slow decay as well as a high-frequency content (i.e., order of hundreds of kilohertzs).

12.3.1 Characterization of Earthing Systems Under Impulse Conditions

The grounding system used to safely drain to earth the fault currents at the power frequency is also employed to dissipate to ground the lightning current. Earthing electrodes, in fact, may be connected, by means of down-conductors, to lightning protection systems (LPS), such as masts, installed on the roof of buildings being protected.

Grounding systems through which high-frequency currents circulate, though, do not behave in the same fashion as examined in the previous chapters for fault currents at the network frequency. Earth electrodes, in fact, cannot be assumed to be purely resistive in the presence of pulse currents, as the inductance of their metal parts cannot be neglected.[4] Thus, the earth electrode must be modeled as an ohmic-inductive *pi* circuit, and we will use the term *earth impedance* instead of earth resistance.

To clarify the concept, let us consider, as an earth electrode, a buried horizontal wire (Fig. 12.10).

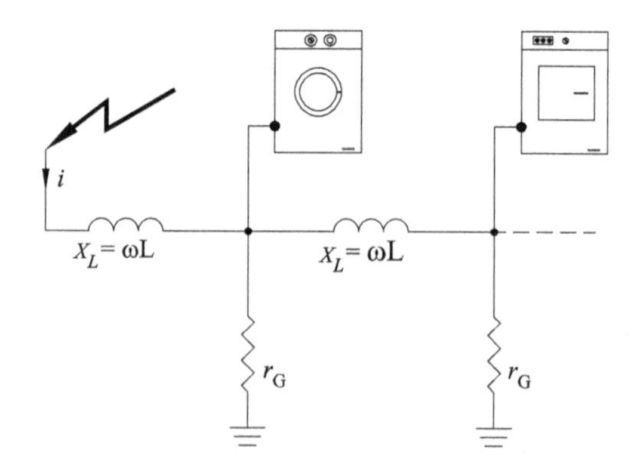

Figure 12.10 High-frequency equivalent circuit of a buried horizontal wire earth electrode.

L, r_G, and x_L are, respectively, the inductance, the earth resistance, and the inductive reactance per unit of length of the wire. The voltage drops across x_L due to pulse currents causes nonzero potential differences along the wire, which becomes no longer equipotential. This behavior may cause two ECPs connected to the same earthing electrode, but in two different locations, to be at different potentials with risk for persons.

High currents in the soil cause its resistivity to decrease, as small voids in the earth are "shorted" by the intense electrical field. Consequently, during lightning impulse conditions, the earth resistance of the electrode reduces. On the other hand, the high frequency of the current causes the reactance to increase. Thus, the earth impedance is the result of the combination of these two opposite effects.

12.3.2 Induced Overvoltages

The lightning current flows to the earth through the down-conductors connecting the LPS to the ground electrode of the building. As it is known, the circulation of currents in conductors creates magnetic fields. If the field is variable with time, which is our case, overvoltages will be induced in any metal loops present in the building.

The down-conductor and any linear metal parts with vertical path within the structure, such as EXCPs, power, and telecommunication circuits, etc., can form metal loops (Fig. 12.11).

The induced overvoltage can be expressed through Eq. (12.2):

$$V_i = L_l l \frac{di}{dt} \tag{12.2}$$

With reference to Fig. 12.11, l is the vertical length of the natural gas pipe from the main equipotential bonding connection (MEB); i is the lightning current flowing at the point of strike. L_l is the inductance,

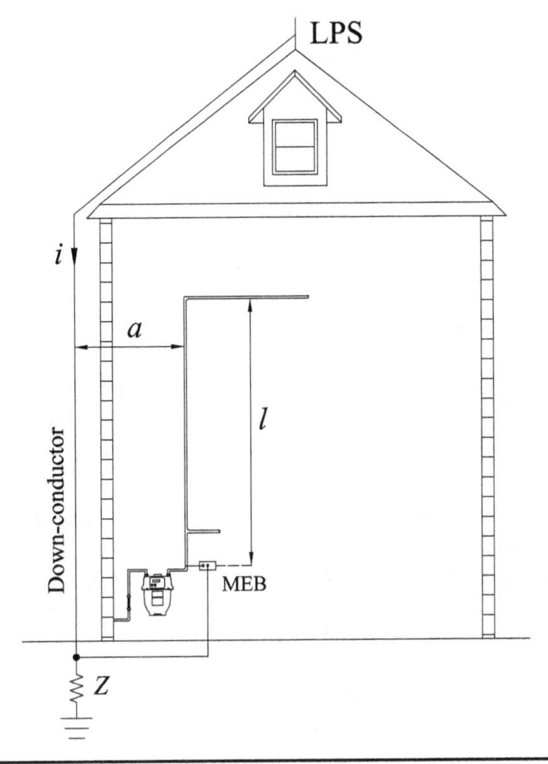

FIGURE 12.11 Closed metal loop formed by the down-conductor and the gas pipe.

in air, of the loop, per unit of length given by Eq. (12.3):

$$L_l = \frac{\mu_0}{2\pi} \ln \frac{a}{r} \qquad (12.3)$$

where μ_0 is the magnetic permeability of the air,[5] r is the radius of the down-conductor, and a is the distance between the surface of the vertically developing metal part (i.e., the natural gas pipe in Fig. 12.11) and the center of the cross section of the down-conductor.

If the induced overvoltages exceed the impulse withstand capability of the separation means between the down-conductor and the natural gas pipe, which we can consider as electrodes, a lateral spark may occur across the gap a that separates them. This sideflash may be extremely dangerous, as it can have sufficient energy to trigger fires or ignite explosive atmospheres eventually present.

The dangerous discharge between the aforementioned electrodes can be prevented if they are separated from each other by a safety distance s. In fact, the intensity of electric fields existing between two electrodes depends on both the voltage applied across them and their reciprocal distance [Eq. (12.4)]:

$$V_i = k_m E s \quad (12.4)$$

where V_i is the overvoltage induced by the lightning strike[6]; E is the dielectric strength of the air in dry conditions; k_m is a multiplier that takes into account separation materials more "dense" than air, for which it assumes values greater than one. s can be calculated from Eq. (12.3) as the minimum space between the two conductive parts between which no dangerous electrical sparking can occur in lightning conditions.

If metal parts cannot be separated by at least the safety distance (i.e., $a < s$) due, for example, to structural constraints, protective measures must be assumed to prevent the adverse effects of the lightning current.

A possible alternative solution is to provide equipotential bonding connections to bridge the gap where $a < s$. In this case, though, part of the lightning current flows to ground through the metal part within the building, which results in parallel to the down-conductor. This amount of current is reduced if multiple down-conductors are installed.

Alternatively, the adoption of separation materials with higher impulse dielectric capability, by increasing the multiplier k_m, reduces the value of the safety distance, thereby allowing a closer distance between down-conductors and metal parts.

Endnotes

1. "*Protection for Safety—Protection Against Voltage Disturbances and Electromagnetic Disturbances*," IEC 60364-4-44, 2001.
2. IEEE Standard 100 "*The Authoritative Dictionary of IEEE Standard Terms*," 7th ed.
3. The dielectric strength of the air in standard dry conditions is 3000 kV/m, but this value can be greatly reduced by the presence of humidity and/or atmospheric motes.
4. At 50/60 Hz, the inductance of ground electrodes can be estimated as 1 μH/m.
5. $\mu_0 = 1.257 \times 10^{-6}$ H/m.
6. For more details see: M. Mitolo, "*Shall Masts and Metal Structures Supporting Antennae Be Grounded?*," Proceedings of the IEEE-IAS Industrial & Commercial Power Systems Technical Conference, Edmonton, Canada, May 6–10, 2007.

CHAPTER 13

Safety Against Static Electricity and Residual Voltages

If knowledge can create problems, it is not through ignorance that we can solve them.
ISAAC ASIMOV (1920–1992)

13.1 Introduction

Static charge can accumulate on different types of objects such as conductive and insulated parts, high-resistivity liquids, and also gasses. The static charge creates potential differences between the aforementioned elements, as well as between them and the earth, which may reach 100 kV.[1] As a consequence, an electric field possibly exceeding the dielectric strength of the interposed matter among charged parts (e.g., the air) is created and a spark accompanied by release of energy (i.e., the heat associated with the spark) can occur. Static charge can, therefore, be the source of ignition for flammable materials if they are in the right concentration in air, and can even cause their explosion.[2] Flammable materials can explode due to their very rapid combustion, which causes a sudden increase in their temperature and pressure. Explosive substances, instead, do not need to be in the right proportion with the oxygen, and can explode even without it.

The risk of fire or explosion may be present in occupancies when all the following circumstances occur:

1. A process of generation and accumulation of electric charge is present.
2. The gap between parts with opposite charge is small enough to allow a spark discharge.
3. Flammable materials, such as gasses, vapors, and dusts, are present.
4. Flammable atmospheres are present. Such atmospheres are created by flammable materials in the optimum concentration in air, ranging between their lower explosive limit (LEL) and upper explosive limit (UEL). At concentrations in air above the UEL, the mixture air–flammable material is too poor in oxygen to start combustion; at concentrations in air below the LEL, there is not enough flammable material to sustain combustion. Flammable atmospheres can be continuously present in the area during normal operations, or momentarily due to accidents (e.g., rupture of a tank).
5. The electrostatic energy stored and available to be discharged exceeds the minimum ignition energy (E_{MIE}) of the flammable atmosphere, which is typically in the order of millijoule. Note that dusts require much more energy to ignite than gasses and vapors.

All the above elements can be represented in the *fire tetrahedron* of Fig. 13.1.

Previous point 3 is represented by the *fuel* leg of the tetrahedron, whereas points 1, 2, and 5 are the *heat* leg. Point 4, that is, the presence of the optimum flammable atmosphere, is the *chemical reaction* leg. The risk of fire or explosion is eliminated if at least one of the above legs is removed.

Additional risk offered by charged objects is the impulsive current that can possibly flow through persons upon touch and discharging to ground. This current is generally well below the threshold of danger, but in some cases might be above the tingling sensation and, therefore, cause sudden shock and induce accidental falls.

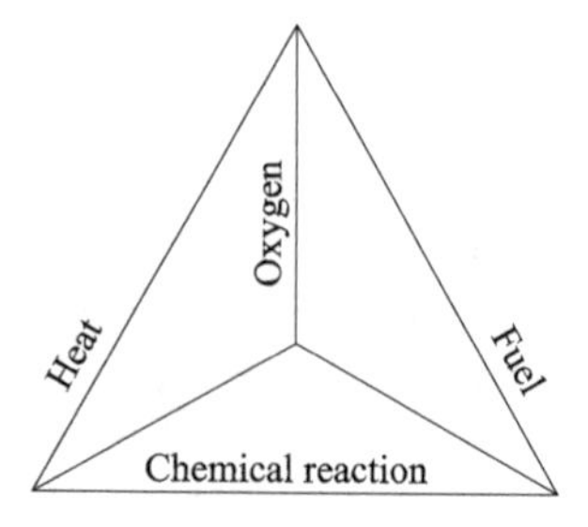

FIGURE 13.1 The fire tetrahedron.

For all of the above reasons, strategies to avoid or mitigate the formation of static charge must be taken into consideration in the design of installations, primarily to guarantee persons' safety and, secondarily, to prevent economic losses. In the following sections, we will examine the causes of generation and accumulation of static charge and possible mitigation strategies.

13.2 Generation of Static Electricity

The common cause of the generation of static charge is the friction between unlike materials, be solid or fluid. The relative motion of the two materials, made of contacts and subsequent separations, allows the transfer of electrons from one surface to the other. If either, or both, materials are insulators, when they separate, some electrons may not be able to return to their original location due to their elevated resistance. Therefore, the electrons' counterparts in the atoms, the protons, are left behind not neutralized, originating ions. This process causes a charge within the insulating material, which may even take days to be spontaneously neutralized. Among other factors such as material characteristics, areas of contact, etc., the speed of separation of the parts in contact plays an important role in determining the magnitude of their electrification.

As an example, let us consider a liquid flowing through a pipe and filling a tank. In correspondence of the conduit's surface, the fluid tends to develop a positively charged layer, while negative ions will be present in its inner part. The system is, therefore, electrically neutral, but when the liquid abandons the tube the negative charge may be carried out with it. This effect is particularly accentuated in fluids with high electrical resistivity (i.e., orders of hundreds of megaohms) due to the charges' difficulties at moving within it. In return, the negatively charged liquid will electrify the container that collects it (Fig. 13.2).

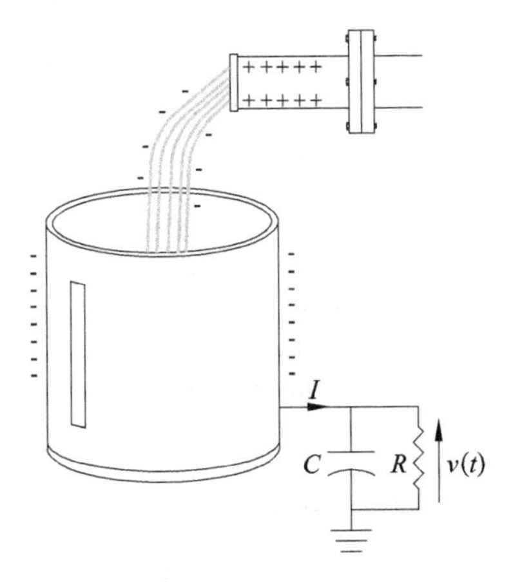

FIGURE 13.2 Charging of a container by flow of liquids.

The circulation of the fluid, by moving charges, creates a stream current. If the liquid is characterized by volumetric flow rate F (m^3/s), density d (kg/m^3), and specific charge density δ (C/kg), the resulting stream current will have a magnitude that can be expressed in amperes as

$$I = Fd\delta \tag{13.1}$$

It is important to stress that, as represented in Fig. 13.2, any object may have either, or both, "natural" capacitance C and resistance-to-ground R (e.g., C for a 3.6-m-diameter tank with insulating lining equals 100 nF, as per IEEE 142–1991). The capacitance-to-ground can store the static charge even after the process of electrification is over, and then discharge the energy, for example, upon person's touch.

13.3 Static Charge Energy

The accumulation of static charge over an object consists of the "deposit" of the charge Q, during the time t_Q, therefore, the electrification process can be analyzed by studying the stream current $I = Q/t_Q$.

Once the charge deposits, the object's earth potential $v(t)$ elevates, and coincides with the voltage to which the capacitor is charged. To calculate such earth potential, reference is made to Fig. 13.2. The charge current I flowing to ground divides through C and R, hence, by applying Kirchhoff's first law, we can write

$$I = i_C(t) + i_R(t) = C\frac{dv(t)}{dt} + \frac{v(t)}{R} \tag{13.2}$$

where $i_C(t)$ and $i_R(t)$ are, respectively, the currents charging the earth capacitance and the current through the earth resistance of the tank.

By assuming as the initial condition that the object is not charged [i.e., $v(0) = 0$], the solution $v(t)$ of the above differential equation is

$$v(t) = RI(1 - RIe^{-\frac{t}{RC}}) \tag{13.3}$$

The product RC is dimensionally a time and is defined as the *time constant* of the charging process. The time constant is the time the earth voltage takes to reach 63.2% of its final magnitude (i.e., the product RI). To better clarify this concept, let us evaluate the earth potential given in Eq. (13.3) for $t = nRC$, for $n = 0$ to 3. The results are shown in Table 13.1.

Obviously, the greater the time constant, the longer the capacitance-to-ground takes to reach its full charge. In theory, because of the exponential function of Eq. (13.3), such capacitance reaches the voltage RI after an infinite time. In practice, after three or four times the time constant, we can deem the capacitance-to-ground fully charged.

If the body is completely insulated from ground (i.e., $R = \infty$), the electric charge, as it forms, cannot be drained to earth. Equation (13.2)

t	$v(t)$
RC	63.2% RI
$2RC$	86.4% RI
$3RC$	95.0% RI

TABLE 13.1 Evaluation of the Earth Potential as a Function of the Time Constant RC

then becomes

$$I = i_C(t) = C\frac{dv(t)}{dt} \tag{13.4}$$

By assuming again $v(0^+) = 0$, we solve the above differential equation for $v(t)$ and obtain

$$v(t) = \frac{I}{C}t \tag{13.5}$$

In this case, the voltage to earth linearly increases without having the upper limit of RI. Therefore, when the magnitude of the potential difference exceeds the dielectric strength of the air, a discharge of energy occurs in the form of heat (i.e., a spark). The charge process restarts again and repeats itself.

The graph of $v(t)$ as a function of time is qualitatively shown in Fig. 13.3 for both the above cases.

The energy being stored in the charged object as a function of time is given by

$$E = \frac{1}{2}Cv(t)^2 = \frac{1}{2}CR^2I^2\left(1 - e^{-\frac{t}{RC}}\right)^2 \tag{13.6}$$

If the process of accumulation of charges lasts at least three times the time constant (i.e., $t_Q > 3RC$), the exponential function in Eq. (13.6)

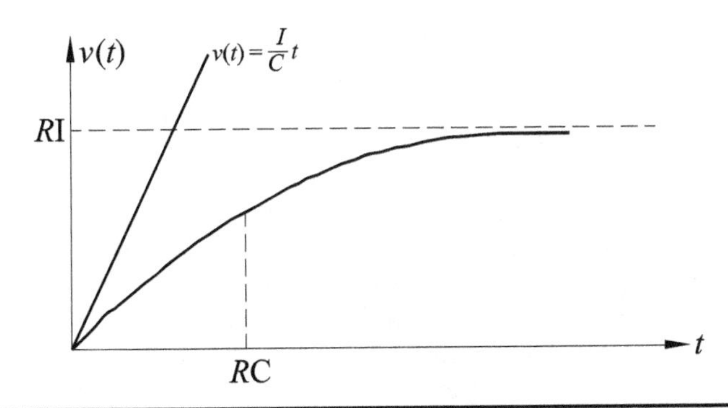

FIGURE 13.3 Graph of $v(t)$ as a function of time.

becomes insignificant, as it approaches zero when t approaches infinity. In this case, the expression of the stored energy can be simplified by neglecting it and results proportional to the capacitance-to-ground.

13.4 Mitigation Strategies

In flammable atmospheres, bonding and earthing can be effectively employed to prevent the accumulation of static charge over conductive bodies.

Equipotential bonding conductors bridge the gap existing between two metal parts likely to be charged, keeping them at the same potential. In doing so, the risk of spark discharge is eliminated or greatly reduced.

Earthing, instead, by solidly connecting metal bodies likely to be charged to ground, allows the draining of the static charge, preventing its accumulation. This connection to ground must have a resistance low enough so that the energy possibly being accumulated is lower than the minimum ignition energy E_{MIE} of the flammable atmosphere. To this end, we can calculate the maximum permissible value for R, by assuming $t_Q > 3RC$, thereby neglecting the exponential and solving Eq. (13.6) for R. Thus, we obtain the following inequality:

$$R < \frac{1}{I}\sqrt{\frac{2E_{\text{MIE}}}{C}} \tag{13.7}$$

For instance, let us calculate the minimum safe resistance-to-ground of a tank, whose capacitance-to-ground is $C = 100$ nF, bleeding a current I of 50 μA, and surrounded by a flammable atmosphere of minimum ignition energy $E_{\text{MIE}} = 10$ mJ. We obtain

$$R < \frac{1}{50 \times 10^{-6}}\sqrt{\frac{2 \times 10^{-3}}{100 \times 10^{-9}}} = 2.8\ \text{M}\Omega \tag{13.8}$$

For static charge purposes, therefore, even high values of ground resistance are suitable to prevent dangerous accumulation. Applicable standards conservatively recommend that the earth resistance should not exceed 1 MΩ.

Equipotential bonding conductors and earthing conductors used to bleed-off the charge will only carry currents in the order of the microamperes, therefore, there are no issues regarding their current-carrying capability. The mechanical strength of these conductors, though, must be guaranteed by using minimum cross-sectional areas of at least 4 mm^2, if aluminum, and 2.5 mm^2, if copper.

Needless to say that for static electricity purposes we must only earth metal objects isolated from ground, or whose resistance-to-ground is greater than 1 MΩ. We have previously substantiated, in fact, that by connecting to the main grounding system metal objects other than these ones, we make them prone to transferred potentials generated in faulty equipment.

13.5 Residual Voltages

Residual voltages are potentials caused by static charge accumulated in capacitors within equipment during its normal operation. Residual voltages may persist after the supply has been turned off, even for hours, and may expose maintenance personnel to the risk of electrocution.

During normal operations of electrical systems, voltages across capacitors are sinusoidal, and after disconnection of supply they remain charged at the value V_0 the sinusoid had at the instant of the interruption.

Upon direct contact with one, or both terminals, of the capacitor, the discharge process will initiate and its potential will decay with exponential law. Such potential will cause the circulation of an impulse current through the person's body of duration in the order of a few milliseconds. If we assume constant the person's body resistance R_B, such current, as a function of time, can be expressed by using the Ohm's law:

$$i(t) = \frac{V_0}{R_B} \times e^{-\frac{t}{R_B C}} = I_{peak} e^{-\frac{t}{\tau}} \tag{13.9}$$

where τ indicates the time constant of the discharge process.

The r.m.s. value of the discharge current can be calculated as:

$$I_{rms} = \frac{I_{peak}}{\sqrt{6}} = \frac{V_0}{R_B\sqrt{6}} \tag{13.10}$$

We can assume that the above impulse current will have practically transferred almost the whole static energy accumulated in the capacitor in a period of time equal to 3τ. At this time, the current will be at 5% of its initial value (Fig. 13.4).

The energy E_R released during the discharge and dissipated in the person's body is the quantity that determines the probability of ventricular fibrillation.

E_R can be so expressed as:

$$E_R = R_B \int_0^{\infty} i^2(t)\,dt \cong R_B \int_0^{3\tau} i^2(t)\,dt \cong R_B I_{rms}^2 3\tau \tag{13.11}$$

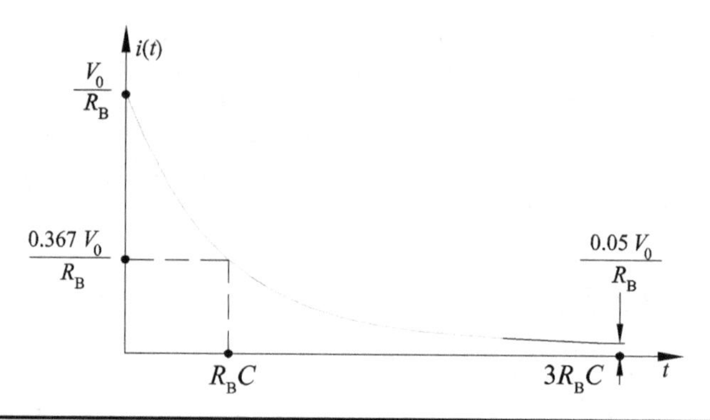

FIGURE 13.4 Graph of $i(t)$ as a function of time.

Hence,

$$\int_0^{3\tau} i^2(t)\,\mathrm{d}t \cong I_{\mathrm{rms}}^2 3\tau \qquad (13.12)$$

The left-hand side of Eq. (13.12) is the Joule integral as referred to the impulse discharge current. The Joule integral can be used to express the risk of ventricular fibrillation in the case of exposure to capacitor discharges, or to unidirectional single impulse currents of short duration.

Equation (13.12) allows the evaluation of the Joule Integral, which must be below the thresholds deemed likely to cause ventricular fibrillation. Conventional time–body current curves, which describe the probability of ventricular fibrillation in the case of impulse currents, have been elaborated for the left-hand-to-foot path.[3] These curves identify current–time zones of increasing hazard for persons (Fig. 13.5).

All the pairs (t_c, I_{Brms}) below curve A cause no ventricular fibrillation; between curves A and C, the risk of fibrillation is increasingly higher up to 50%; beyond curve C, the risk of fibrillation is greater than 50%.

To mitigate the risk of electrocution "bleeders," that is, resistors are connected in parallel to capacitors embedded in the equipment. Therefore, the static charge can be drained off in the specific time and to the harmless values as assigned by applicable standards, before any maintenance. For example, IEC 60335–1[4] establishes that 1 s after disconnection of appliances, the voltage between the pins of the plug must not exceed 34 V.

Example 13.1 To put things in perspective, let us calculate the Joule integral latent in a capacitor $C = 1\ \mu\mathrm{F}$, initially charged at $V_0 = 10^4$ V. We assume the person's body resistance $R_B = 1\ \mathrm{k}\Omega$ and a contact duration equal to $3\tau = 3$ ms.

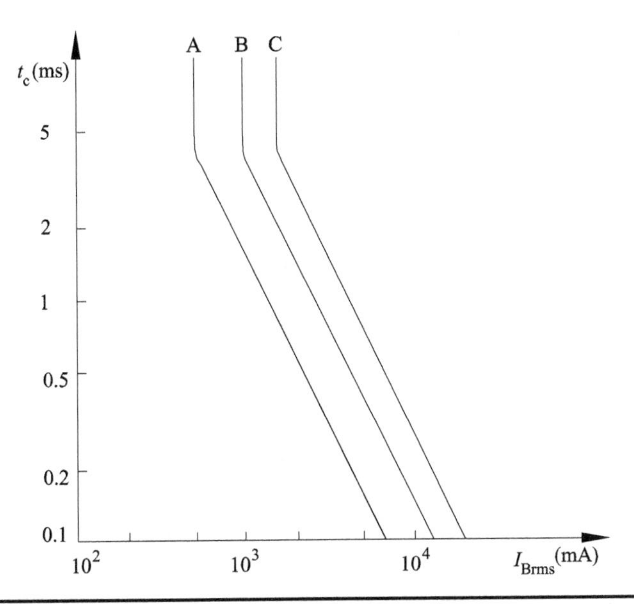

FIGURE 13.5 Impulse current thresholds for ventricular fibrillation as function of the contact duration (left-hand-to-foot path).

Solution The initial energy W_C stored in the capacitor is:

$$W_C = \frac{1}{2}CV_0^2 = 50 \text{ J}.$$

The peak of the impulse current is given by:

$$I_{peak} = \frac{V_0}{R_B} = \frac{10^4}{10^3} = 10 \text{ A}$$

From Eq. (13.10), we calculate the root mean square value of the impulse current circulating through the body:

$$I_{Brms} = \frac{10}{\sqrt{6}} = 4000 \text{ mA}$$

From Eq. (13.11), we calculate the value of the Joule integral:

$$\int_0^{3R_BC} i^2(t)\,dt = I_{Brms}^2\,(3R_BC) = 48 \times 10^{-3} \text{ A}^2\text{s}$$

The calculated value of the body current of 4000 mA, sustained for a contact time of 3 ms, causes ventricular fibrillation with a probability greater than 50%, as shown in the chart in Fig. 13.5.

Endnotes

1. IEEE Std. 142–1991 "*Recommended Practice for Grounding of Industrial and Commercial Power Systems.*"

2. Many solid substances (e.g., coal, grain, sugar, etc.), normally not explosive, may become explosive if reduced to fine dust with particles not exceeding the size of 420 μm (NFPA Journal, Nov/Dec 2008).

3. Italian Standard CEI 64–4985R "*Effects of Current Passing Through the Human Body,*" 1999–2001.

4. IEC 60335–1 "*Household and Similar Electrical Appliances—Safety, Part 1: General Requirements.*" Ed. 4.1, 2004–07.

CHAPTER 14

Testing the Electrical Safety

Zeal without knowledge is fire without light.
THOMAS FULLER (1608–1661)

14.1 Introduction

The parameters that ensure electrical safety in installations can, and must, be tested to positively assure that the risk of electric shock is below the limit deemed acceptable by the electrical design.

Some measurements must be carried out during the design phase (e.g., soil resistivity test), in order to have objective input data, and for others (e.g., earth resistance test) after the installation of the electrical system, and prior to being put into service. This allows a comparison between test results and the relevant design data, which must be in possession of the verifier.

Electrical safety obtained due to the proper deployment of protective measures against direct and indirect contact tends to decrease in time due to aging of the electrical system; therefore, cyclic testing, together with preventive maintenance, are necessary to assess, and eventually restore, the level of safety. Test procedures, of course, must not endanger persons or damage the property, and must be carried out with procedures and instruments in compliance with relevant technical standards.

14.2 Soil Resistivity Measurement

As we already know, the resistivity of soils play a crucial role in determining the performance of ground electrodes, as it is a major factor in influencing their resistance to earth. For this reason, prior to designing

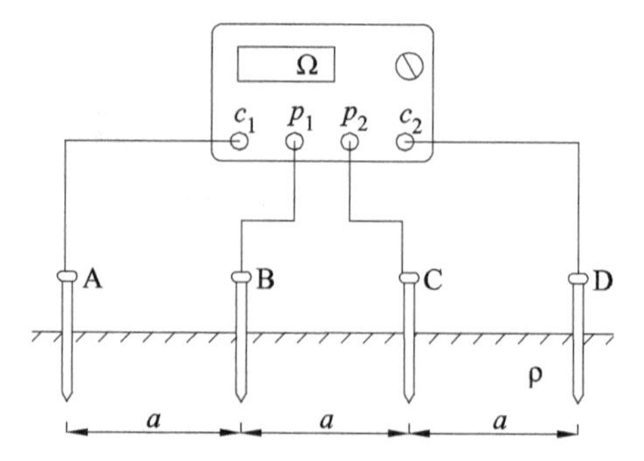

FIGURE 14.1 The Wenner method.

earthing systems, the actual site of its installation should be tested in order to obtain this very important parameter.

The Wenner method is probably the most widely used technique to create soil models. With this method, four small rods are driven into the earth in a straight line and uniformly spaced by the distance a. The rods are driven to a depth much smaller than a, so that we can consider them as "point" electrodes.

With reference to Fig. 14.1, a known constant current I is then injected between the outer rods A and D, and the resulting potential difference V_{BC} between the two inners pins B and C is measured.

Let us now consider the four rods as identical hemispherical electrodes separated by the distance a in a uniform soil of resistivity ρ. We will assume that the length of their radii is very small compared to a, so that the soil can be considered undisturbed by the presence of the hemispheres (Fig. 14.2).

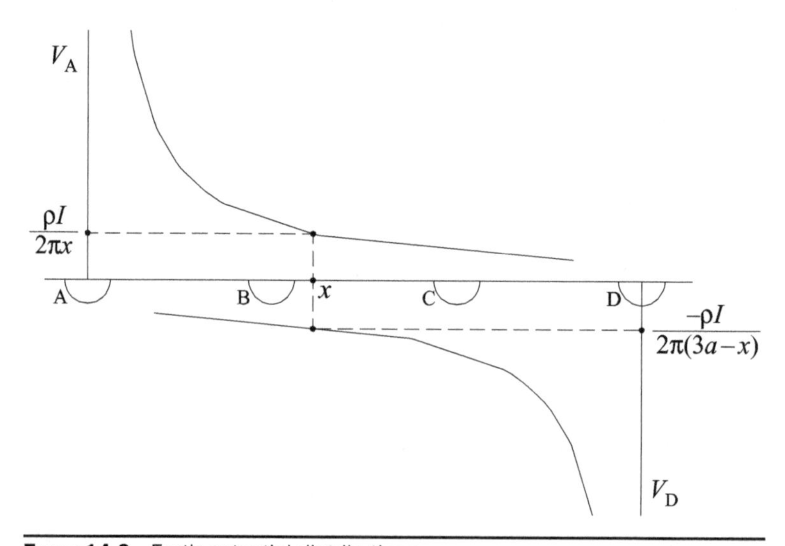

FIGURE 14.2 Earth potential distributions.

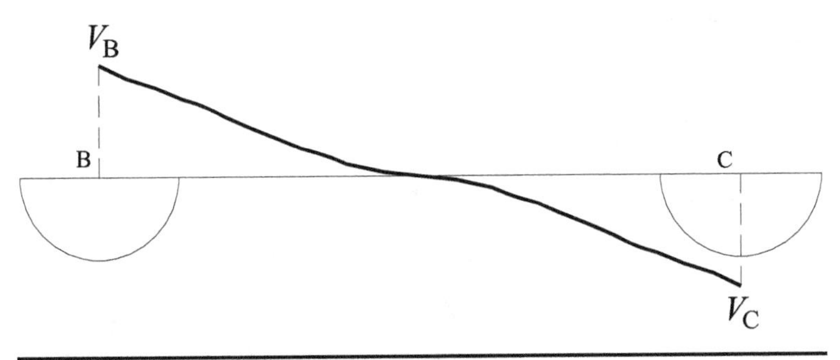

FIGURE 14.3 Total earth potential distribution.

Since the current I leaves the electrode A and enters the electrode D, supposed independent of each other, the earth potentials generated in the soil by A and D are identical but of opposite sign, as shown in Fig. 14.2. The value of the total earth potential at any value can be obtained by superposing the contributions due to the two electrodes. The result for x varying between the electrodes B and C is given in Eq. (14.1) and graphed in Fig. 14.3.

$$V(x) = V_A(x) + V_D(3a - x) = \frac{\rho I}{2\pi x} - \frac{\rho I}{2\pi(3a - x)} \tag{14.1}$$

As a consequence, the potential difference V_{BC} across the rods B and C measured by the tester is calculated by substituting the variable x in Eq. (14.1) with a and $2a$, as follows:

$$V_{BC} = V(a) - V(2a) = \frac{\rho I}{4\pi a} + \frac{\rho I}{4\pi a} = \frac{\rho I}{2\pi a} \tag{14.2}$$

Solving the above equation for the earth resistivity ρ, we obtain

$$\rho = \frac{2\pi a V_{BC}}{I} = 2\pi a R \tag{14.3}$$

where R is the value directly provided by the instrument, since I is known and V_{BC} is measured. Thus, the soil resistivity can be obtained by multiplying the reading of the tester by $2\pi a$.

If the soil is uniform, the above method will provide the same result, regardless of the separation distance a between the electrodes. In reality, the soil is typically not homogeneous, as it is composed of layers of different nature laid one upon the other; therefore, its resistivity changes with the depth. Thus, the resistivity that is measured can be considered as the average value found in the volume of soil shown in Fig. 14.4, as per theoretical considerations not herein reported.

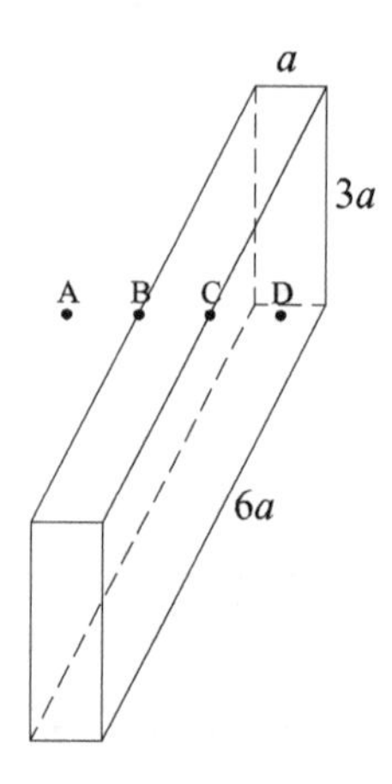

FIGURE 14.4 The resistivity that is measured can be considered as the average value found in the volume of soil of volume $18a^3$.

It is apparent that by increasing/decreasing a it is possible to "prospect" the soil at various depths, thereby allowing the determination of a multilayer soil model in terms of its resistivity.

14.3 Earth Resistance Measurement

Recall from Eq. (4.6) that the earth potential V_G depends, among other parameters, on the total ground resistance R_G of the electrode system (which, in turn, depends on the soil resistivity).

The voltage exposure upon ground faults is, therefore, dependent on R_G, whose value must be investigated after the system has been installed to assure its correspondence with the design data.

The method of the *fall of potential* (also referred to as *3-point measurement*), which is based on Ohm's law, can be employed to determine R_G (Fig. 14.5).

With this method, an a.c. current I is injected into the soil between the electrode X under test and the auxiliary current electrode Z, and is measured by the ammeter A. As discussed in the previous section, because of the circulation of this current, an earth potential between the outer electrodes will be originated. The earth potential V_Z of the auxiliary current electrode is generally greater than V_T, as Z is usually a rod of small dimensions, while the electrode under test may be an entire grounding system. For this reason, V_Z may reach dangerous potentials, and therefore must be kept inaccessible to persons during the test.

The potential difference V_{XY} between X and Y is measured by the voltmeter V.[1] By applying the Ohm's law, the earth resistance R_G is given by the ratio of V_{XY} to I, which is automatically calculated by the tester.

The precision of this test depends on the mutual position of potential and current rods with respect to the electrode under test. The

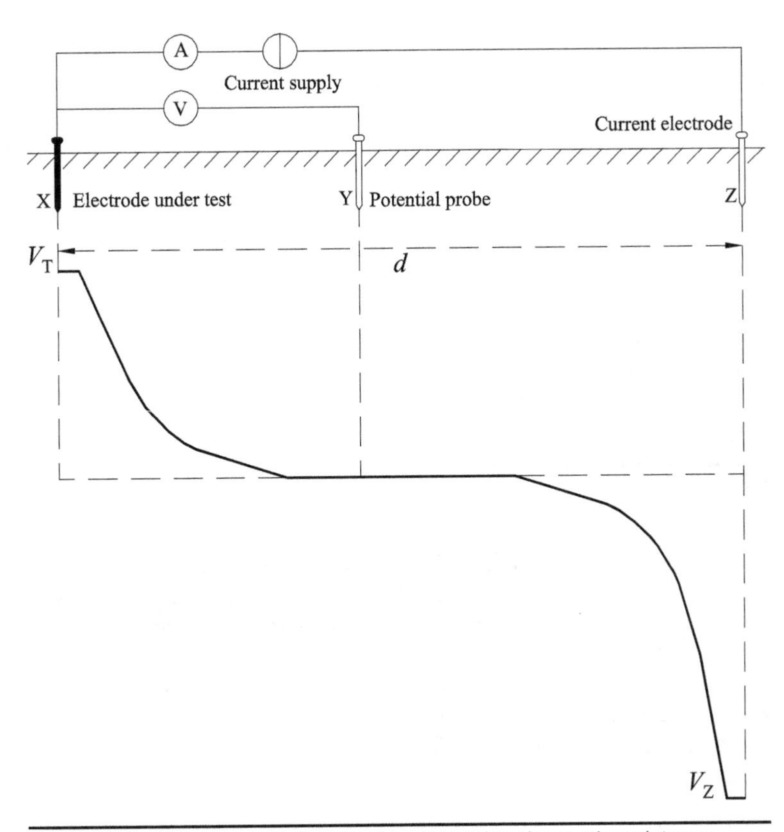

FIGURE 14.5 Fall of potential method to determine the earth resistance.

current electrode Z must be away from X by a distance *d* large enough so that they are independent of each other, and the total earth potential curve will be characterized by a flat region. Based on what was studied in Chap. 4, if the earth electrode under test is a rod, the independency will be guaranteed if *d* is at least five times the rod's length. If the electrode being examined has a more complex structure (e.g., an earthing grid), the minimum distance to be considered is five times its maximum diagonal, or five times the diameter of a circle of equivalent area.

The potential probe Y must be driven in a point at zero potential (flat portion of the earth potential distribution in Fig. 14.5), that is, outside of both the influence areas of X and Z. Erroneous values for R_G will be obtained if Y is placed too close to X, or too far from X.

For example, let us consider Fig. 14.6 where Y is too close to X. The potential difference measured by the tester is lower than V_T, thus the instrument will return an incorrect lower value for R_G. If Y and X coincided, the tester would measure zero.

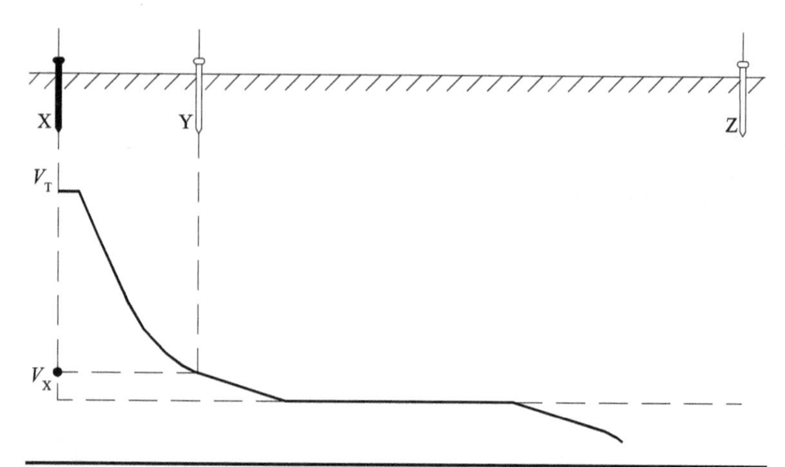

Figure 14.6 The potential probe Y is placed too close to X.

On the other hand, if Y and Z coincided, the tester would provide the summation of the earth resistances of both the electrode under test and the auxiliary electrode, hence presenting an incorrect higher value for R_G.

To locate the point at zero potential, the operator must take successive readings after moving the potential probe Y toward Z. The flat region of the earth potential is found when subsequent readings of R_G do not appreciably change.

14.4 Earth Resistance Measurements in Industrial Facilities

In industrial facilities generally characterized by grounding grids, the earth resistance measurement can be facilitated in the presence of metal bodies embedded in the earth (e.g., EXCPs such as underground cold water pipes). If such metalwork extends well beyond the facility's area and has an estimated negligible resistance-to-ground (i.e., $R_{EXCP} \ll R_G$), it can be used at the same time both as a current and as a potential electrode (Fig. 14.7).

The tester will measure the sum of the earth resistances of both the electrode under test and the EXCP, which, in the above assumptions, will practically coincide with the resistance-to-ground of the electrode under test. Alternatively, the low-voltage neutral of a neighboring utility substation, or even of a dwelling unit, can be used as a simultaneous current and potential electrode. In this case, the neutral must be checked to assure its de-energization.

Grounding grids in industrial facilities may have a very large extension (e.g., perimeters of several hundreds of meters). Therefore,

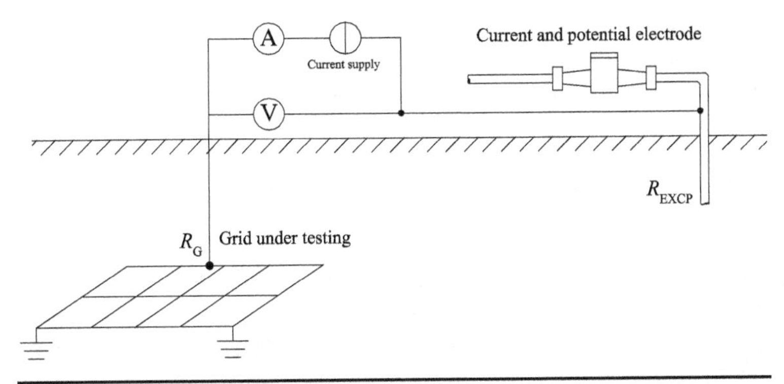

FIGURE 14.7 Current and potential electrodes coincide with an EXCP of negligible resistance-to-ground.

placing the auxiliary current electrode at a distance of five times the grid's maximum diagonal (e.g., distance of the order of kilometers) may be challenging, especially in urban areas.

In such cases, the grounding grid of the power station feeding the facility (Fig. 14.8) might conveniently constitute the current electrode.

The test current (order of tens of amperes) is injected through a de-energized power conductor running between the station and the facility.

In overhead distribution lines, an overhead ground wire is usually present as a protection against lightning strikes. As previously seen, users can conveniently use this overhead wire, or suitable metal armor/sheath of the incoming medium voltage cables, as a conductor to connect their earthing grid to the utility's grid. This connection

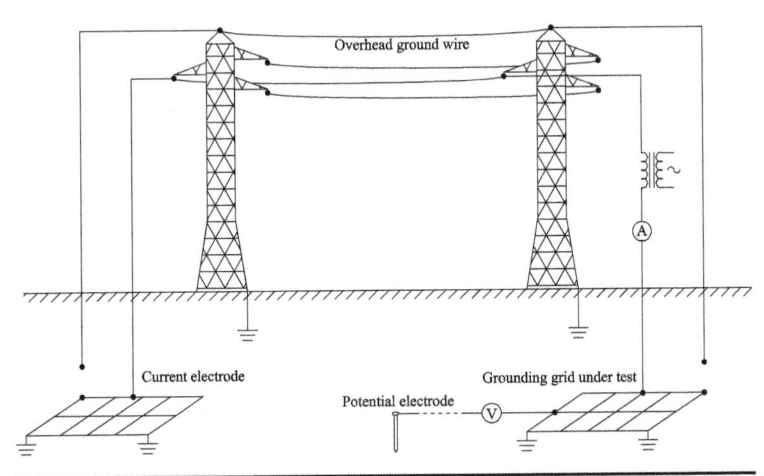

FIGURE 14.8 Grounding grid of the power station feeding the facility as the current electrode.

is effective in preventing parts of fault currents from flowing through ground, thereby lowering touch and step voltages at the user's premises. During the measurement, this connection must be removed at both ends, so that the entire test current will return to the source through the earth, allowing the proper determination of the unknown ground grid resistance.

14.5 Earth Resistance Measurement in TT Systems

As discussed in Chap. 6, the fault-loop in TT systems includes both the ground resistance R_G of the consumer's electrode and the utility's electrode earth resistance R_N, which are independent of each other (Fig. 6.1). In the assumption that $R_G \gg R_N$, which holds true especially in urban areas, a simplified method to measure the earth resistance of the user's electrode may be adopted.

With the aforementioned method, which does not require any auxiliary electrode, the total fault-loop resistance is measured, and the result will basically coincide with R_G (Fig. 14.9).

The voltmeter V in Fig. 14.9 measures the phase voltage V_{ph} (switch in open position) and then after closing the switch the fall of potential V_R across the resistance of the potentiometer R, of known value, caused by the test current I_G. I_G must not exceed the operating threshold of the RCD in order to prevent this protective device from tripping during the test.

Thus, by applying Ohm's law, we obtain

$$V_R = \frac{V_{ph}}{R_N + R_G + R} \times R \tag{14.4}$$

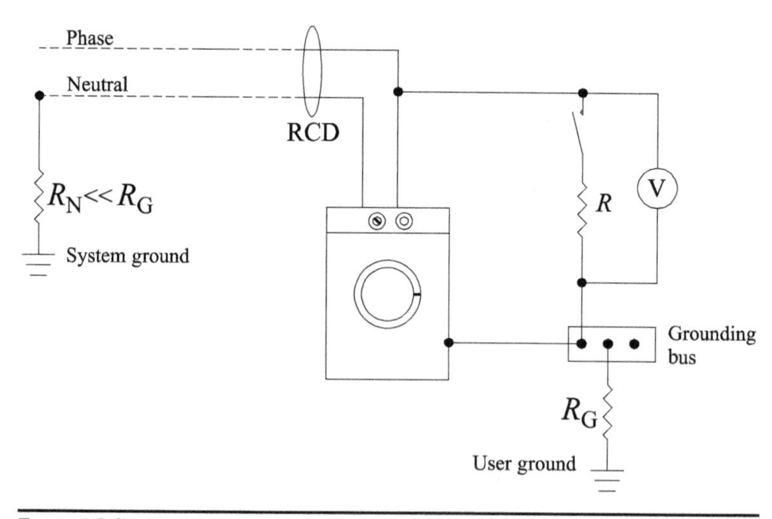

FIGURE 14.9 Fault-loop resistance test in TT systems.

Hence,

$$R_N + R_G \cong R_G = R\left(\frac{V_{ph}}{V_R} - 1\right) \tag{14.5}$$

R_G is automatically provided by the tester because all the quantities at the right-hand side of Eq. (14.5) are known.

It is important to note that during the automatic operations of the tester, all the enclosures will become energized at the voltage $R_G I_G$. The duration of such energization must not exceed the maximum permissible times given in Table 6.1.

As examined in Chap. 6, in TT systems RCDs are the most effective way to protect against indirect contact, as long as the following equation, already studied in Chap. 6, which ties together residual operating currents and earth resistance R_G, is fulfilled:

$$R_G \leq \frac{50\ \text{V}}{I_{dn}} \tag{14.6}$$

Equation (14.6) calls for the maximum value of 50 V as the touch voltage.

The correct operation of the RCD, within the times as per the chart in Fig. 2.6, as well as the fulfillment of Eq. (14.6) can, and must, be instrumentally verified.

With reference to Fig. 14.9, after closing the switch, the resistance of the potentiometer R, which connects the phase conductor to ground, is decreased until the RCD operates. The ammeter A will measure the actual ground current I_d in correspondence with the tripping of the RCD, thereby allowing the verification that the operating threshold I_{dn} of the device (e.g., 30 mA) is not exceeded. The voltmeter V measures the phase voltage V_{ph} (switch in open position) and then, after closing the switch, the voltage drops V_R across the known resistance R in correspondence with the tripping. By assuming R_N negligible with respect to R_G, we can write

$$V_{ph} - V_R \cong V_G = R_G I_d \tag{14.7}$$

Hence, by imposing the safety condition of Eq. (14.6), we will obtain

$$V_{ph} - \frac{50}{I_{dn}} I_d \leq V_R \tag{14.8}$$

Thus, the system is protected against indirect contact if the reading of the voltmeter exceeds the right-hand side of the previous inequality.

The tester will also automatically measure the tripping time in correspondence with I_{dn}, which must not exceed the permissible clearing times of 300 ms, as given in Fig. 2.6.

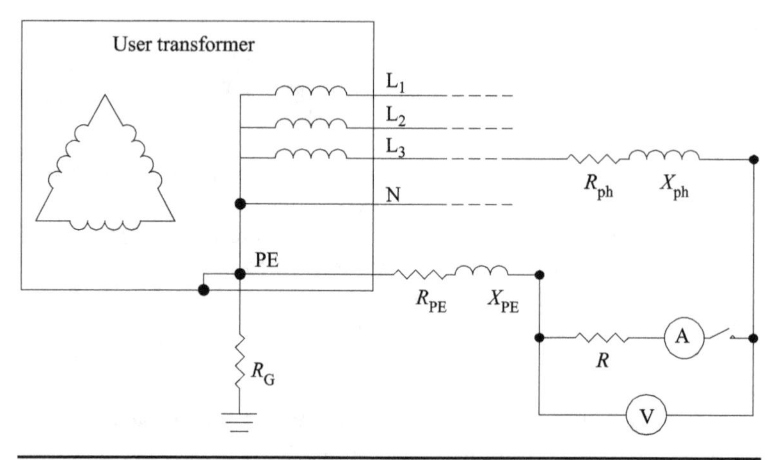

FIGURE 14.10 Test circuit to measure the fault-loop impedance in energized TN systems.

14.6 Measurement of the Fault-Loop Impedance in TN Systems

As discussed in Chap. 7, in TN systems, under low-voltage ground-fault conditions, $\underline{Z}_{\text{Loop}}$ is constituted by the series of the following impedances: phase conductor, protective conductor (PE), and secondary winding of the transformer. The magnitude of $\underline{Z}_{\text{Loop}}$ must be measured at the farthest point of each circuit being protected (e.g., at receptacles) to verify that it matches the value calculated in the design phase.

In Fig. 14.10, a diagrammatic representation of a typical loop-tester and the test circuit is provided.

As in the previous test arrangements, the voltmeter V will measure both the phase voltage V_{ph} when the switch is in the open position, and, after closing it, the voltage drops V caused by the test current I across the resistance R of known value.

If we neglect the impedance of the secondary winding of the transformer, the fault-loop impedance can be written as: $\underline{Z}_{\text{Loop}} = (R_{\text{ph}} + R_{\text{PE}}) + j(X_{\text{ph}} + X_{\text{PE}}) = R_{\text{Loop}} + jX_{\text{Loop}}$. As it is known, the underline quantities conventionally represent complex numbers, also referred to as phasors.[2]

Thus, the theoretical value of $\underline{V}$ is given by

$$\underline{V} = R\underline{I} = R \times \frac{\underline{V}_{\text{ph}}}{R + \underline{Z}_{\text{Loop}}} \tag{14.9}$$

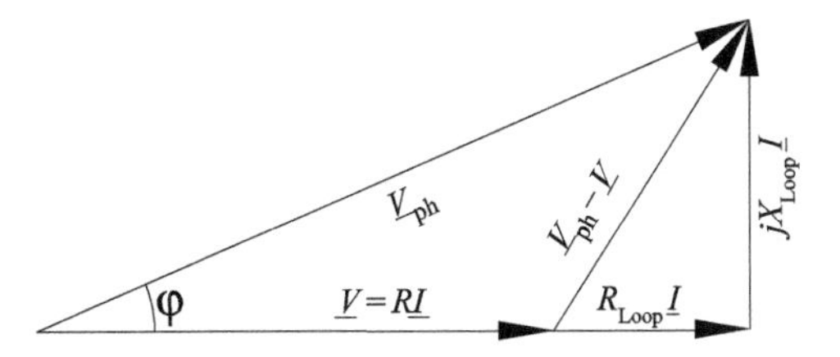

FIGURE 14.11 Phasor diagram of voltages and currents of the fault-loop test circuit in TN systems.

Thus, by solving Eq. (14.9) for $\underline{Z}_{Loop}$, we obtain

$$\underline{Z}_{Loop} = R \times \frac{\underline{V}_{ph} - \underline{V}}{\underline{V}} = \frac{\underline{V}_{ph} - \underline{V}}{\underline{I}} \tag{14.10}$$

where $\underline{I}$ is the test current flowing through R.

The above Eq. (14.10) is a complex number in which voltages and currents are also symbolized via phasors. Each phasor has a magnitude (i.e., the r.m.s. values of the quantity) and an angular phase (i.e., the argument of the complex number), and both must be considered in the determination of $\underline{Z}_{Loop}$.

To clarify this concept, let us represent all the phasor quantities in Eq. (14.10) by applying the Kirchhoff's voltage law to the fault-loop, as follows:

$$\underline{V}_{ph} - R\underline{I} = \underline{Z}_{Loop}\underline{I} = R_{Loop}\underline{I} + jX_{Loop}\underline{I} \tag{14.11}$$

The phasor diagram in Fig. 14.11 graphically represents Eq. (14.11).

It can be noted that $\underline{V}_{ph}$ and $\underline{V}$ are not in phase, but displaced by the angle φ. Thus, the loop-tester must return the following value:

$$|\underline{Z}_{Loop}| = \frac{|\underline{V}_{ph} - \underline{V}|}{|\underline{I}|} \tag{14.12}$$

where the numerator is the magnitude of the vectorial difference between the complex numbers representing the voltages, and the denominator represents the magnitude of the current phasor.

If X_{Loop} were negligible with respect to the resistance (e.g., $X_{Loop} \leq 0.1|Z_{Loop}|$), the fault-loop would essentially be resistive and the test might just assess the fault-loop resistance. This may happen when the fault-loop is made of conductors having small cross-sectional area (i.e., $S \leq 95\ \text{mm}^2$), or the circuit is far from large transformers or generators. $\underline{V}_{ph}$ and $\underline{V}$ would then be practically in phase with each other, thereby, allowing the simplification of Eq. (14.12) as follows:

$$|\underline{Z}_{Loop}| \cong R_{Loop} = \frac{|\underline{V}_{ph}| - |\underline{V}|}{|\underline{I}|} \tag{14.13}$$

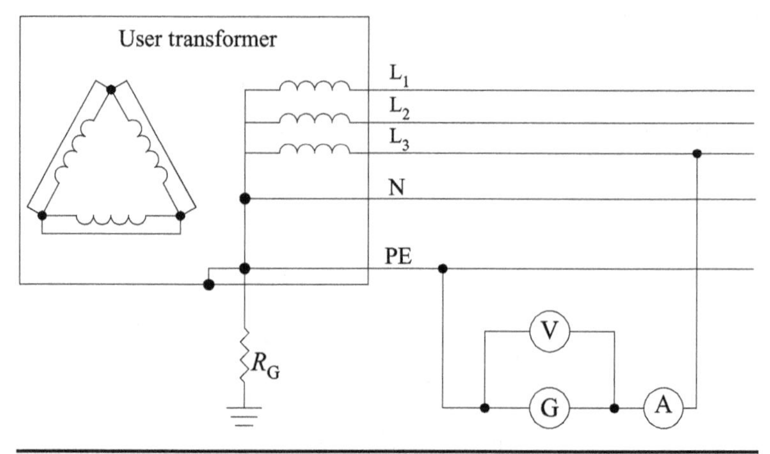

Figure 14.12 Test circuit to measure the fault-loop impedance in de-energized TN systems.

The numerator represents the much simpler algebraic difference between the magnitudes of the two voltages.

In the case of a prevalently resistive fault-loop, ohmmeters, performing the measurement as per Eq. (14.13), may be used for the test. If the loop is also reactive, for instance near large transformers (i.e., >100 kVA), loop-testers performing the test as per Eq. (14.12) should be employed in order to prevent errors in the measurement.

A more rigorous method of measurement of the fault-loop impedance employs loop-testers with a built-in independent generator G at the power system frequency. In this case, one must disconnect the supply to the primary side of the transformer and short circuit its windings (Fig. 14.12).

The fault-loop impedance is directly given by the ratio of the readings of voltage and current. A major drawback of this method is the necessity to put the substation out of service.

14.7 Touch Voltage Measurement in TN Systems (Low-Voltage Earth Faults)

The measurement of the actual touch voltage for the worst-case scenario of hand-to-hand contacts occurring between ECPs (e.g., low-voltage panels) and EXCPs (e.g., cold water pipes) can be carried out with the method of the fall of potential (Fig. 14.13). If the measurement's result does not exceed 50 V, the installation can still be considered safe against indirect contact, even if $|\underline{Z}_{\mathrm{Loop}}|$ is not in compliance with Eq. (7.5).

With this arrangement, the test current $\underline{I}$ will circulate through the protective conductor and the loop impedance will be measured in

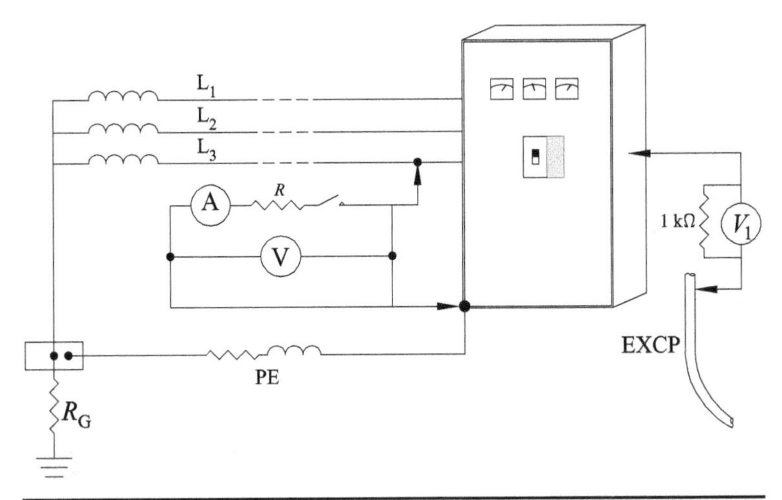

Figure 14.13 Test circuit to measure touch voltages in TN systems.

the same way as discussed in Sec. 14.6. At the same time, the voltage between the ECP and the EXCP is measured by the voltmeter V_1. As already explained in Chap. 4, the voltmeter V_1 has a 1-kΩ resistance connected in parallel to its leads to simulate the conventional body resistance of a standard person.

By knowing $|\underline{Z}_{\text{Loop}}|$, the actual fault current $|\underline{I}_{\text{G}}|$, which is greater than $|\underline{I}|$, can be so determined:

$$|\underline{I}_{\text{G}}| = \frac{|\underline{V}_{\text{ph}}|}{|\underline{Z}_{\text{Loop}}|} \tag{14.14}$$

Assuming a linear relationship between currents and voltages, the magnitude of the touch voltage V_{T} at the measurement point can be calculated as:

$$V_{\text{T}} = V_1 \frac{|\underline{I}_{\text{G}}|}{|\underline{I}|} \tag{14.15}$$

14.8 Step and Touch Voltage Measurements in TN Systems

As already substantiated in Chap. 7, ground faults that occur on the primary side of substations may cause circulation of high currents in their earthing systems, thereby causing touch and step voltages. The magnitude of such quantities must be ascertained via field tests and compared against values considered safe by applicable standards.

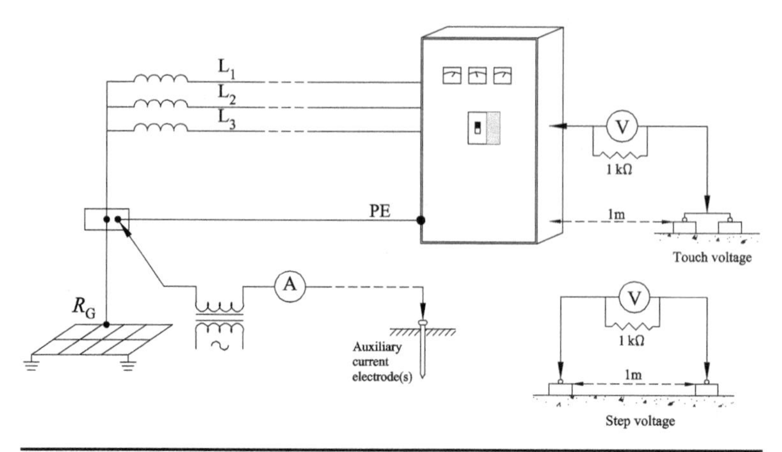

Figure 14.14 Test circuit to measure touch and step voltages in TN systems.

The touch and step voltage measurements are carried out by simulating ground faults by means of a test current I_t of low magnitude, but of at least 1% of the actual ground-fault current I_G impressed across the facility grounding system R_G and auxiliary current electrode(s). The auxiliary current electrode(s) will be buried at a sufficient distance from R_G to be considered independent of it (Fig. 14.14).

For the touch voltage measurement, a voltmeter will measure the potential differences between any accessible ECPs/EXCPs and two electrodes with a contact surface of 200 cm^2, pressed against the soil with a force of 250 N and placed at the distance of 1 m from each other. For the step voltage measurement, a voltmeter will measure the potential difference across the same two test electrodes in various locations within the facility.

The test electrodes must be placed radially with respect to the grid, so that they will not sit on the equipotential lines of the electric field, thereby measuring zero potential difference.

Also in this case, we assume a linear relationship between test current and touch and step potentials and, therefore, obtain their values in correspondence of the actual fault current, as follows:

$$V_{T,S} = V_1 \frac{I_G}{I_t} \tag{14.16}$$

where V_1 is the reading of the voltmeter.

The reading V_1, though, may include the unavoidable and unwanted contribution of "stray" potentials existing across the earth and not produced by the test circuit. The subsequent disturbance voltage $\underline{V}_d$ vectorially adds itself to the true values of touch and step potentials

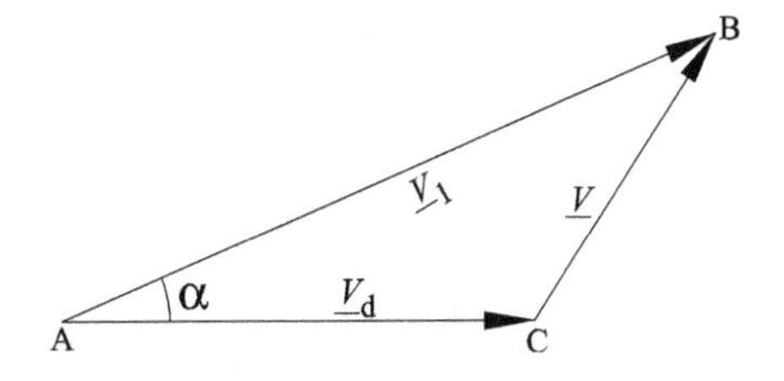

Figure 14.15 Erroneous reading $\underline{V}_1$ due to the disturbance voltage $\underline{V}_\mathrm{d}$.

$\underline{V}$, as follows:

$$\underline{V}_1 = \underline{V}_\mathrm{d} + \underline{V} \tag{14.17}$$

Figure 14.15 shows the phasor diagram of Eq. (14.17).

The voltmeter then returns an erroneous result, which may invalidate the measurement.

Thus, in order to determine the correct values for touch and step potentials, a set of three measurements should be performed:

1. Measurement of the r.m.s. value of $\underline{V}_\mathrm{d}$, obtained in the absence of the test current.
2. Measurement of the r.m.s. value of $\underline{V}_1$, by impressing the test current I_t.
3. Measurement of the r.m.s. value of $\underline{V}_2$, by impressing the test current I_t of same intensity as in the previous measurement, but with inverted polarity.

The measurement with inverted polarity is described by the vector diagram in Fig. 14.16.

We assume that the disturbance remains constant during the two measurements, therefore $\underline{V}_\mathrm{d}$ in Figs. 14.16 and 14.17 will have the same length. However, $\underline{V}$, although having a different angular displacement with reference to $\underline{V}_\mathrm{d}$, will not change its magnitude even if we change the direction of the test current.

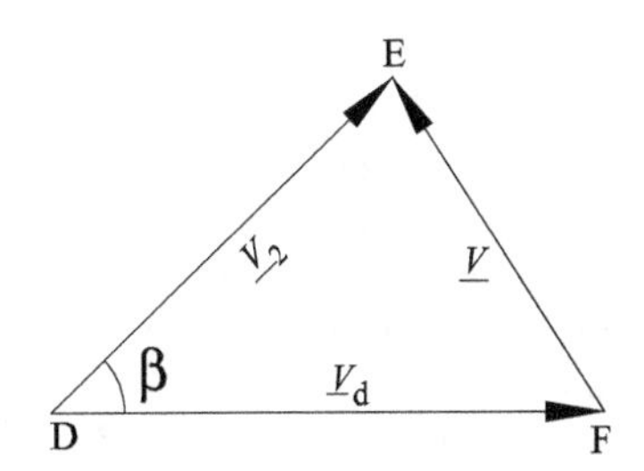

Figure 14.16 Erroneous reading $\underline{V}_2$ (test current with inverted polarity) due to the disturbance voltage $\underline{V}_\mathrm{d}$.

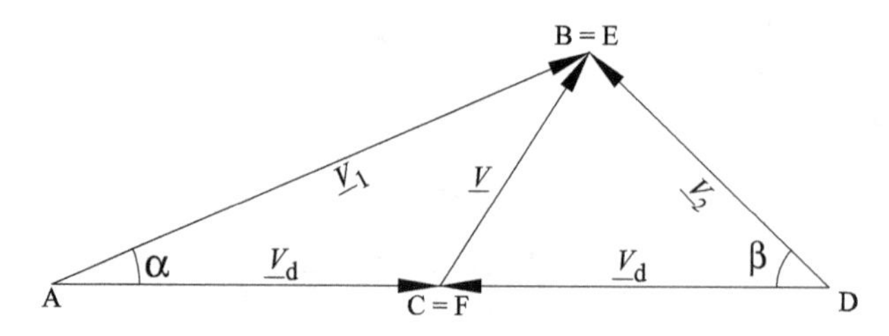

FIGURE 14.17 Triangle obtained by composing together the phasor diagrams of Figs. 14.16 and 14.17.

By composing together the two previous phasor diagrams, which have in common the magnitudes of $\underline{V}_d$ and $\underline{V}$, we obtain the following triangle, whose base equals $2|\underline{V}_d|$.

$|\underline{V}|$ will coincide with the triangle's median, whose length is given by

$$V = \sqrt{\frac{V_1^2}{2} + \frac{V_2^2}{2} - V_d^2} \tag{14.18}$$

Equation (14.18) provides the correct value that should be used in Eq. (14.16).

14.9 Fundamental Measurements in IT Systems

As examined in Chap. 9, under first-fault conditions, safety against indirect contact in IT systems is assured if

$$R_G \leq \frac{50}{I_G} \tag{14.19}$$

where R_G is the earth resistance of the grounding system and I_G is the first-fault current to ground.

To guarantee the fulfillment of Eq. (14.18), I_G must be measured. A typical testing circuit using a clamp-on current meter is shown in Fig. 14.18.

During this measurement, a line conductor is gradually connected to earth via a rheostat[3] R. When the rheostat is fully disengaged (i.e., zero resistance), the clamp-on current meter reads the first-fault current. The presence of the rheostat is advisable in order to prevent the inception of a short circuit, in the eventuality that a nonresolved first fault involving another phase conductor should still exist in the system at the time of the test.

As previously explained in Chap. 9, under first-fault conditions, the IT system "evolves" into a TT (ECPs earthed individually, or in

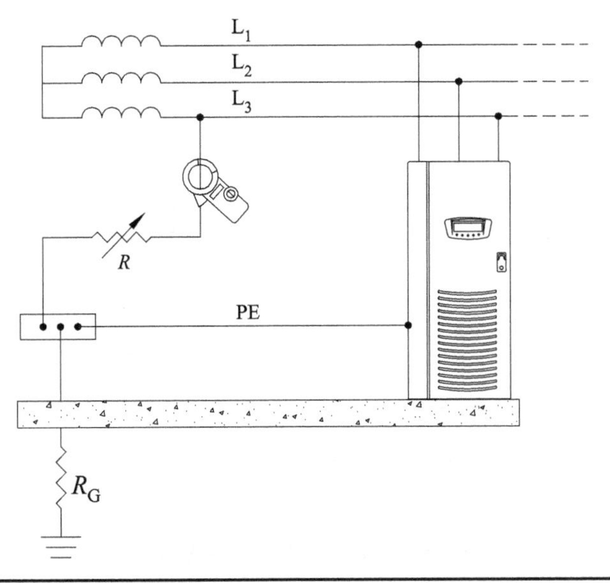

Figure 14.18 Test circuit to measure the first-fault current with clamp-on current probe in IT systems.

groups) or into a TN (ECPs collectively earthed). In the first case, the fulfillment of Eq. (9.22) must be verified by measuring the earth resistance of the local grounding system, as explained in Sec. 14.5.

In the most common case of ECPs collectively earthed, the fulfillment of Eq. (9.24) (i.e., neutral wire not shipped to loads) or Eq. (9.26) (i.e., neutral wire shipped to loads) must be verified. In either case, the fault-loop impedances Z_S and Z'_S must be measured.

A typical testing arrangement for IT systems with no neutral conductor is shown in Fig. 14.19.

For the duration of the test, a temporary connection between the neutral point and the earth must be carefully realized in order to close the fault-loop otherwise open by definition.

14.10 Protective Conductor Continuity Test

As already substantiated in previous chapters, protection against indirect contact for Class I equipment requires both the basic insulation of live parts and an effective link to the means of earthing, via the protective conductor (PE) in the connecting cable and plug, as well as in the socket outlet. In addition, main (MEB) and supplementary (SB) bonding conductors must be present to ensure equipotentiality between ECPs and EXCPs in fault conditions.

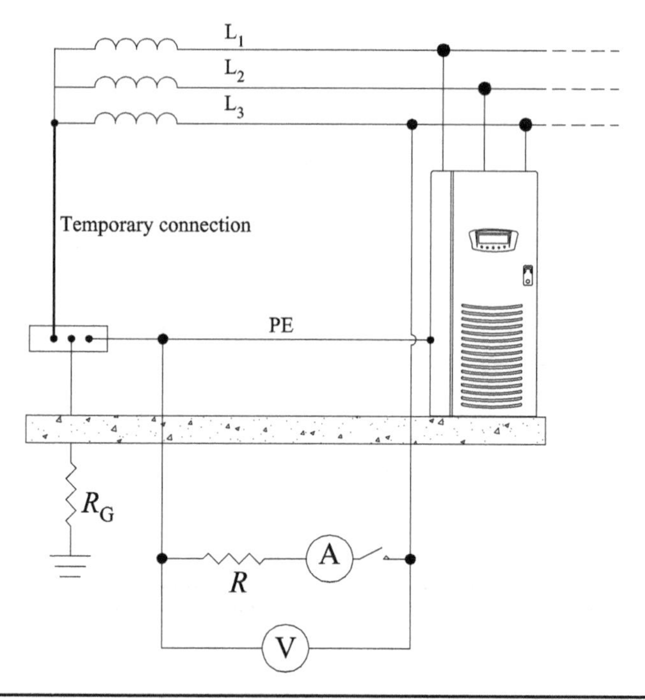

Figure 14.19 Test circuit to measure Z_S in IT systems with no neutral.

The continuity of PEs between all the ECPs and the ground bus, and the continuity of MEBs and SBs between all the EXCPs and the ground bus must be tested (Fig. 14.20).

The continuity test is not aimed at measuring the resistance of the PE, but only its electrical integrity. Thus, an ohmmeter equipped with a generator with a no-load voltage ranging between 4 and 24 V will inject a current (a.c. or d.c.) of at least 200 mA between the aforementioned metal parts. This test is also referred to as the *soft* test. A low value for the PE's resistance will indicate its electrical continuity.

14.11 Insulation Resistance Test

High insulation levels between circuits are of paramount importance in preventing faults. During the normal operation of the electrical system, though, thermal and mechanical stresses, as well as the aging of equipment, can decrease the effectiveness of such insulation, making the system prone to faults. Verifications, therefore, must be carried out.

An ohmmeter will determine the resistance existing between circuit conductors, including the neutral, connected together or taken

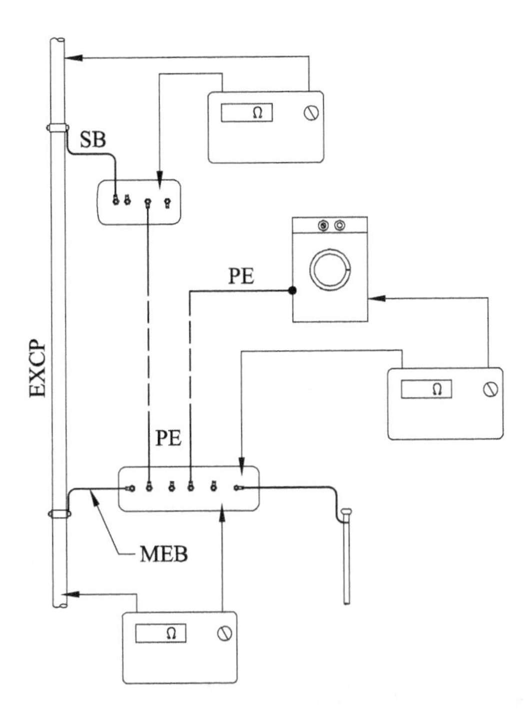

FIGURE 14.20 Protective conductor continuity test.

singularly, and the earth, by applying d.c. test voltages, for up to a minute, of the magnitudes shown in Table 14.1.[4]

We need to employ d.c. voltages as opposed to a.c., in order to prevent eventual capacitive reactances-to-ground from being taken into account during the measurement. We want to determine, in fact, the resistance of the insulation, and not its impedance. At 0 Hz such reactances are open circuits, and, therefore, will have no influence on the measurement.

It is clear that in the presence of voltages of these magnitudes, precautions should be taken to safeguard both the person's safety and the functionality of the equipment. For this purpose, the supply

Nominal Voltage of Circuit (V a.c.)	Test Voltage (V d.c.)	Minimum Insulation Resistance (MΩ)
SELV and PELV	250	0.25
Low-voltage circuits up to 500 V	500	0.5
Low-voltage circuits above 500 V	1000	1

TABLE 14.1 Test Voltages and Minimum Insulation Resistances

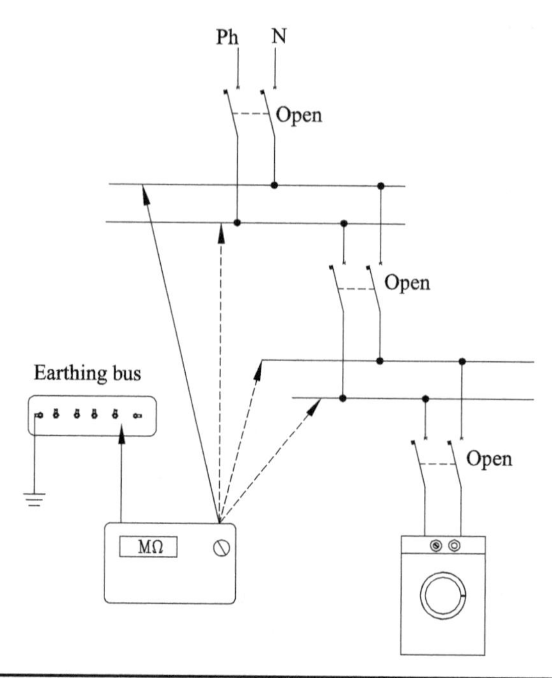

FIGURE 14.21 Insulation resistance test.

will be turned off by opening circuit breakers, as well as loads will be disconnected (Fig. 14.21).

Endnotes

1. The voltmeter must have very high internal resistance so as to drain a negligible test current.
2. See App. A for further details.
3. A rheostat is an adjustable resistor, which allows variation in its resistance without breaking the electrical circuit of which it is a part.
4. Italian Standard CEI 64–14, 2d ed., 2007.

CHAPTER 15

Applications of Electrical Safety in Special Locations and Installations

I studied with the masters long ago,
And long ago did master all they know;
Here now the end and issue of it all,
From earth I came, and like the wind I go!

OMAR KHAYYAM

15.1 Introduction

In this final chapter we will analyze the safety requirements against indirect contact employed in special installations or locations, where environmental conditions may increase the risk of indirect contact defined in Eq. (3.5) as $r(t) = [1 - S(t)]k(t)v(t)$.

Following are examples of special locations and installations:

Locations where the presence of water or moisture decreases the resistance-to-ground of the person's body by lowering his/her skin resistance. This does increase the probability $v(t)$ that the superficial touch voltage appearing over a faulty equipment is harmful to the person.

Wet locations where the humidity can increase the failure rate of equipment by compromising the integrity of its basic insulation, thereby raising the probability $1-S(t)$ that the metal enclosure is energized.

Publicly exposed installations (e.g., lighting systems for public places, roads, etc.) with which persons may come in contact.

In the above installations and locations, additional or tighter requirements for safety must be met.

15.2 Electrical Safety in Marinas

Marinas are facilities for the mooring of pleasure craft[1] with fixed wharves. They are equipped with a.c. receptacles to feed the boats, installed in marine-style pedestals located as close as possible to the berth. As per IEC standards, the nominal supply voltage must not exceed 230 V single phase, or 400 V three phase. Mechanically supported flexible cables will directly connect the receptacles to the boats (Fig. 15.1).

The metal hull acts as a grounding electrode to the shipboard's electrical system, both during navigation, when it is a TN system, and when the vessel is berthed. The seawater, in fact, acts as the earth to a land-based installation.[2] Thus, an "earthing" conductor (EC) is employed to link the shipboard's main grounding bus to the hull, which, therefore, becomes an EXCP (i.e., at zero potential). Supplementary equipotential bonding connections may be additionally required between hull and onboard equipment to reduce further potential differences caused by faults.

Marinas are "unfriendly" environments, electrically speaking, as the person's body resistance because of salt and moisture is lower than the standard values shown in Fig. 5.15 in dry conditions. For this reason, international standards do not permit in marinas the protection against indirect contact by nonconducting location or earth-free local equipotential bonding; for the same reason, protection against direct

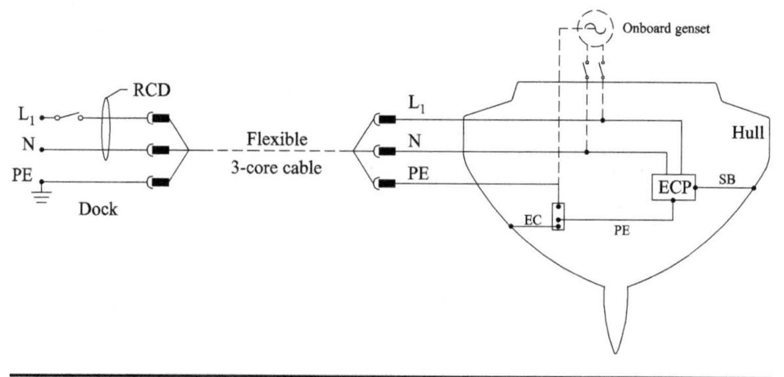

FIGURE 15.1 Direct single-phase connection of the vessel to the dockside outlet equipped with RCD.

contact cannot be carried out by obstacles or by placing live parts out of reach.

An effective protection against indirect contact can be obtained by disconnection of supply carried out by residual-current protective devices with a residual rating not exceeding 30 mA (Fig. 15.1). In this arrangement, a protective conductor (PE) connects the dockside earthing electrode to the boat's metal hull, the underwater gear, and/or the d.c. negative bus, allowing the RCD to trip upon both boat- and dock-originated ground faults.

A drawback of this arrangement is the energization of the hull under fault conditions because of its equipotentialization with the boat's ECPs. Consequently, across hull and land a potential difference arises, which causes part of the fault current to flow through the parallel path eventually constituted by the seawater. This potential difference is present during the RCD's clearing time and may be dangerous for swimmers. The residual device, in fact, is designed to operate in standard conditions (i.e., dry skin), and may not act fast enough if the person's body impedance is lowered due to wet conditions. In addition, if the boat is moored in fresh water, which is a poor conductor, nearly no stray current would circulate, unless swimmers, by entering the water, increase its conductivity and become themselves a return path to the source.

Other negative aspects concerning the presence of the PE between berth and boat is the possibility to trigger the electrolytic corrosion of the hull. The dockside earth electrode and the hull are, in fact, unlike materials immersed in electrolytes (i.e., seawater and earth). The two metals electrically connected by the protective conductor in the 3-core flexible cable, as shown in Fig. 15.1, constitute a galvanic cell, which causes circulation of direct current. If the hull is anodic to the earth electrode, corrosion will occur at its expenses. The same phenomenon may occur to vessels docked alongside one another when they plug in at the same pedestal. Their respective protective conductors electrically link the boats' hulls to each other and, if they are made of dissimilar metals, the less noble metal will corrode.

To prevent corrosion, protection against indirect contact can be carried out by electrical separation (Fig. 15.2).

The isolating transformer separates the shipboard electrical system from the shore supply and can be either shore-mounted or installed onboard. The effectiveness of the electrical separation is ensured by not connecting the protective conductor serving the vessel's loads to the grounding system ashore. As a consequence, since there is no longer a metal connection between hull and earthing electrode, corrosion cannot occur.

In addition, even if the hull becomes energized, there will be no circulation of fault current through the seawater due to the galvanic decoupling with evident benefit for the swimmers safety.

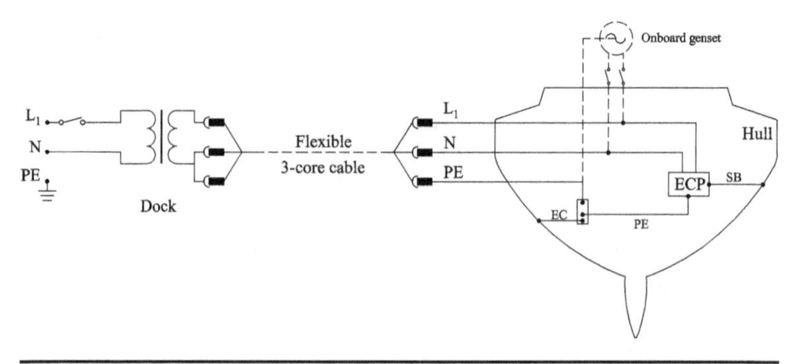

Figure 15.2 Connection of the vessel to a single-phase, shore-mounted isolating transformer.

15.3 Electrical Safety Requirements for Equipment Having High Protective Conductor Currents

Electronics equipment (e.g., computers, telecommunication equipment, etc.) may be sensitive to electromagnetic interferences irradiated by "disturbing" loads. For this reason, radio frequency input filters constituted by capacitors connected between supply conductors and enclosures of equipment are employed to enhance the electronic systems' immunity. Such filters may cause continuous leakage currents through the protective conductors in excess of 3.5 A (Fig. 15.3).

International standard IEC 60950–1[3] limits the maximum value of leakage currents of Class I, stationary, or pluggable, equipment to 5% of the input current. The presence of high currents on protective conductors might cause nuisance trippings of RCDs, when their residual settings are exceeded even in the absence of ground faults.

Serious hazard is caused by the accidental loss of the protective conductor serving the equipment (Fig. 15.4).

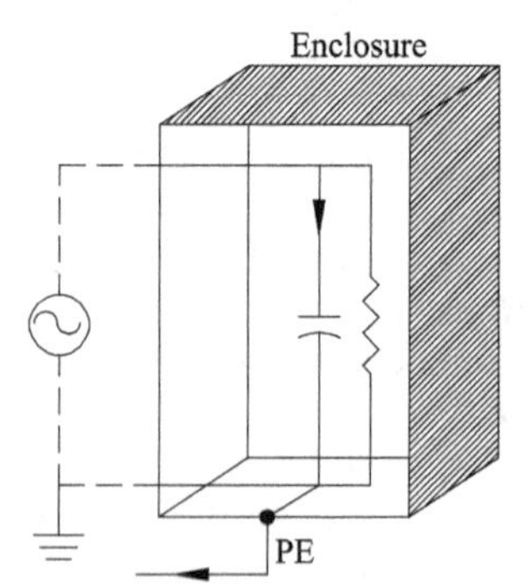

Figure 15.3 Earth leakage currents due to radio frequency filters.

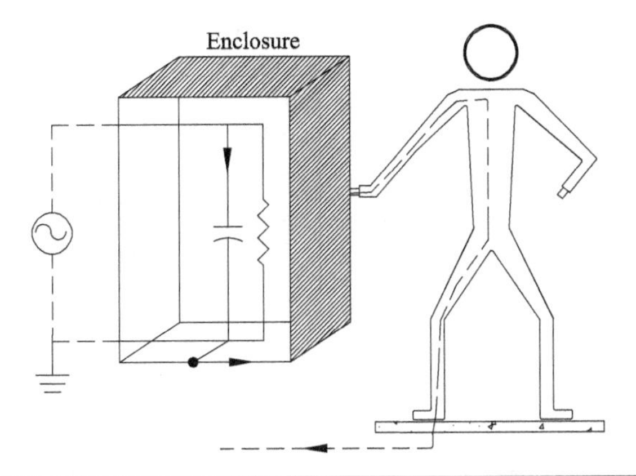

Figure 15.4 Hazardous situation caused by the accidental loss of the PE.

The leakage current, in fact, will circulate through the person's body if she/he comes into contact with the enclosure. Thus, for safety reasons, the bonding of high leakage current equipment must be assured by enhancing the reliability of the PE. This can be obtained by doubling its cross-sectional area, with respect to minimum permissible values, by using more than one PE conductor in parallel, and/or monitoring its electrical continuity.

If the leakage current exceeds 3.5 A, the aforementioned IEC standard requires a warning label to be affixed adjacent to the equipment power connection, indicating the necessity of connecting the protective conductor before switching on the supply.

To prevent electrical noise[4] from interfering with sensitive electronic equipment, manufacturers may require a dedicated grounding system. An erroneous interpretation of this condition may lead the designer to install in the same building one separate ground for the high-frequency electronic apparatus and one for the 50/60-Hz equipment (Fig. 15.5).

The above arrangement is extremely unsafe and must be avoided, because it does not assure equal potential between equipment under fault conditions. In the case of ground faults on either ECP, persons in simultaneous contact with both pieces of equipment are subject to the whole earth potential. Each ECP becomes, in fact, an EXCP to the other one.

In addition, separate earthing points, possibly energized at different potentials under fault conditions, can cause circulation of ground currents, and be the source of the electrical noise one wants to eliminate.

Another issue pertaining to continuous high leakage currents may be the inevitable nuisance trippings of protective residual current devices. In this situation, the general protection provided by RCDs

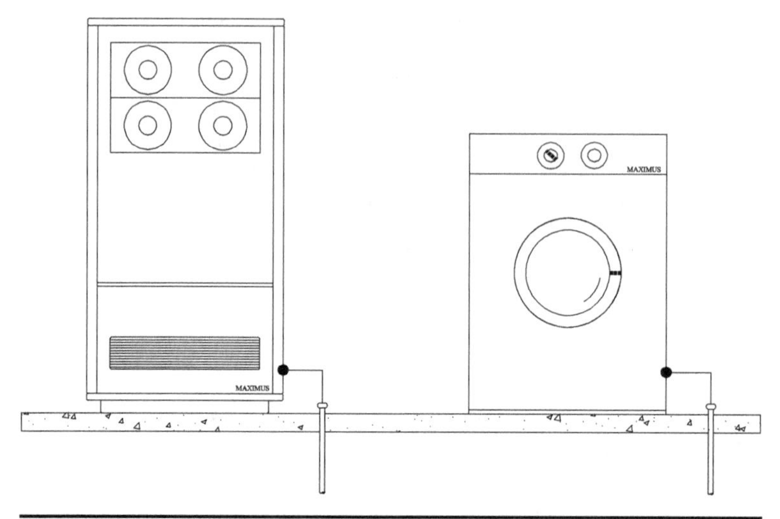

Figure 15.5 Schematic representation of distinct grounding electrodes for sensitive electronic equipment and for 50/60-Hz equipment.

becomes ineffective, circumstance extremely dangerous in TT systems. The solution to this problem might be to supply the "offender" equipment with a dedicated RCD with residual operating threshold set to a sufficiently high value. Alternatively, a separation transformer with RCDs on the primary side can be used to supply the load, as RCDs would not sense any leakage currents on its secondary side.

15.4 Electrical Safety in Train Stations

The typical traction electrification system is composed of overhead contact lines, a.c. traction power substations, d.c. substations (e.g., at 3 kV d.c.), and a.c./d.c feeders. Both the rails, at a negative potential, constitute the return path of the d.c. train current to the source (Fig. 15.6). In some cases, instead of the overhead line a third rail is employed (e.g., New York City subway, 600 V d.c.).

Should the overhead contact line break and fall, or the pantograph dewire,[5] metal structures (e.g., fences) and publicly exposed equipment (e.g., light poles) in their vicinity may become dangerously energized. If these items are insulated from ground, or have a high earth resistance, the fault current might not be large enough to promptly and safely trip the protective device of the traction line.

Thus, for safety reasons, the European Norm EN 50122–1[6] prescribes that all the ECPs likely to become live due to faults of the train electrification system must be directly connected to the traction's earth, usually the running rails (Fig. 15.7).

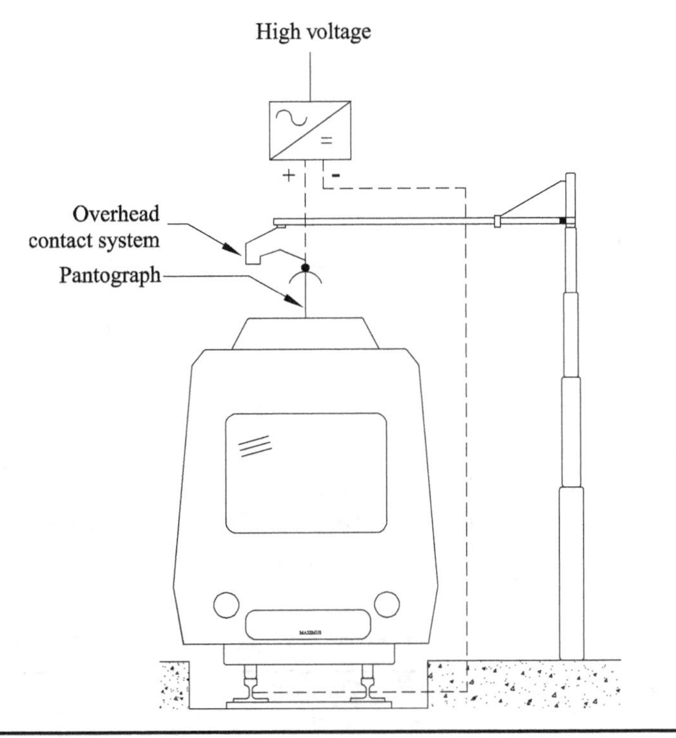

FIGURE 15.6 Typical traction electrification system.

In TT systems, the bond to the running rails prevents the circulation of the d.c. fault current through the train station's grounding system R_G, which earths all the alternating current ECPs. Because of this link, most of the fault current is returned to the source via the rails, with the result to lower both the perspective touch voltage appearing over the ECPs and the fault clearing time of the direct current protective devices.

A negative aspect of this arrangement is the establishment of permanent d.c. stray currents through the soil, which, as we already know, may cause corrosion of metal buried parts. The rails, in fact, are earthed through the ECPs grounding system of the station. The return current, in the ordinary operation of the train, in fact, will go back to the source not only through the rails, the legitimate return circuit, but also via the parallel path constituted by the actual earth.

In order to minimize the stray currents in a d.c. traction system, therefore, the direct bonding connection ECP, rails should be avoided and voltage-limiting device (e.g., diodes) should be used (Fig. 15.8).

In the regular operation of the train, the diode is an open circuit and prevents the d.c. traction current to circulate through the earth.

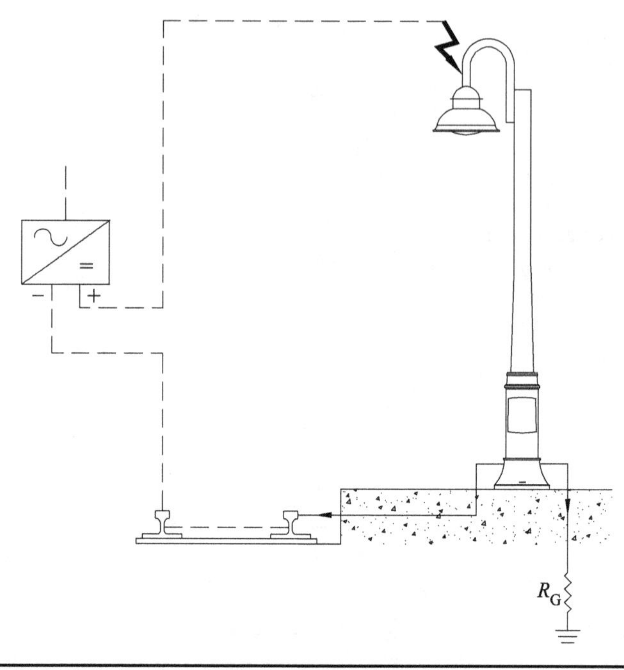

Figure 15.7 Connection of ECPs at the running rails.

In fault conditions, the diode is directly polarized, becomes a short circuit, and links the ECPs to the rails.

15.5 Electrical Safety in Swimming Pools

Swimming pools may contain submersed electrical items (e.g., light fixtures), which may break down. The underwater faulted equipment can be modeled as a spherical electrode radially leaking current toward the source both in the water and in TT systems, in the surrounding earth. Thus, a current field is generated and is characterized by a current density vector $\underline{J}$, defined as the current passing through the unit area of the conductive medium (e.g., the water). Thus, through the swimmer, which can be thought of as a submersed conductive cylinder of radius k, a dangerous current $\underline{I}_s$ may circulate. To calculate this current, let us assume that the equivalent cross-sectional area of the swimmer/cylinder S_k is oriented perpendicularly to $\underline{J}$ (Fig. 15.9) and that the swimmer is at a distance r from the faulty item.

In the above conditions, the current $\underline{I}_s$ flowing through the swimmer is a scalar quantity given by Eq. (15.1):

$$I_s = JS_k = \frac{I}{4\pi r^2}\pi k^2 = \frac{Ik^2}{4r^2} \tag{15.1}$$

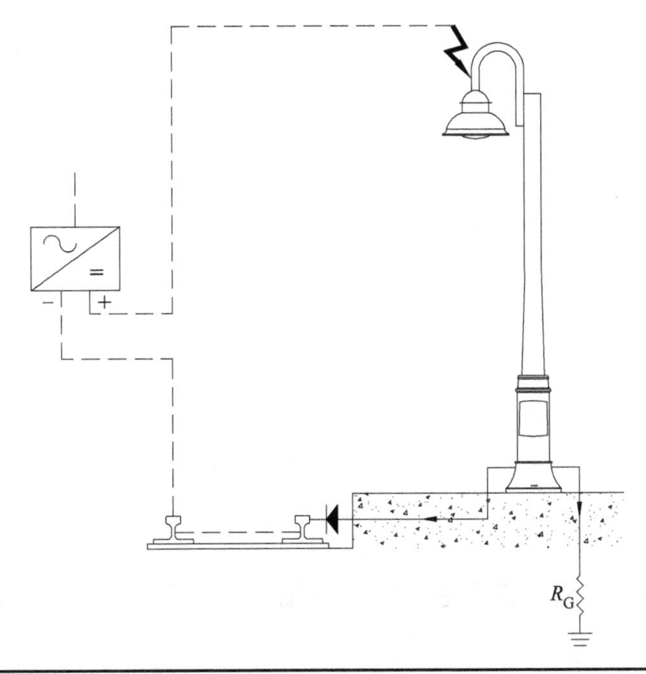

Figure 15.8 Connection of ECPs at the running rails through a diode.

where I is the leakage current and r corresponds to the radius of the generic spherical equipotential surface.

Equation (15.1) shows that the electrocution hazard decreases with the square of the swimmer's distance from the faulty item, and increases with the square of the radius k, representing the person's exposed surface to the leakage current, which depends on her/his size.[7]

Persons not in contact with the water may still be exposed to the risk of electrocution when ground faults occur into the pool, as the earth potential also develops outside of it. The hazard is greater if persons, who may likely have minimal clothing and be dripping wet, are in contact with EXCPs (Fig. 15.10).

In the case of Fig. 15.10, the contact with the shower water pipe, which is at zero potential, exposes the person to a larger touch

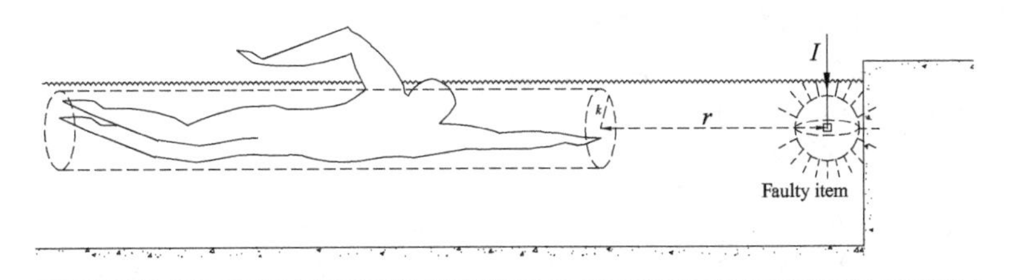

Figure 15.9 Swimmer exposed to a current field in the water.

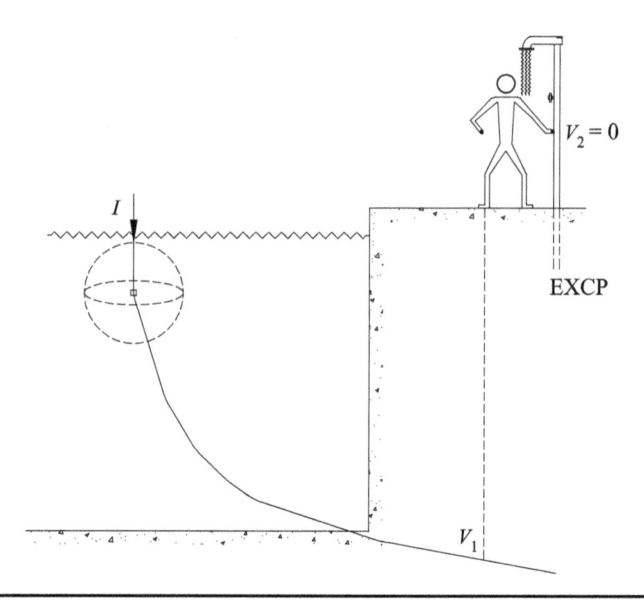

Figure 15.10 Person in contact with an EXCP during faults of submersed items.

voltage, with respect to the absence of contact, with potentially harmful consequences.

Thus, to reduce the risk of electrocution, supplementary equipotential bonding must be employed in pool areas as per IEC 60364–7-702.[8] All the ECPs and the EXCPs (e.g., metal ladder, diving-board's metal supports, hand railings, re-bars in concrete, etc.) located within Zone 0, 1, and 2 (where Zone 1 is in the pool itself, Zone 2 extends 2 m beyond Zone 1 and 2.5 m above it, and Zone 3 extends a further 1.5 m beyond Zone 2) must be connected together. This equipotential bonding connection must be realized by using an insulated conductor of adequate cross-sectional area. The supplementary equipotential bonding, which is in addition to the main equipotential bonding, will greatly reduce the voltage gradients in fault conditions.

15.6 Electrical Safety in Restrictive Conductive Locations

IEC 60364–7-706[9] defines restrictive conductive locations (RCLs) as the locations in contact with the earth and where workers may come into bodily contact with large areas of their conductive constituting material (e.g., metal tanks, wet tunnels, transmission towers, etc.). The extended bodily contact may be due not only to the RCLs' reduced dimensions, which also restricts the freedom of movement, but also possibly to the nature of the task workers must perform. The hazard is constituted by the presence, within the RCL, of electrical equipment, either fixed and/or hand-held, which may break down.

Extended contact with large earthed conductive surfaces greatly reduces the person's body resistance to ground. In these conditions, the threshold of ventricular fibrillation is lowered and the restrictions in movement makes more difficult to let-go of energized parts. Workers are in a hazardous situation as they do not have the benefit of the standard body resistance to ground, which would limit the flow of the current through the person.

In order to ensure protection against indirect contacts, IEC 60364–7-706 requires equipment used in RCL be supplied by isolating transformers (i.e., electrically separated systems; see Chap. 2) or through SELV systems (see Chap. 10). Class II hand-held pieces of equipment are also advisable, although not required.

15.7 Electrical Safety in External Lighting Installations

As per IEC 60364–7-714,[10] external lighting installations comprise lighting fixtures, along with their wiring and accessories, located outside buildings. Accessories may include transformers, breakers, reclosers, switches, manholes, poles, and whatever is functional to the performance of the system.

External lighting installations may expose the general public to touch potentials caused by faults.

In TT systems, an important safety requirement is the prohibition of earthing lighting poles, whose circuits are protected by the same RCD by means of independent ground electrodes (e.g., one rod for each pole).

To better understand the reasons behind this restriction, let us consider Fig. 15.11.

Let us assume that the basic insulation of the neutral wire of pole A fails and it comes in contact with the metal structure. This is a fault situation; however, the residual current device cannot pick up because the neutral conductor is not energized, and, therefore, there is no current leakage to ground. If the phase conductor of pole B also fails, the ground current $\underline{I}_2$ will be impressed to earth, as the fault-loop in TT system comprises it, and we expect the RCD to trip promptly.

However, part of the fault current, indicated in Fig. 15.11 with $\underline{I}_1$, will return to the source through the ground rod R_{G1} and the undue neutral-to-enclosure connection occurred at the first pole. $\underline{I}_1$ will flow also through the RCD, thereby, desensitizing it. The RCD, in fact, will not sense the actual fault current $\underline{I}_2$, but only the portion $\underline{I}_3$ not circulating through it, whose magnitude may be lower than its residual operating current. The fault might not be cleared by overcurrent

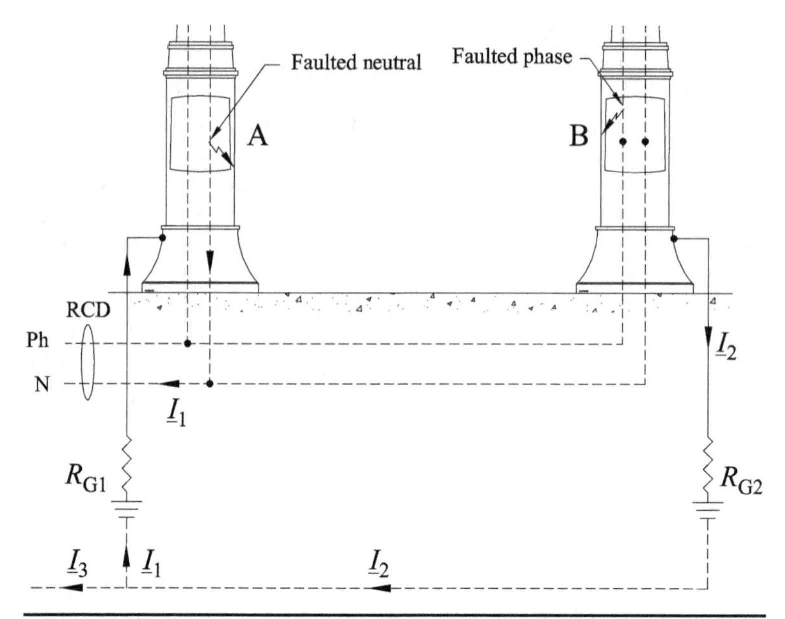

FIGURE 15.11 Lighting poles in TT systems independently earthed.

protective devices either, as $\underline{I}_2$, limited by the ground resistance R_{G2}, might be too low for their instantaneous pick up.

As a result, the ground fault can permanently energize pole B and cause stray voltage over it.

Alternatively, earthing the lighting poles collectively would allow overcurrent devices to trip, in the case of double faults (Fig. 15.12).

The fault current, in fact, corresponds to a short circuit phase-neutral, and the overcurrent device can promptly disconnect the supply. Electrical safety is, thus, assured, even if the RCD may still not intervene due to its desensitization caused by the neutral-to-enclosure fault.

15.8 Electrical Safety in Medical Locations

15.8.1 Microshock

In medical locations (e.g., hospitals, medical, and dental practices, etc.), patients are exposed to increased hazard of electric shock due to their particular conditions. Patients, in fact, may be unconscious, or anaesthetized and, therefore, unable to let-go of an energized part. In addition, patients may be connected to medical equipment either through applied parts to the skin (e.g., sensors, electrodes, etc.), often

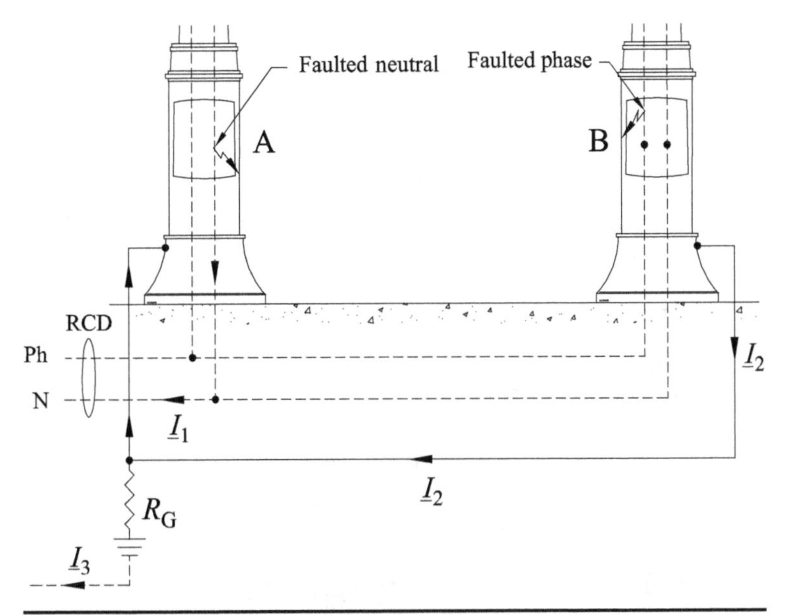

FIGURE 15.12 Lighting poles in TT systems collectively earthed.

locally treated to lower its resistance, and/or through the introduction of catheters[11] directly into their body's organs (e.g., the heart).

The use of conductive intracardiac probes, electrically connecting the heart to medical equipment, makes the patient extremely vulnerable to electric currents, because it lowers the threshold of danger. In fact, in a catheterized patient subject to touch voltages, leakage, or fault, currents will entirely flow through his/her heart,[12] and leave the body via the catheter. In these conditions, the current is no longer limited by the body resistance-to-ground, because this resistance does not form part with the fault-loop. Patients become particularly susceptible to the adverse effects of electricity, and currents of magnitude of a few tens of microamperes can trigger ventricular fibrillation. This phenomenon is defined as *microshock*.

15.8.2 Leakage Currents

Ordinary Class I equipment, medical or not, during its use may leak current through the insulation, and into the protective conductor. As discussed in Sec. 15.3, in the case of interruption of the protective conductor, the leakage current may circulate through the persons in contact with the enclosure. For ordinary equipment (i.e., equipment having low protective conductor currents), the magnitude of such current is so low that it does not constitute a hazard for persons. In medical locations, though, the interruption of the PE, defined as *single*

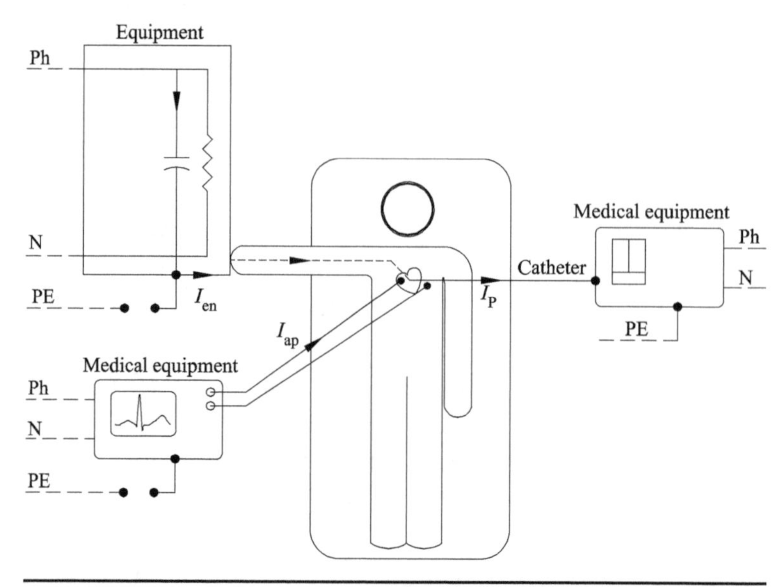

FIGURE 15.13 Patient leakage current caused by the interruption of the PE.

fault condition, is very dangerous because the same low amount of leakage current flowing through the catatherized patient, defined as *patient leakage current* I_P, can cause the microshock.

In Fig. 15.13, it is shown how in a single fault condition, the patient leakage current I_P can directly circulate through his/her heart due to contacts with any Class I equipment (I_{en}) and/or due to applied parts to the body (I_{ap}).

In the above situation, the person is at great risk of microshock, as the patient current may exceed the fibrillation threshold.

15.8.3 Local Equipotential Earthing Connection

In addition to the "chronic" problem of leakage currents from regularly operating Class I equipment emphasized by the interruption of their protective conductors, the safety of the patient can also be endangered by actual ground faults.

Figure 15.14 exemplifies a ground fault in TT systems because of the failure of an electrical component in the *patient vicinity*. The *patient vicinity* is defined as the space with enclosures likely to be touched by the patient, which extends 1.83 m beyond the perimeter of the bed and 2.29 m above the floor.[13]

In the above situation, a potential difference V_A, caused by the voltage drop on the protective conductor of resistance R_{PEA}, appears between the two pieces of equipment the patient may be in simultaneous contact with by touch and via catheters.

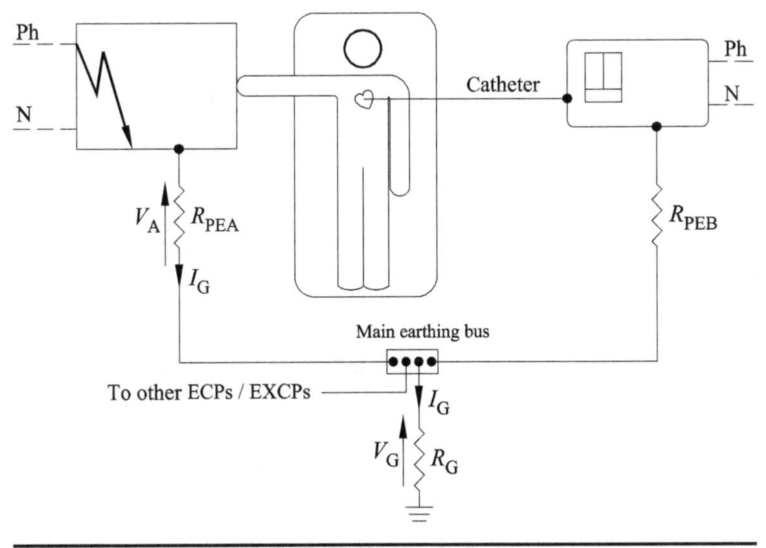

FIGURE 15.14 Hazardous condition in the presence of a sound PE.

The reduction of such voltage drop can be obtained by connecting together all the ECPs and the EXCPs to a local equipotential earthing bus located within the patient vicinity (Fig. 15.15).

This supplementary equipotential bonding connection lowers the resistance of the protective conductor serving the faulty ECP (i.e., $R_{PEA1} < R_{PEA}$), thereby decreasing the touch voltage (i.e., $V_{A1} < V_A$).

It is important to note that due to the patient's enhanced sensitivity to electric currents, EXCPs in medical locations need to be redefined, with respect to the standard definition that we gave in ordinary locations. In medical locations, in fact, IEC standards assume the threshold of 25 V as the maximum permissible touch voltage, and a lethal current for catheterized patients of 50 μA. In these assumptions, the resulting resistance-to-ground of any metal part in the patient vicinity must exceed 500 kΩ in order not to be an EXCP, and therefore, not bonded to the local earthing bus.

15.8.4 Electrical Separation

The local supplementary bonding connection, even though within the patient vicinity, cannot always sufficiently decrease the resistance of the PE and therefore limit the touch voltage to safe values. This is true especially in TN systems, where the ground-fault current may be rather high and so may the voltage drop on the PEs. In nonfault conditions, instead, currents leaking from pieces of equipment connected to the same local earthing bus are virtually identical and in phase, and

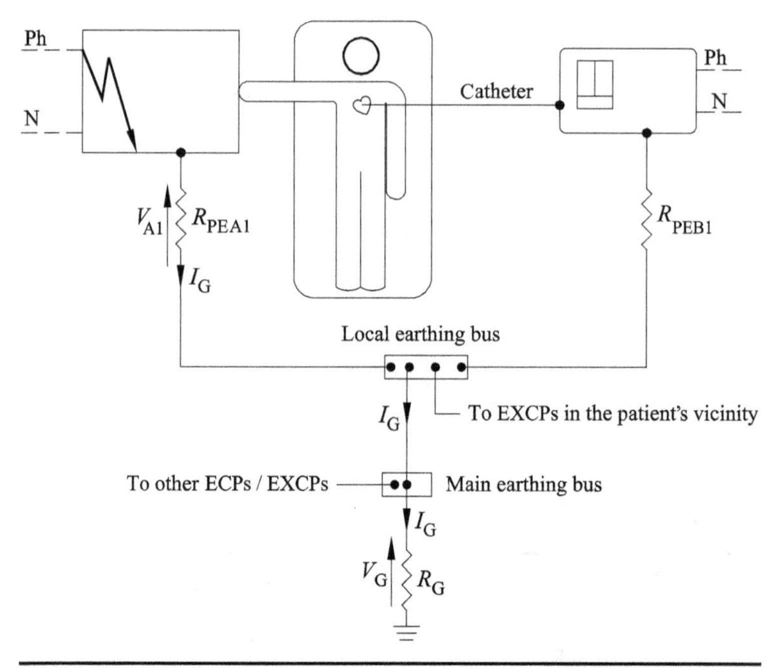

FIGURE 15.15 A local equipotential earthing bus reduces the voltage drop on the PE.

cause almost identical voltage drops over their PE, thereby determining no appreciable potential differences.

To improve the protection against indirect contact under fault conditions, in addition to the local equipotential earthing bus, an isolating transformer supplying the circuits in the patient vicinity may be adopted (Fig. 15.16).

First faults occurring in electrically separated systems, in compliance with the definition given in Chap. 2, cause the flow of capacitive currents I_G of low magnitude (i.e., order of milliamperes) through the PEs. As a consequence, the touch voltage the patient might be exposed to is well within safe limits.

Problems may arise at the occurrence of a subsequent second fault involving the other pole in another piece of equipment in contact with the patient. In that case, the resulting short circuit current circulating through the protective conductors might cause dangerous potential differences between the faulty ECPs, even in the presence of the supplementary equipotential bonding. For this reason, in medical locations, the first fault must be promptly traced by means of an insulation-monitoring device and then cleared.[14]

The earthing connection of the enclosures of the separated system, shown in Fig. 15.16, makes this system resemble the IT system. In the

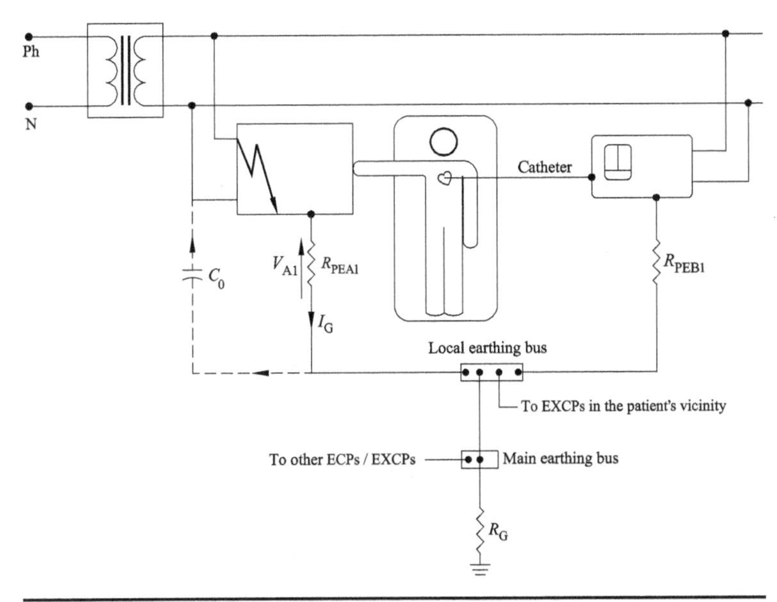

Figure 15.16 Isolating transformer supplying the circuits in the patient vicinity.

patient vicinity, in fact, there may be small equipment, not supplied by the isolating transformer, requiring the ground connection available at the local earthing bus. The above arrangement is a violation of the general rule, which prohibits the ECPs of separated systems to share the earth with nonseparated systems (prohibition also applicable to Class II equipment). As seen, this rule intends to prevent enclosures of separated systems from becoming dangerously "live" due to potentials transferred by means of earthing connections.

In the presence of the supplementary equipotential bonding in medical locations, this risk is, indeed, very low and deemed acceptable. In fact, even if the earthing bus attains a certain potential under fault conditions, all the ECPs in the patient vicinity will simultaneously reach this same value, as Fig. 15.16 shows; ergo, no potential differences can appear among them and the patient is safe. As a consequence, the grounding connection in separated systems adopted in medical locations is deemed safe in the presence of the local equipotential bonding.

15.8.4.1 Interruption of the Protective Conductor in Separated Systems

The interruption of the PE is dangerous even in separated systems because the resulting capacitive current through the patient may exceed the fibrillation limits (Fig. 15.17).

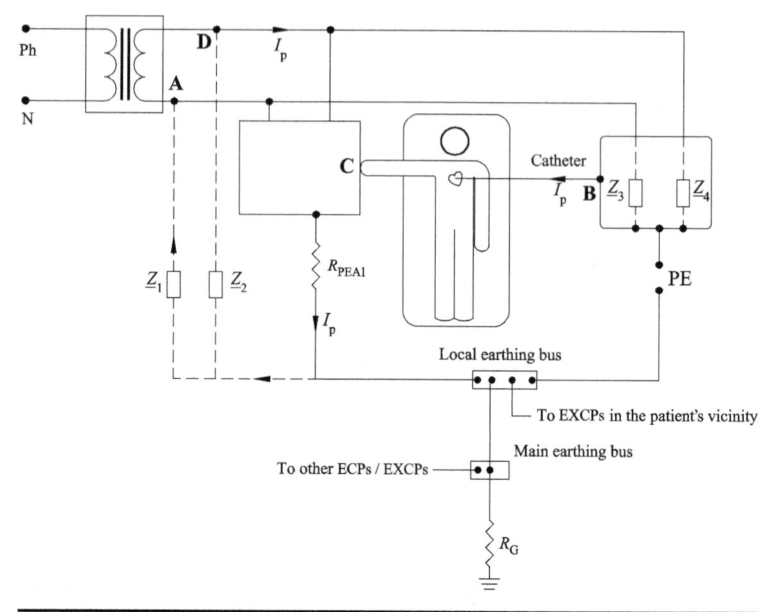

FIGURE 15.17 Separated system with interrupted PE.

Let $\underline{Z}_1$, $\underline{Z}_2$, $\underline{Z}_3$, and $\underline{Z}_4$ be the capacitive impedances-to-ground of supply and equipment, as indicated in Fig. 15.17. Upon loss of the protective conductor, the leakage current impressed by pole D through $\underline{Z}_4$ will reclose toward pole A by circulating through the patient.

The above impedances are connected in a "bridge" configuration across whose diagonal BC patients may find themselves linked (Fig. 15.18).

If the impedance bridge is balanced, that is, $\underline{Z}_1\underline{Z}_4 = \underline{Z}_2\underline{Z}_3$,[15] the patient is safe, as $\underline{V}_{BC} = \underline{I}_P = 0$. If the bridge is not balanced, the patient current $\underline{I}_P$ can be obtained by deducing the Thevenin equivalent circuit as seen at the points B and C (Fig. 15.19).

The equivalent Thevenin voltage $\underline{V}_{th}$ is calculated by applying Kirchhoff's second law to the loop DBAD and the voltage divider rule:

$$\begin{aligned}\underline{V}_{th} = \underline{V}_{BC} &= \underline{V}\left(\frac{\underline{Z}_3}{\underline{Z}_3+\underline{Z}_4}\right) - V\left(\frac{\underline{Z}_1}{\underline{Z}_1+\underline{Z}_2}\right)\\ &= \underline{V}\left(\frac{1}{1+(\underline{Z}_4/\underline{Z}_3)} - \frac{1}{1+(\underline{Z}_2/\underline{Z}_1)}\right)\end{aligned} \tag{15.2}$$

The Thevenin impedance $\underline{Z}_{th}$ is given by

$$\underline{Z}_{th} = \left(\frac{\underline{Z}_1\underline{Z}_2}{\underline{Z}_1+\underline{Z}_2} + \frac{\underline{Z}_3\underline{Z}_4}{\underline{Z}_3+\underline{Z}_4}\right) \tag{15.3}$$

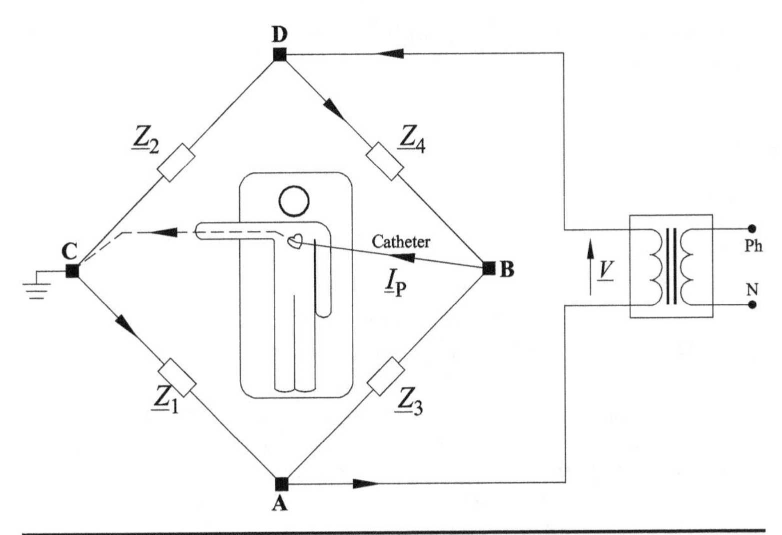

FIGURE 15.18 The patient may be connected across the impedance bridge's diagonal.

Thus, the patient current $\underline{I}_P$ is

$$\underline{I}_P = \frac{\underline{V}_{th}}{\underline{Z}_{th} + R_B} \tag{15.4}$$

where R_B is the patient's body resistance.

In balance conditions of the bridge (i.e., $\underline{Z}_1\underline{Z}_4 = \underline{Z}_2\underline{Z}_3$), it appears clear from Eq. (15.2) that $\underline{V}_{th}$ equals zero; therefore, the patient is safe even if the protective conductor of the medical equipment is interrupted.

In practice, the balance condition is rather challenging to both achieve and maintain in time. Although results have been achieved in manufacturing separation transformers with identical capacitance-to-ground, that is, $\underline{Z}_1 = \underline{Z}_2$, it is rather difficult to obtain medical equipment with symmetrical impedance-to-ground, that is, $\underline{Z}_3 \neq \underline{Z}_3$. Thus,

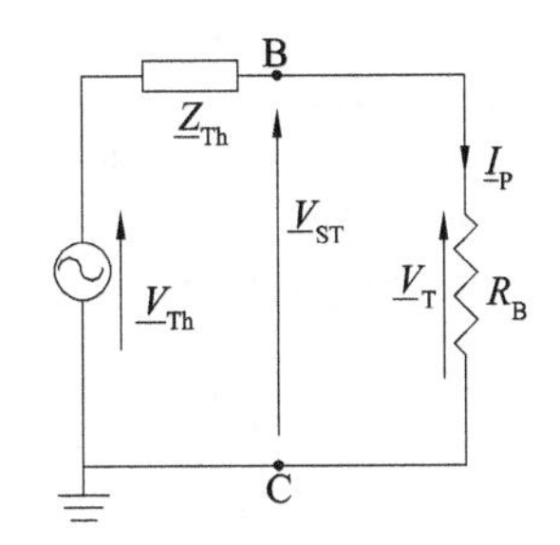

FIGURE 15.19 Thevenin equivalent circuit as seen at the points B and C of the bridge.

designers cannot rely upon balanced bridges to achieve safety in medical locations.

Endnotes

1. The international standard IEC 60364–7-709; 2007–05, 2d ed., "*Electrical Installations of Buildings—Part 7: Requirements for Special Installations or Locations—Section 709: Marinas and Pleasure Craft*" defines pleasure craft as any boat used exclusively for sport or leisure.
2. We can assume the seawater to have a resistivity of 10 $\Omega \cdot$ m.
3. IEC 60950–1 "*Information Technology Equipment—Safety—Part 1: General Requirements*," Ec1:2006–08.
4. Electrical noise is caused by unintentional, and unpredictable, high-frequency potential differences between electronic units. These voltage differences may be caused by electromagnetic interferences radiated and received by electronic equipment. The noise can cause the failure of components, as well as errors in data/signal processing.
5. Broken lines or dewired pantographs must be considered energized, as they may also be in contact with other neighboring live pantographs or contact lines.
6. "*Railway Applications—Fixed Installations Part 1: Protective Provisions Relating to Electrical Safety and Earthing*," EN 50122–1:1997–06.
7. In fish tanks, failures of electrical items (e.g., oxygenators, light fixtures, etc.) are generally not dangerous to the fish, as their exposed surface to the leakage current is very small.
8. IEC 60364–7-702:1997–11, "*Electrical Installations of Buildings—Part 7: Requirements for Special Installations or Locations—Section 702: Swimming Pools and Other Basins.*"
9. IEC 60364–7-706:2005, "*Electrical Installations of Buildings—Part 7: Requirements for Special Installations or Locations—Section 706: Conducting Locations with Restricted Movement.*"
10. IEC 60364–7-714:1996, "*Electrical Installations of Buildings—Part 7: Requirements for Special Installations or Locations—Section 714: External Lighting.*"
11. A catheter is a tube, flexible or rigid, which allows both drainage of physiological fluids and injection of solutions for therapeutic purposes (e.g., saline solution). Both liquids must be considered conductive.
12. In conditions of macroshock, upon touch of an energized part, less than 10% of the body current circulates through the cardiac muscle (see Sec. 5.4.2).
13. IEC 60601–1-SER, 1st ed. , 2008–01-22 "*Medical Electrical Equipment.*"
14. In ordinary locations, the second fault is not dangerous, as the nongrounded equipotential bonding conductors reduce the potential difference between enclosures to harmless values for persons (Fig. 2.18).
15. See App. B for more details on balancing bridges of impedances.

APPENDIX A

Sinusoids and Phasors

A.1 Sinusoids

Sinusoids are fundamental in power systems. After the invention of the transformer in 1885 and the induction motor in 1888 by Ferraris and Tesla, a.c. systems were preferred to d.c. systems. Sinusoids were chosen both due to their relatively simple generation and due to a well-established mathematical theory to analyze them. In this appendix, we will provide the basic concepts of sinusoids and phasors, as they apply to the previous chapters.

Sinusoids are periodic signals that have the form of a sine, or a cosine, function. They are described as a function of time t by the following mathematical expression:

$$v(t) = V_{\mathrm{M}}\sin(\omega t + \theta) \tag{A.1}$$

where V_{M} is the maximum value (or peak) of the sinusoid, ω is the angular frequency (radian/s), θ is the initial phase (radian), and $\omega t + \theta$ is the argument of the sinusoid (radian), which is also referred to as instantaneous phase; V_{M}, ω, and θ are real constants (i.e., time-invariant) and t is the independent variable. When $\theta \neq 0$, the sinusoid is nonzero at $t = 0$, but its magnitude depends on the initial phase. In this case, the sinusoid looks as if it were shifted along the time-axis, toward left or right, by the amount of time of θ/ω seconds. If the phase is negative, the sinusoid is defined as "delayed," while if positive, it is defined as "ahead of time." In Fig. A.1, the sinusoid with a generic phase $\theta > 0$ is graphically represented as a function of time.

Sinusoids are periodic functions, that is, they repeat themselves every T seconds. In formulas:

$$v(t) = v(t + nT) \tag{A.2}$$

for each integer n and for all t.

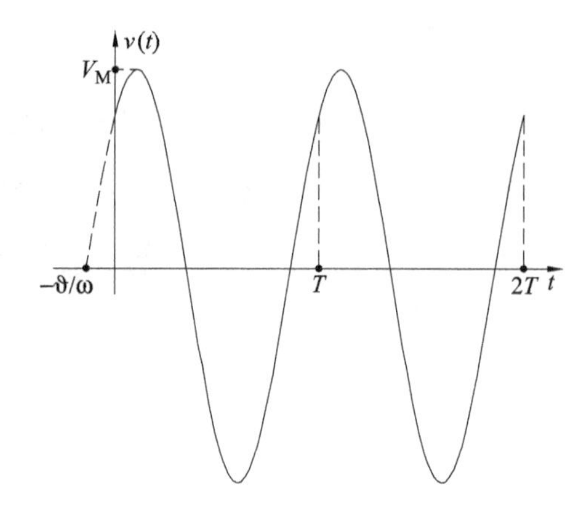

FIGURE A.1 Sinusoid $v(t)$ as a function of time t.

T is defined as the *period* of the sinusoid, that is, the time (second; s) the sinusoid takes to complete an entire cycle corresponding to 2π radians before repeating itself. The reciprocal of the *period* is the *frequency* f defined as the number of cycles the sinusoid completes in the unit of time (hertz; Hz).

Due to the periodicity of the sinusoid, and assuming $\theta = 0$, we can write

$$v(0) = v\left(\frac{T}{2}\right) = V_M \sin\left(\omega \frac{T}{2}\right) = 0 \tag{A.3}$$

By knowing that the sine function equals zero when its argument is a multiple of π, we obtain:

$$\omega \frac{T}{2} = \pi \Rightarrow \omega = \frac{2\pi}{T} = 2\pi f \tag{A.4}$$

Sinusoids are characterized by a zero *mean* or *average value* in one period, as per its mathematical definition given in Eq. (A.5):

$$g_m = \frac{1}{T} \int_0^T v(t)\, dt \tag{A.5}$$

To clarify this concept, let us apply Eq. (A.5) to the generic periodic signals $g(t)$ in Fig. A.2.

The integral in Eq. (A.5) represents the measure of the area enclosed between $g(t)$ and the t-axis (the dotted area in Fig. A.2). The area above the t-axis is conventionally considered positive, whereas the area below it is negative. The positive area is greater than the negative one, thus, the net area is nonzero. We can also interpret the

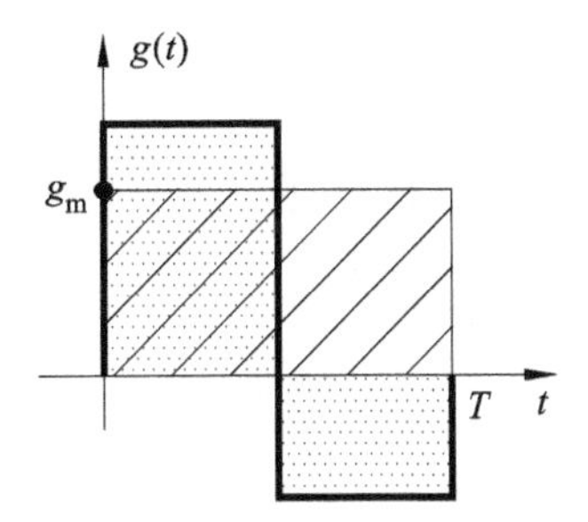

FIGURE A.2 Generic periodic signals $g(t)$ as a function of time t.

previous definition of mean value as the height of a rectangle of base T and net area as defined above (i.e., the hatched area). As a result, $g(t)$ has a nonzero mean value.

As to sinusoids, in a period they are perfectly symmetrical with respect to the t-axis. The areas underneath both half-periods are equal, but are of opposite sign regardless of the value of the initial phase. Ergo, the integral in Eq. (A.5) is zero, and so is their mean value.

For this reason, we define their *mean value* in a half-period,[1] and we can calculate it, assuming for simplicity $\theta = 0$, as follows:

$$\begin{aligned} V_{\mathrm{m}} &= \frac{2}{T}\int_0^{T/2} V_{\mathrm{M}} \sin \omega t \, \mathrm{d}t = \frac{2V_{\mathrm{M}}}{T}\left| -\frac{\cos \omega t}{\omega}\right|_0^{T/2} \\ &= \frac{2V_{\mathrm{M}}}{\omega T}\left(-\cos \omega \frac{T}{2} + 1\right) = \frac{2V_{\mathrm{M}}}{2\pi}(-\cos \pi + 1) = \frac{2V_{\mathrm{M}}}{\pi}. \end{aligned} \tag{A.6}$$

Another important parameter for periodic signals is the root mean square (r.m.s.) value, mathematically defined as follows:

$$V_{\mathrm{rms}} = \sqrt{\frac{1}{T}\int_0^T v^2(t)\, \mathrm{d}t} \tag{A.7}$$

Under the sign of square root, the mean value of the squared sinusoid appears. As squared signals are always positive, their mean value during a period is always nonzero. By applying Eq. (A.7) to the case of sinusoids and knowing the following trigonometry identity:

$$\sin^2 \varphi = \frac{1 - \cos 2\varphi}{2} \tag{A.8}$$

we obtain

$$V_{\mathrm{rms}} = \sqrt{\frac{1}{T}\int_0^T (V_{\mathrm{M}} \sin \omega t)^2\, dt} = \sqrt{\frac{1}{T}\int_0^T V_{\mathrm{M}}^2 \frac{1-\cos 2\omega t}{2}\, dt}$$

$$= V_{\mathrm{M}}\sqrt{\frac{1}{2T}\int_0^T (1-\cos 2\omega t)\, dt}$$

$$= V_{\mathrm{M}}\sqrt{\frac{1}{2T}\int_0^T dt - \frac{1}{2T}\int_0^T \cos 2\omega t\, dt}$$

$$= V_{\mathrm{M}}\sqrt{\frac{1}{2T}T} = \frac{V_{\mathrm{M}}}{\sqrt{2}} \tag{A.9}$$

The r.m.s. of a sine wave is the value that a constant signal (i.e., d.c.) should have to yield the same average power dissipation.

Although sinusoids are univocally identified once their maximum value, frequency, and initial phase are known, other parameters are of interest.

The *form factor* of a sinusoidal wave is defined as the ratio of the r.m.s. value to the mean value in one half-time:

$$k_f = \frac{V_{\mathrm{rms}}}{V_{\mathrm{m}}} = \frac{(V_{\mathrm{M}}/\sqrt{2})}{(2V_{\mathrm{M}}/\pi)} \cong 1.11 \tag{A.10}$$

The *crest factor* (also referred to as peak-to-r.m.s. ratio) of a sinusoidal wave is defined as the ratio of the peak value V_{M} to the r.m.s. value:

$$k_f = \frac{V_{\mathrm{M}}}{V_{\mathrm{rms}}} = \sqrt{2} \cong 1.41 \tag{A.11}$$

It is possible to execute operations between sine waves of the same frequency in the time-domain, even though it is not very convenient to operate in this domain. As an example, let us consider the addition of two sinusoids:

$$\begin{aligned} v_{\mathrm{T}}(t) &= v_1(t) + v_2(t) = V_{1\mathrm{M}} \sin(\omega t + \alpha) + V_{2\mathrm{M}} \sin(\omega t + \beta) \\ &= V_{\mathrm{TM}} \sin(\omega t + \gamma) \end{aligned} \tag{A.12}$$

The result is still a sinusoid of the same frequency, and we can prove that V_{TM} and γ are given by Eqs. (A.13) and (A.14):

$$V_{\mathrm{TM}} = \sqrt{V_{1\mathrm{M}}^2 + V_{2\mathrm{M}}^2 + 2V_{1\mathrm{M}}V_{2\mathrm{M}} \cos(\alpha - \beta)} \tag{A.13}$$

$$\tan \gamma = \frac{V_{1\mathrm{M}} \sin \alpha + V_{2\mathrm{M}} \sin \beta}{V_{1\mathrm{M}} \cos \alpha + V_{2\mathrm{M}} \cos \beta} \tag{A.14}$$

Let us consider the product of two sinusoids:

$$v_T(t) = v_1(t)v_2(t) = V_{1M}\sin(\omega t + \alpha)V_{2M}\sin(\omega t + \beta)$$
$$= \frac{1}{2}V_{1M}V_{2M}[\cos(\alpha - \beta) - \cos(2\omega t + \alpha + \beta)] \qquad (A.15)$$

The product of sine waves, such as the instantaneous electrical power, is the summation of a constant term plus a sinusoid of angular frequency 2ω.

Now, let us consider the derivative of a sinusoid:

$$\frac{d}{dt}a(t) = \frac{d}{dt}[A_M\sin(\omega t + \alpha)] = \omega A_M\cos(\omega t + \alpha)$$
$$= \omega A_M\sin\left(\omega t + \alpha + \frac{\pi}{2}\right) \qquad (A.16)$$

The derivative of a sinusoid has the same angular frequency of the original sinusoid and leads it by a quarter cycle.

As said, these methods of calculation, although possible, are inconvenient as they are time-consuming. An alternative procedure, the symbolic method that we will examine in the following sections, has been developed.[2]

Example A.1 If $v(t)$ is a sinusoidal voltage, whose peak is 10 V and frequency 50 Hz. Calculate the following:

1. The period
2. The angular frequency
3. The mean value in one-half cycle
4. The root mean square value

Solution

1. $T = \frac{1}{f} = 0.02$ s.
2. From Eq. (A.4), we obtain $\omega = 2\pi f = 314$ radian/s.
3. From Eq. (A.6), we obtain $V_m = \frac{2V_M}{\pi} = 6.37$ V.
4. From Eq. (A.9), we obtain: $V_{rms} = \frac{V_M}{\sqrt{2}} = 7.1$ V.

A.2 Phasors

Let us consider a vector $\underline{A}$, of magnitude A_M, rotating counterclockwise with constant angular velocity ω in the Gauss chart (Fig. A.3).

Let this revolving vector, also referred to as *phasor*, be initially displaced by an angle θ with respect to the *real*-axis. During its revolution

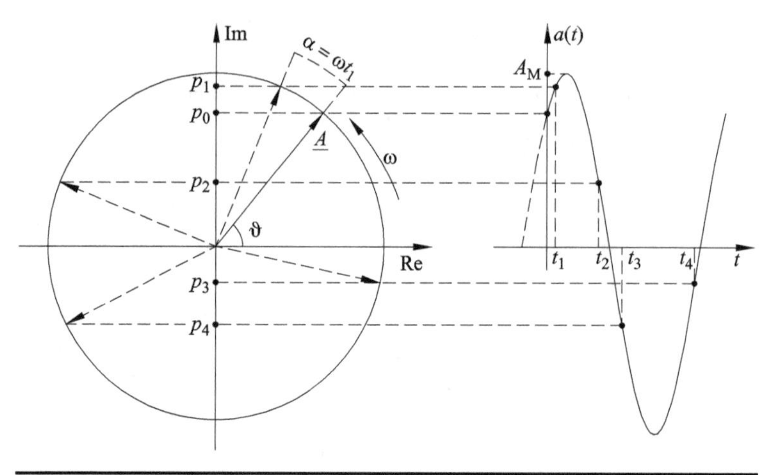

FIGURE A.3 The rotating vector $\underline{V}$ (*phasor*) in the Gauss chart.

the phasor traces a circumference of radius equal to its length. After the time t_1, the *phasor* will have "swept" the angle $\alpha = \omega t_1$ on the circumference, positioning itself at the $\omega t_1 + \theta$ on it.

Let us project the phasor on the *imaginary*-axis as it rotates, obtaining the segments $0p_0 = V_M \sin\theta$, $0p_1 = V_M \sin(\omega t_1 + \theta)$, $0p_2 = \sin(\omega t_2 + \theta)$, $0p_3 = \sin(\omega t_3 + \theta)$, etc., in correspondence with the times $0, t_1, t_2, t_3$, etc. If we graph these pairs $(0p_i, t_i)$ in a chart, we will univocally obtain a sine wave of phase θ as a function of time (Fig. A.3). Vice versa, once a sinusoid is defined in the time-domain, that is, peak value, phase, and angular frequency are known, only one corresponding phasor can be univocally associated.

Thus, a biunivocal relation[3] exclusively links together sinusoids and phasors, allowing us to advantageously use the latter elements in lieu of the former ones for our calculations. For example, instead of carrying out the addition of sinusoids in the time-domain, one can add the corresponding phasors up, by conveniently using the simpler rule of addition of vectors.

In addition, another biunivocal relation between phasors and complex numbers does exist: any phasor is univocally determined by a complex number and vice versa (Fig. A.4).

Thus, we can conveniently represent sinusoids by means of complex numbers rather than with phasors, knowing that two biunivocal relations do exist: sinusoids $\leftrightarrow$ phasors $\leftrightarrow$ complex numbers.

Complex numbers can be written in two different forms as

$$\underline{A} = a + jb \qquad \text{(rectangular form)} \tag{A.17}$$

where a and b are, respectively, the real and the imaginary parts of the complex number; and as

$$\underline{A} = |\underline{A}|\, e^{j\theta} \qquad \text{(exponential form)} \tag{A.18}$$

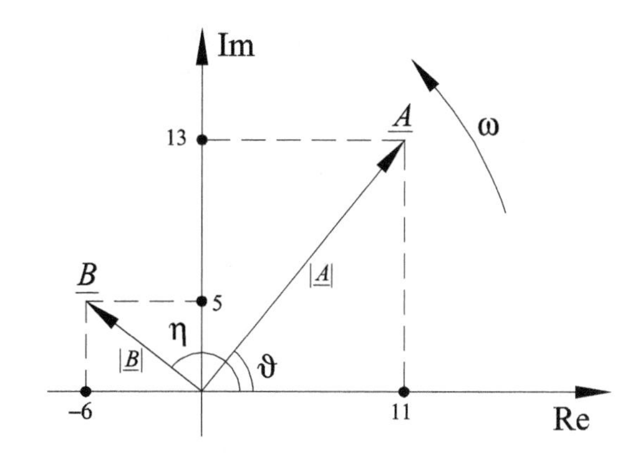

FIGURE A.4 Biunivocal relation between phasors and complex numbers.

where $|\underline{A}|$ and θ are, respectively, the magnitude and the argument of the complex number.

The relations between the terms of the two forms are

$$|\underline{A}| = \sqrt{a^2 + b^2} \tag{A.19}$$

$$\vartheta = \tan^{-1}\frac{b}{a} \tag{A.20}$$

$\underline{A}$ can be rewritten by using the Euler's identity as follows:

$$\underline{A} = |\underline{A}|(\cos\theta + \mathrm{j}\sin\theta) \tag{A.21}$$

By using the previous equation, we can write

$$\begin{aligned} a(t) &= A_{\mathrm{M}}\sin(\omega t + \theta) = \mathrm{Im}\{A_{\mathrm{M}}[\cos(\omega t + \theta) + \mathrm{j}\sin(\omega t + \theta)]\} \\ &= \mathrm{Im}[A_{\mathrm{M}}e^{\mathrm{j}(\omega t + \theta)}] \end{aligned} \tag{A.22}$$

If we assume to consider only sinusoids of the same angular frequency, and, therefore, same period, their representative phasors revolve at the same speed, which makes their phase relation constant at any given time. We can focus, therefore, just on the instant $t = 0$, and "suppress" the time variable in Eq. (A.21). By doing so we can study sinusoids not in the time-domain, but in the more convenient frequency-domain of the phasors/complex numbers characterized by a constant ω. Vice versa, given a phasor/complex number, we can obtain the sinusoid in the time-domain by knowing its magnitude and phase.

This process is summarized in Eq. (A.23):

$$a(t) = A_{\mathrm{M}}\sin(\omega t + \theta) \leftrightarrow \underline{A} = A_{\mathrm{M}}e^{\mathrm{j}\theta} \tag{A.23}$$

In sum, $a(t)$ is the instantaneous value of the sinusoid, which depends on the time t and is always a real number; $\underline{A}$ is a phasor, which is time-independent and is generally a complex number (i.e., with both a real and an imaginary part).

Fundamental operations between complex numbers are listed below:

$$\underline{A} \pm \underline{B} = (a_1 + ja_2) \pm (b_1 + jb_2) = (a_1 \pm b_1) + j(a_2 \pm b_2) \qquad \text{(A.24)}$$

$$\underline{A} \times \underline{B} = |\underline{A}|e^{j\theta} \times |\underline{B}|e^{j\eta} = |\underline{A}| \times |\underline{B}|e^{j(\theta+\eta)} \qquad \text{(A.25)}$$

$$\frac{A}{B} = \frac{|\underline{A}|}{|\underline{B}|}e^{j(\theta-\eta)} \qquad \text{(A.26)}$$

As to the derivative of sinusoids in the frequency-domain, by considering Eqs. (A.16) and (A.22), we can write

$$\frac{d}{dt}a(t) = \omega A_M \sin\left(\omega t + \alpha + \frac{\pi}{2}\right) = \mathrm{Im}[\omega A_M e^{j\omega t} e^{j\alpha} e^{j(\pi/2)}]$$
$$= \mathrm{Im}\left[j\omega A_M e^{j\omega t} e^{j\alpha}\right] = \mathrm{Im}\left[j\omega \underline{A} e^{j\omega t}\right] \qquad \text{(A.27)}$$

By "suppressing" the time variable in Eq. (A.27), we can write

$$\frac{d}{dt}a(t) \leftrightarrow j\omega \underline{A}. \qquad \text{(A.28)}$$

Equation (A.26) shows that the derivative of a sinusoid is obtained in the frequency-domain by multiplying the original phasor $\underline{A}$ by $j\omega$.

It is important to note that operations between phasors are possible only if the sinusoids have the same angular frequency. For example, let us consider the phasors $\underline{A}$ and $\underline{B}$, respectively, rotating at ω and 2ω radian/s and displaced by α radians (Fig. A.5).

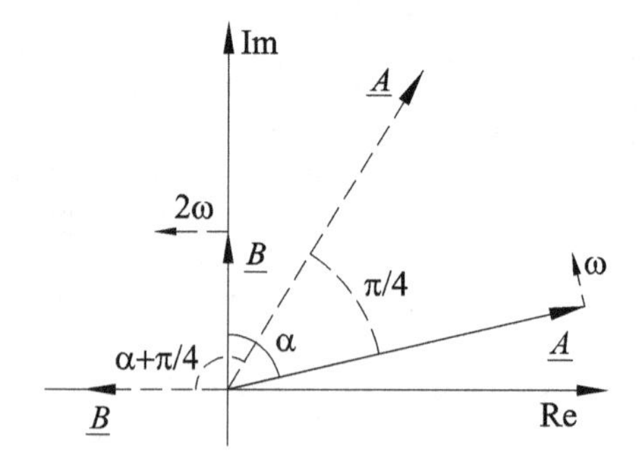

FIGURE A.5 Phasors rotating at different angular velocity.

After the time t_1, if the phasor $\underline{A}$ rotates $\pi/4$ radian, the phasor $\underline{B}$ rotates instead $\pi/2$ radians. As a consequence, the angular displacement of the two phasors has increased to $(\alpha + \pi/4)$. The variability with time of the phase relation between the two phasors renders meaningless their representation on the same chart.

Example A.2 With reference to Fig. A.4, let us write the expressions of the two sinusoids $a(t)$ and $b(t)$ in the time-domain, represented by the complex numbers $\underline{A} = 11 + j13$ and $\underline{B} = -6 + j5$. Let us also add up the sinusoids, and obtain the corresponding sinusoids in the time-domain.

From Eq. (A.18), $|\underline{A}| = 17$ and $|B| = 7.8$.

From Eq. (A.19), $\theta = 0.86$ radian (49.7°), $\eta = 2.44$ radians (140°).

Note that there are always two different angles, differing by multiple of π radians, whose tangent equals b/a. In our case, the alternate angles are $\theta = 4$ radians (229.7°) and $\eta = -0.7$ radian (−39.8°). However, even if the alternate angles are mathematically correct, we must decide the right values according to which quadrant the phasors are located in.

The expression of the sinusoids in the time-domain are

$$\alpha(t) = 17 \sin(\omega t + 0.86) \quad \text{and} \quad b(t) = 7.8 \sin(\omega t + 2.44)$$

To write the expression of the addition of the sinusoids in the time-domain, it is more convenient to operate in the frequency-domain by adding up the two corresponding complex numbers and then transform. Thus, $\underline{S} = \underline{A} + \mathrm{B} = 5 + j18$ (Fig. A.6).

From Eqs. (A.19) and (A.20), $|\underline{S}| = 18.7$ and $\delta = 1.3$ radians (74.5°). Thus, the addition of the two sinusoids in the time-domain is given by $s(t) = 18.7 \sin(\omega t + 1.3)$.

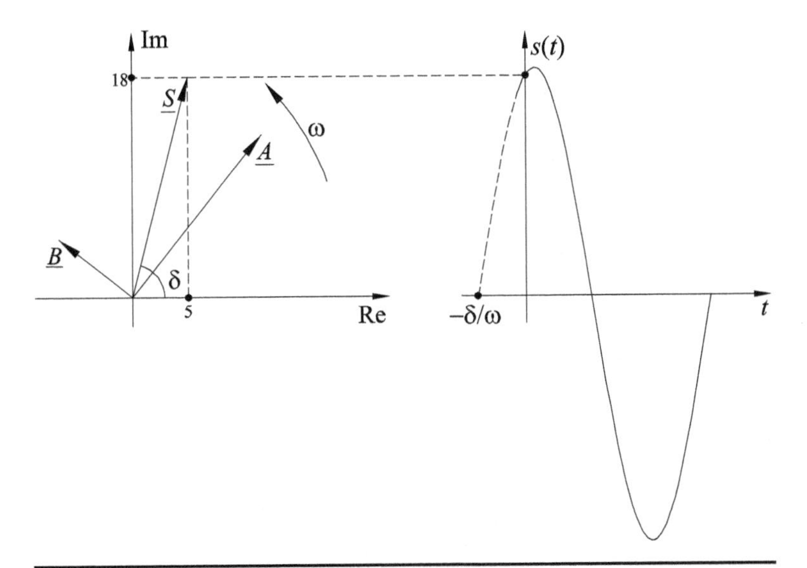

FIGURE A.6 Summing complex numbers.

It is also possible to represent sinusoids with phasors of length equals the sinusoids' r.m.s. value, instead of their maximum value. This is due to the greater significance of this parameter. The phasors then are only scaled down by $1/\sqrt{2}$, while their phase relation is kept the same.

Endnotes

1. We could calculate the *mean value* in reference to any length of time within the half-period (e.g., quarter-period, etc.), as the result does not change.
2. Charles Proteus Steinmetz, *Theory and Calculation of Alternating Current Phenomena*, Electrical World and Engineer Incorporated, New York, 1897.
3. A biunivocal relation is usually indicated with the mathematical symbol $\leftrightarrow$.

APPENDIX B

Fundamental Conventions and Electric Circuit Theorems

B.1 Introduction

In this appendix, we will examine the fundamental conventions and the electric circuit theorems that have been extensively used throughout the text. The purpose of this appendix is to give the reader basic theoretical support for the comprehension of the technical methodologies profusely applied previously.

B.2 Fundamental Electrical Conventions

In solving electric circuits, we need to determine magnitude and direction of currents and magnitude and polarity of voltages. We approach the problem by establishing an initial arbitrary direction for currents, and we consequently establish the polarity of voltages by observing the *passive sign* convention (Fig. B.1*a*) and the *active sign* convention (Fig. B.1*b*).

The initial direction of currents is arbitrary in the sense that the actual direction can only be ascertained after the calculation has been performed. If the calculated current in Fig. B.1*a* results positive, its direction is correct and the component is absorbing power (i.e., it is an electric load). If the calculated current is, instead, negative, its direction must be reversed.

The same logic can be applied to the active sign convention in Fig. B.1*b*. We keep the direction of the calculated current if positive; otherwise we reverse it.

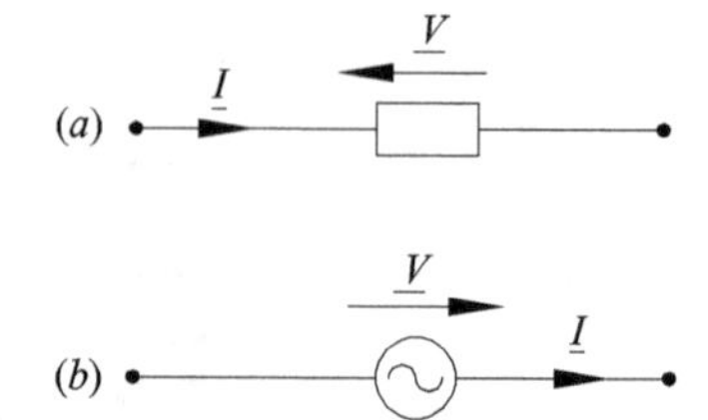

FIGURE B.1 (*a*) Passive sign convention and (*b*) active sign convention.

B.3 Kirchhoff's Laws

Kirchhoff's laws can be applied both to d.c. and a.c. electrical systems.

Kirchhoff's first law, based on the law of conservation of charge, states that the sum of the n currents entering a node equals the sum of the m currents leaving it:

$$\sum_{i=1}^{n} \underline{I}_i = \sum_{j=1}^{m} \underline{I}_j \tag{B.1}$$

It is important to note that current directions must be arbitrarily decided before applying the first law.

Kirchhoff's second law, based on the principle of conservation of electrostatic field, states that the algebraic sum of n voltages around a loop is zero:

$$\sum_{i=1}^{n} \underline{V}_i = 0 \tag{B.2}$$

It is important to note that voltage polarities must be conventionally established before applying the second law by applying the conventions of Fig. B.1*a* and *b*.

To illustrate the two Kirchhoff's laws, let us consider the circuit in Fig. B.2.

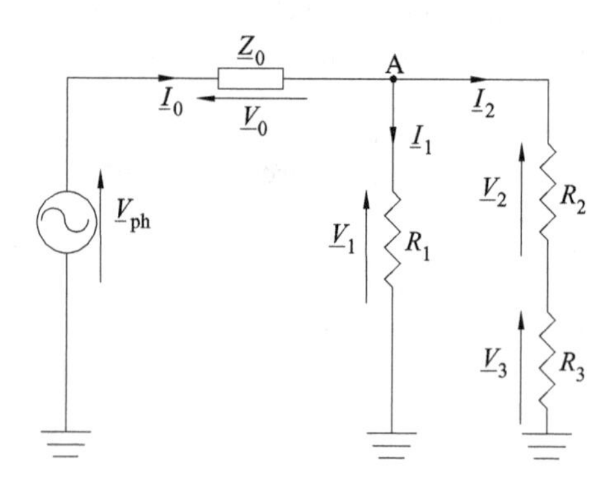

FIGURE B.2 A circuit illustrating the Kirchhoff's laws.

Directions of currents and polarities of voltages conform to the aforementioned passive and active sign conventions.

At node A, we can write as per Eq. (B.1):

$$\underline{I}_0 = \underline{I}_1 + \underline{I}_2 \tag{B.3}$$

To apply the second law, we are free to take either a clockwise or a counterclockwise trip around the loop and can start at any component of the circuit. In applying Eq. (B.2), the sign of each voltage is assumed negative, if the conventional arrow opposes the direction of the trip, and positive if it is in favor. In formulas, by starting with the generator and taking a clockwise trip:

$$\begin{aligned} \underline{V}_{\text{ph}} - \underline{V}_0 - \underline{V}_1 &= 0 \\ \underline{V}_{\text{ph}} - \underline{V}_0 - \underline{V}_2 - \underline{V}_3 &= 0 \\ \underline{V}_1 - \underline{V}_2 - \underline{V}_3 &= 0 \end{aligned} \tag{B.4}$$

B.4 Voltage and Current Dividers

If two impedances are in series, we may want to know the voltage across each component. With reference to Fig. B.3, we can write:

$$\underline{I} = \frac{\underline{V}_{\text{ph}}}{\underline{Z}_1 + R_2} \tag{B.5}$$

$$\underline{V}_1 = \underline{Z}_1 \underline{I}_1 \tag{B.6}$$

$$\underline{V}_2 = \underline{R}_2 \underline{I}_1 \tag{B.7}$$

By combining the previous equations, we obtain the expressions for the voltages across $\underline{Z}_1$ and R_2:

$$\underline{V}_1 = \underline{V}_{\text{ph}} \times \frac{\underline{Z}_1}{\underline{Z}_1 + R_2} \tag{B.8}$$

$$\underline{V}_2 = \underline{V}_{\text{ph}} \times \frac{R_2}{\underline{Z}_1 + R_2} \tag{B.9}$$

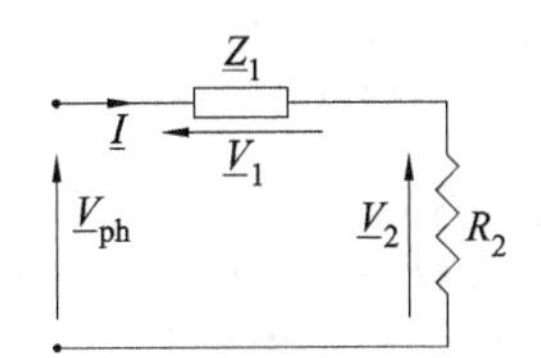

FIGURE B.3 A circuit illustrating the voltage division.

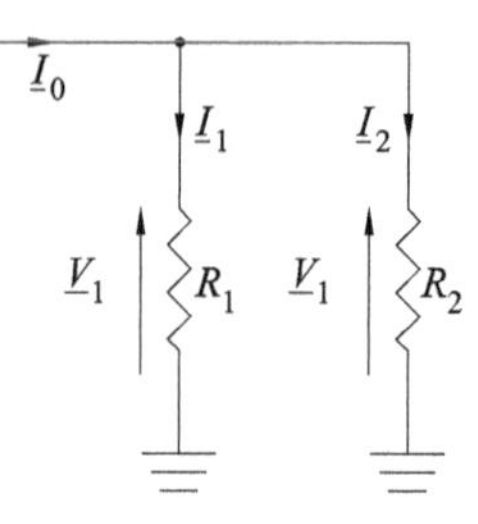

FIGURE B.4 A circuit illustrating the current division.

If two impedances are in parallel, we may want to know the currents flowing in each component. With reference to Fig. B.4, we can write

$$\underline{V}_1 = \underline{R}_1\underline{I}_1 \tag{B.10}$$

$$\underline{V}_1 = R_1\underline{I}_2 \tag{B.11}$$

$$\underline{V}_1 = R_T I_0 = \frac{R_1 R_2}{R_1 + R_2} \times I_0 \tag{B.12}$$

By combining the previous equations, we obtain the expressions for the current circulating through R_1 and R_2:

$$I_1 = \frac{R_2}{R_1 + R_2} \times I_0 \tag{B.13}$$

$$I_2 = \frac{R_1}{R_1 + R_2} \times I_0 \tag{B.14}$$

B.5 Superposition Principle

The superposition principle is applicable only to linear circuits having at least two independent sources. With this principle, to establish the value of a variable (i.e., voltage across or current through a component) we determine the contribution to the variable due to each independent source acting alone by turning off all the others.[1] One source at a time must be active. The results will be, then, algebraically summed. Each single contribution is obtained by short circuiting the other voltage generators and opening the other current sources.

This methodology, which provides simpler circuits, is diagrammatically shown in Fig. B.5, where $\underline{V}$ and $\underline{J}$ are, respectively, voltage and current generators, and $\underline{I}_1 = \underline{I}_1' + \underline{I}_1''$.

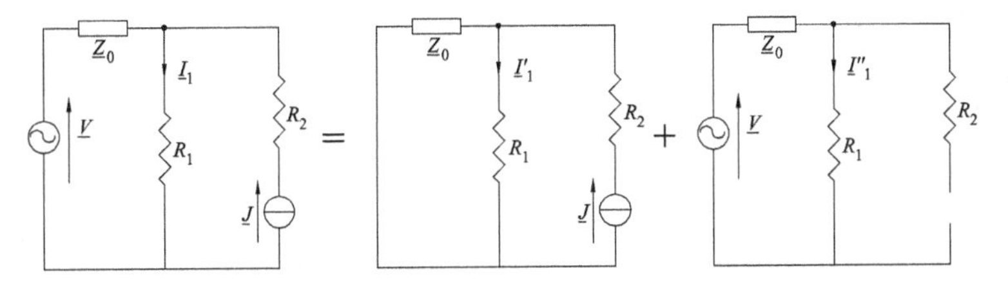

FIGURE B.5 Superposition principle.

B.6 Thevenin's Theorem

Thevenin's theorem states that any linear network, as seen from any pair of terminals A and B, can be replaced by an equivalent circuit having the same voltage–current relation at those terminals (Fig. B.6), as long as $\underline{V}_{\mathrm{Th}}$ and $\underline{Z}_{\mathrm{Th}}$ are properly calculated.

Thevenin's theorem can be proved by using the superposition principle as follows.

Let us add to the network of Fig. B.6 two identical voltage generators $\underline{V}_{\mathrm{Th}}$ with opposite polarities so that their presence has no effect on the original network (Fig. B.7).

Let us now apply the superposition principle and decompose the previous system into two, so that $\underline{I} = \underline{I}' + \underline{I}''$ (Fig. B.8).

The first system is constituted by the original network, whereas the second one is composed of the original network, once all the voltage generators have been short circuited and all the current generators have been opened. The magnitude of $\underline{V}_{\mathrm{Th}}$ is chosen equal to the no-load voltage $\underline{V}_{\mathrm{AB}}$ as seen at terminals A and B, so that $\underline{I}' = 0$. The no-load voltage is the potential difference between A and B obtained by disconnecting the load (i.e., R in our case). As a consequence, we will have $\underline{I}'' = \underline{I}$, that is, the same current as in the original system will flow in the second circuit of Fig. B.8. Thus, the two circuits combined together cause the original current $\underline{I}$ to flow. Thus, the equivalence at the terminals A and B between the original system and the Thevenin circuit is assured if $\underline{V}_{\mathrm{Th}} = \underline{V}_{\mathrm{AB}}$ (no-load voltage) and $\underline{Z}_{\mathrm{Th}}$ equals the impedance of the original system, once the independent sources are made inactive.

FIGURE B.6 Thevenin equivalent circuit.

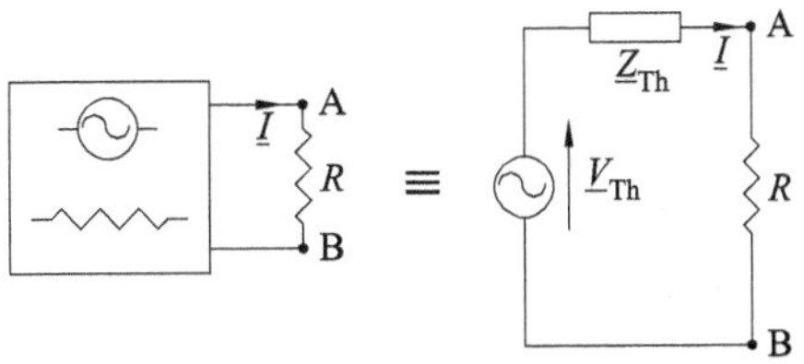

FIGURE B.7 The two voltage generators do not affect the original network.

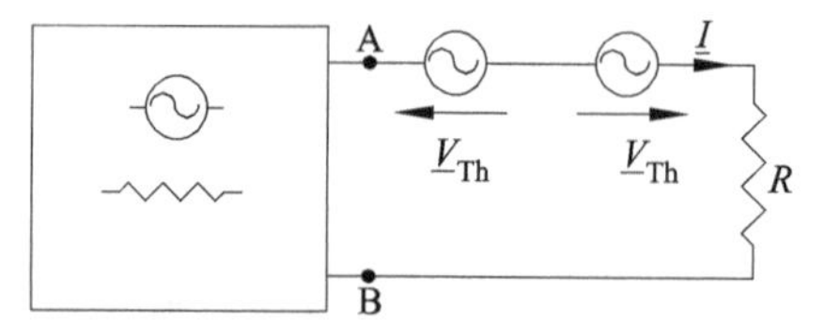

B.7 Millman's Theorem

Millman's theorem applies to circuits made up of branches in parallel with respect to two nodes (Fig. B.9).

This theorem allows the quick calculation of the voltage V_{AB} across all branches through the following formula:

$$V_{AB} = \frac{\sum_i (\pm V_i / R_i) + \sum_i \pm I_i}{\sum_i (1/R_i)} \tag{B.15}$$

where V_i and I_i are, respectively, voltage and current generators.

Each term at the numerator represents the current that would flow through the ith branch, if the nodes A and B were short circuited. If a branch contains no generator (i.e., $V_i = 0$), but only a passive components (e.g., R_i), its contribution will not appear into the summations at the numerator. The algebraic signs of the short circuit currents at the numerator are chosen according to the polarity of V_{AB}: positive if the currents enter the node A, and negative if the currents leave the node. At the denominator we have the arithmetic sum of the conductances of each branch.

To put things in perspective, let us apply Millman's formula to the circuit in Fig. B.9:

$$V_{AB} = \frac{(V_1/R_1) - (V_2/R_4) - I}{(1/R_1) + (1/R_2) + (1/R_4)} \tag{B.16}$$

B.8 Impedance Bridge

In general, an impedance bridge is defined as balanced when, with reference to Fig. B.10, the potential at nodes B and C are equal, that is, $\underline{V}_B = \underline{V}_C$.

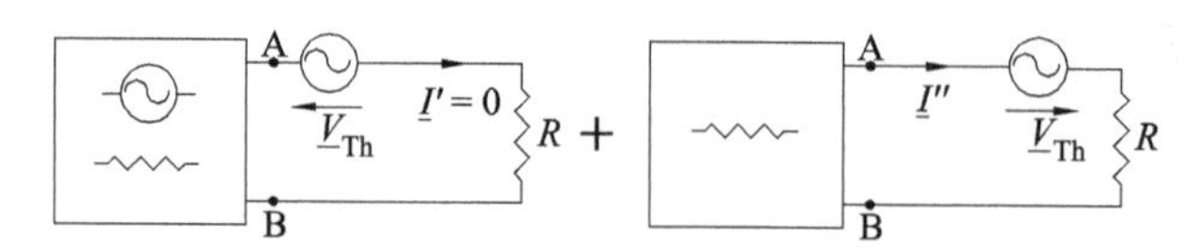

FIGURE B.8 Application of the superposition principle to the network with two identical generators.

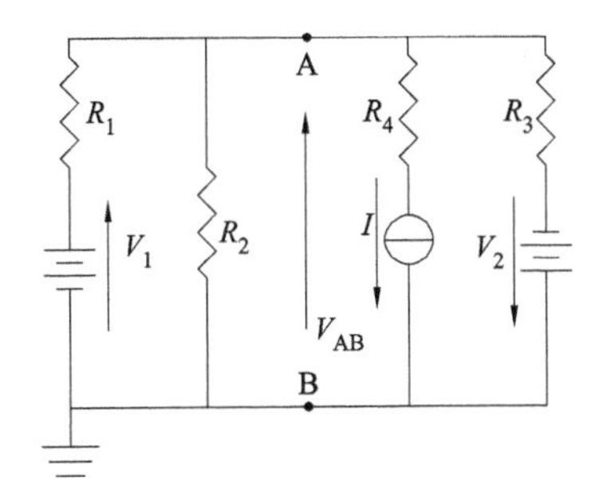

FIGURE B.9 Millman's theorem applies to circuits made up of only branches in parallel.

In practical applications, the "balance" condition can be revealed by a null-detector meter, which can indicate a state of zero volts across its terminals.

If $\underline{V}_{BC} = 0$ and no current flows through the diagonal BC. We then have $\underline{I}_4 = \underline{I}_3$ and $\underline{I}_1 = \underline{I}_2$. In these conditions by applying Kirchhoff's law to the bridge circuit, we can write the following system of equations:

$$\begin{cases} \underline{Z}_2\underline{I}_2 = \underline{Z}_4\underline{I}_4 \\ \underline{Z}_1\underline{I}_1 = \underline{Z}_3\underline{I}_3 \\ \underline{I}_1 = \underline{I}_2 \\ \underline{I}_3 = \underline{I}_4 \end{cases} \tag{B.17}$$

The above system yields as a solution

$$\underline{Z}_1\underline{Z}_4 = \underline{Z}_2\underline{Z}_3 \tag{B.18}$$

which is the condition to balance the bridge.

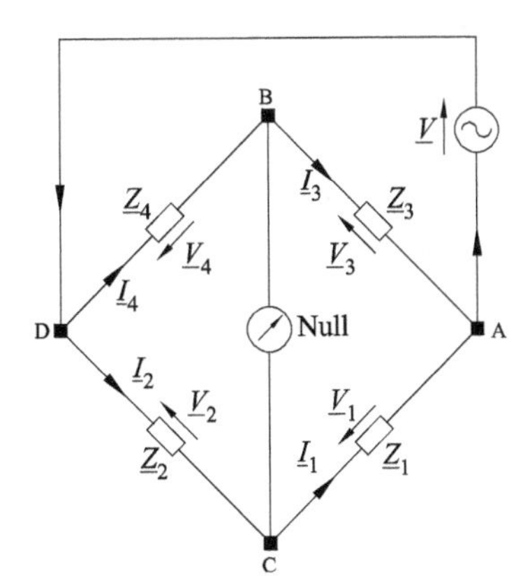

FIGURE B.10 Impedance bridge.

Endnote

1. Dependent sources are left unaltered as they are controlled by circuit variables.

APPENDIX C

Fundamental Units, Symbols, and Correct Spellings

C.1 Synoptic Table

Quantity	Symbol	Unit	Relation to Fundamental MKSA Units	Correct Spelling	Wrong Spelling
Capacitance	C	Farad	m^{-2} kg^{-1} s^4 A^2	10 μF	10 μf
Electric current	I	Ampere	A	10 A; 10 kA	10 a; 10 Ka
Electric current density	J	Am^{-2}	m^{-2} A	10 Am^{-2}; 10 kAm^{-2}	10 am^{-2}; 10 Kam^{-2}
Electric potential	V	Volt	m^2 kg s^{-3} A^{-1}	230 V; 1 kV	230 v; 1 Kv
Electric resistance	R	Ω (ohm)	m^2 kg s^{-3} A^{-2}	10^6 Ω; 1 MΩ	1 mΩ
Electric resistivity	ρ	Ω m	m^3 kg s^{-3} A^{-2}	10 Ωm; 10 Ω · m	10 Ω m
Energy	W	Joule	m^2 kg s^{-2}	10 J; 10 kJ	10 j; 10 kj
Frequency	*f*	Hertz	s^{-1}	50 Hz	50 hz; 0.5 KHZ
Inductance	L	H	m^2 kg s^{-2} A^{-2}	1 μH	1 μh
Power	P	Watt	m^2 kg s^{-3}	10 W; 10 kW	10 w; 10 KW

Index

Note: Page numbers followed by f indicate figures; numbers followed by t indicate tables.

CPSIA information can be obtained
at www.ICGtesting.com
Printed in the USA
BVHW041126230619
551147BV00017B/25/P

9 780071 508186